Vegetation

a

INIVER

Vegetation Description and Analysis

A Practical Approach

Martin Kent and
Paddy Coker

JOHN WILEY & SONS

Chichester · New York · Brisbane · Toronto · Singapore

581.524
KEN

Disk Copied
28.6.96

First published in Great Britain in 1992 by
Bellhaven Press (a division of Pinter Publishers),
25 Floral Street, London WC2E 9DS

First published in the United States of America in 1992 by
the CRC Press, Inc., 2000 Corporate Blvd., N.W.,
Boca Raton, Florida, 33431

Published in 1994 by John Wiley & Sons Ltd,
Baffins Lane, Chichester,
West Sussex PO19 1UD, England

National   (01243) 779777
International (+44) 1243 779777

Reprinted with amendments April 1995.
Reprinted January 1996

Other Wiley Editorial Offices

John Wiley & Sons Inc., 605 Third Avenue,
New York, NY 10158-0012, USA

Jacaranda Wiley Ltd, 33 Park Road, Milton,
Queensland 4064, Australia

John Wiley & Sons (Canada) Ltd, 22 Worcester Road,
Rexdale, Ontario M9W 11.1, Canada

John Wiley & Sons (SEA) Pte Ltd, 37 Julan Pemimpin #05-04,
Block B, Union Industrial Building, Singapore 2057

1105049 7

Learning Resources
Centre

ISBN 0 471 94810 1 (paper)

Filmset by Mayhew Typesetting, Rhayader, Powys
Printed and bound in Great Britain by SRP Ltd, Exeter

17.7.96

Contents

Preface

Our aim in writing this book has been to produce a text which is suitable for the average undergraduate taking a course in vegetation description and analysis within an ecology, geography, environmental science or biological science degree. Having taught such an option for over 15 years, the lack of a textbook which is suitable for the typical undergraduate has always caused both us and our students considerable difficulties. To some ecologists this may seem a strange comment to make, for there are already many textbooks available on this subject (at least 12 on a recent count). However, many of these are not suitable for most undergraduates usually because their treatment is too mathematical and they assume that all students interested in vegetation ecology are equally interested and capable in matrix algebra and multivariate statistical analysis. We have lost count of the number of times we have heard the comment: 'We did this course because we were interested in plants and their ecology, not because we wanted to learn mathematics and statistics!'.

Plant ecology and biogeography are field-based subjects. Most undergraduates come to them because they acquired an interest in botany and ecology at school and also because they are are concerned about their environment and the conservation of ecosystems, plants and animals. While they may feel strongly about the environment and conservation issues, as science students they also appreciate that they should have a rigorous approach to their studies. However, achieving a correct balance between the mathematical and statistical as opposed to the biological, ecological and geographical components of vegetation description and analysis is a difficult and demanding task.

The purpose of this book is to show that vegetation description and analysis can and should be taught in a much more overtly ecological manner, keeping the mathematical and technical material to the minimum necessary for clear understanding. It is quite possible to teach the 'grey box' approach, where students learn primarily through the **application** of the methods to different **ecological situations and problems**. Also, to teach this subject independently of field and practical work, although possible, is rarely satisfactory. For students to identify a problem and then to use the approaches and methods to solve that problem from field through laboratory and computer analysis to final results and interpretation is the optimal way to learn the real value of vegetation description and analysis. It is the **practical approach** that matters.

This book is aimed at a second- or third-year undergraduate taking a course in plant ecology within a biology, geography or environmental science degree and assumes that the student has completed a first-year course in the fundamentals of ecology or biogeography. Many suitable introductory textbooks exist, for example Whittaker (1975), Krebs (1978), Ricklefs (1979), Tivy (1982), Clapham (1983), Putman and Wratten (1984), Kormondy (1984), Cox and Moore (1985), Pears (1985), Odum (1989), Begon *et al.* (1990) and Bradbury (1991). A tropical emphasis is provided in Ewusie (1980) and Deshmukh (1986). Some aptitude for working with numbers is assumed and a grasp of basic statistical concepts would be an advantage.

After reading this text and carrying out some field and practical work within a course, a student should be in a position to take further advanced courses in ecology, perhaps applying the approaches and techniques of this book to applied problems and environmental management. Also, an improved understanding of techniques for vegetation description and analysis should encourage undergraduates to embark on projects and dissertations involving the use of such methods for which there is considerable scope. At this stage, **the case studies** should be particularly useful, as well as **the sources of FORTRAN and BASIC computer programs** listed in Chapter 9. We have also compiled a disk of programs for the analysis of vegetation data which will run on IBM-compatible PCs. This disk is available for a small charge and further details are also given in Chapter 9. Fairly substantial quantities of vegetation data are likely to be generated quite rapidly in student projects and we regard the availability of such user-friendly programs as being extremely important.

We are convinced that too many students and lecturers are put off this subject and shy away from teaching it because of its apparent complexity. Through stressing the ecological and practical approach, we hope that this book will encourage more teaching and application of methods of vegetation description and analysis within terrestrial plant ecology and that both students and teachers will see beyond the academic aspects of the subject into possible applications in biological conservation and environmental management.

Martin Kent   Paddy Coker
Plymouth   Orpington

March 1992

Acknowledgements

Many people have assisted in the writing of this book. In particular we would like to thank Dr. Tom Dargie (formerly of the University of Sheffield, now a private consultant) who has always shared our enthusiasm and interest for the subject, Dr. David Gilbertson (Department of Archaeology and Prehistory, University of Sheffield), Dr. Ken Thompson (Unit of Comparative Ecology, University of Sheffield) and Dr. Peter Wathern (Department of Biological Sciences, University College of Wales, Aberystwyth), all of whom have given support and encouragement.

Martin Kent wrote most of the manuscript and accepts responsibility for any errors contained therein. He would like to offer special thanks to Dr. Ruth Weaver (Geographical Sciences, University of Plymouth) for her patience in carefully reading and correcting the manuscript and to Dr. Dan Charman (Geographical Sciences) and Dr. Pam Dale (Biological Sciences) for additional comments.

Diagrams in the early part of the book were drawn by Jenny Wyatt (now a cartographer at the University of Cambridge) and Sarah Cockerton. However, most of the drawing work has been completed by Tim Absalom and Brian Rogers (Department of Geographical Sciences, University of Plymouth). Tim Absalom, in particular, tolerated the drawing and revision of numerous ordination diagrams with exceptional good humour. We are greatly indebted to them both.

Paddy Coker compiled the package of computer programs for PCs and wrote the accompanying Chapter 9. Most of the programs are in the public domain but both authors wish to thank Dr. Mark Hill (Institute of Terrestrial Ecology) for permission to include the public domain versions of TWINSPAN and DECORANA on the disk of programs. Dr. Warren Kovach (Institute of Earth Studies, University College of Wales, Aberystwyth) kindly gave his consent for the shareware version of his MVSP (MultiVariate Statistical Package) program to be included. Special thanks also go to Dr. Alan Morton (Department of Biology, Imperial College, University of London) for assistance with various programs. At the University of Greenwich, Robert Paynter (Computer Centre) has provided invaluable advice and support.

Both authors would like to thank Iain Stevenson, Sarah Henderson and Jane Evans of Belhaven Press for being so patient while the manuscript was in preparation and for their professionalism during the production of the book.

Finally, both of us owe an enormous debt of gratitude to our families (MK — Gay, Jonathan, Joseph, Holly and Kitty; PC — Rosemary and Bryony) for their infinite patience and tolerance during the writing of the book and the compilation of the computer programs.

Copyright and authorship of all Figures and Tables are acknowledged in the appropriate caption. The authors are grateful to Routledge Publishers for permission to include diagrams from P. Gould and R. White (1986) *Mental Maps* in Chapter 5.

In a small number of cases the authors and publishers have been unable to trace the copyright holders of material reproduced here and would be grateful for any information that would enable them to do so.

Martin Kent   Paddy Coker
Department of Geographical Sciences   School of Environmental Sciences
University of Plymouth   University of Greenwich

Safety in the field

All field work is potentially dangerous, even when carried out in local, well known areas. Precautions should always be taken and local safety codes adhered to. The following recommendations are important:

(1) Always obtain an up-to-date weather forecast.
(2) Take advice from local experts if in doubt.
(3) Be aware of potential health problems of any members of the party.
(4) Collect the address and telephone numbers of family or friends of every member of the party.
(5) Leave this information and details of the route to be followed with a responsible person at the base and an expected time of return.
(6) Never, ever, carry out field work alone: a group of three leaves one person free to go for help while a second person can stay with an injured colleague.
(7) All members of the party should have had a tetanus injection.
(8) Be extra careful in certain habitats such as wetlands, bogs and swamps. Working in the tropics carries special potential dangers.
(9) Be prepared for the worst that can happen in terms of bad weather or an accident. Responsible members of the party must be familiar with basic first aid and safety procedures. The following equipment is essential, depending on environment: suitable footwear (usually stout boots), waterproofs with hood, overtrousers, warm hat and gloves, sunhat, water, first aid kit, torch with batteries, whistle, emergency rations including glucose sweets, spare warm clothing, survival blanket or lightweight tent, map and compass.
(10) The standard SOS signal for torches or whistles is three short signals, three long and three short.

Access

Always obtain permission from landowners, farmers and other relevant agencies before carrying out field work on their land. By far the majority will gladly give permission **provided it is requested before going onto their land**.

Disclaimer

While every reasonable care has been taken, neither the authors nor the publishers accept any liability for any injury, accident, loss or consequent damage, however caused, arising from this book or any information contained therein.

The nature of quantitative plant ecology and vegetation science

The nature of vegetation

Dictionary definitions usually describe vegetation as 'plants collectively' or 'plant growth in the mass'. To the plant ecologist and vegetation scientist, this definition is completely inadequate and perhaps conforms to the view of many students (and teachers and lecturers!), who see vegetation as 'a frightening and unknown mass of green, shrouded in technical terms and Latin names' (Randall, 1978, p. 3). This book is concerned with the techniques for both collecting and analysing data on vegetation, with the primary aim of making sense of the 'frightening and unknown mass of green'. As such, it is a text on **quantitative plant ecology**, which is a clearly recognisable sub-discipline of **ecology** and **biogeography**. The field of quantitative plant ecology is also related to an area of research known as **vegetation science** which, in addition to vegetation description and analysis, also includes plant population biology, species strategies, production ecology and vegetation dynamics (successional processes and vegetation change) (van der Maarel, 1984a). Most researchers and students take the phrase 'vegetation description and analysis' to mean the collection of vegetation data, followed by analysis, usually using complex mathematical methods. However, there has been an increasing tendency for the processes of analysis to become an end in themselves. An important aim of this book is to show that quantitative plant ecology and vegetation description and analysis can and perhaps should be primarily **ecological** rather than **mathematical** in emphasis. The only way variations in vegetation and plant species distributions can be properly understood and explained is within an ecological framework. This introduces the fundamental point that vegetation is always an integral part of an **ecosystem** (Tansley, 1935; Waring, 1989) and can only be studied by fully exploring its role within that ecosystem. Vegetation cannot be isolated as a separate entity from the ecosystem within which it exists.

The building blocks of vegetation are individual plants. Each plant is classified according to a hierarchical system of identification and nomenclature using carefully selected criteria of physiognomy and growth form. The individuals of one species, taken together, form a **species population** and within the local area of a few square metres to perhaps as much as a square kilometre, groups of plant species populations which are found growing together are known as **plant communities**. Much more will be said of plant communities later, but within plant communities, the **presence or absence** of particular species is of primary importance. After this, the amount or **abundance** of each species present is of interest. This book is concerned with reasons and methods for collecting data of these kinds and with techniques for their analysis.

The importance of vegetation within ecology is threefold. First, in most terrestrial parts of the world, with the exception of the hot and cold deserts, vegetation is the most obvious **physical representation** of an ecosystem. When ecologists talk about different ecosystem types, they usually equate these with different vegetation types. Second, most vegetation is the result of **primary production**, where solar energy is transformed through **photosynthesis** by different plant species into green plant tissue. The **net primary production**, which is the amount of green plant tissue accumulated within the area of a particular vegetation type over a given period of time, represents the base of the **trophic pyramid**. All other organisms in both the **grazing** and **detrital food webs** are ultimately dependent on that base for their food supply. Third, vegetation also acts as the **habitat** within which the organisms live, grow, reproduce and die. In the case of the grazing food web, it is among the above ground parts of plants. In the detrital web, it is on the surface and below ground among the roots. Taken together, these three points show the central importance of vegetation to ecology and demonstrate the need for methods to assist with description and analysis (Anderson and Kikkawa, 1986; Diamond and Case, 1986; Cherrett, 1989).

Why study vegetation?

There are many situations where vegetation merits study. The commonest examples of the use of vegetation description are in the recognition and definition of different vegetation types and plant communities known as the science of **phytosociology**; the **mapping of vegetation communities and types**; the study of **relationships between plant species distributions and environmental controls**; and the study of vegetation as a **habitat** for animals, birds and insects. Change in vegetation over time may also need to be described using concepts of **succession and climax**.

Information on vegetation may be required to help to solve an ecological problem: for **biological conservation and management** purposes; as an input to **environmental impact statements**; to monitor **management practices** or to provide the basis for **prediction** of possible future changes.

A useful distinction is into aspects of study which are **academic** as opposed to those which can be termed **applied**. In the academic case, vegetation may be described and analysed largely for its own sake. Applied studies are where vegetation data are collected and analysed with the aim of providing information of relevance to some ecological problem, often to do with environmental conservation and ecosystem management. Many examples of research include elements of both.

Case studies

Throughout this book many different examples of the application of methods for the description and analysis of vegetation will be presented. A brief introduction to four contrasting case studies serves to demonstrate the diversity of situations where vegetation may need to be surveyed and data collected and analysed.

Case Study 1 — The forest vegetation of the Lower Alabama Piedmont, USA (Golden, 1979)

This study is an excellent example of a primarily academic survey of vegetation and its variation in relation to environmental factors. It is described in more detail in Chapter 6. The purpose of the survey is of greatest importance here. The author worked for the Department of Forestry within an agricultural experimental station at Auburn University, Alabama and clearly states that the aim of his research was to examine the forest vegetation of the southern end of the Alabama Piedmont and its relationships to environment. There was a paucity of quantitative vegetation-site information for this region of the United States.

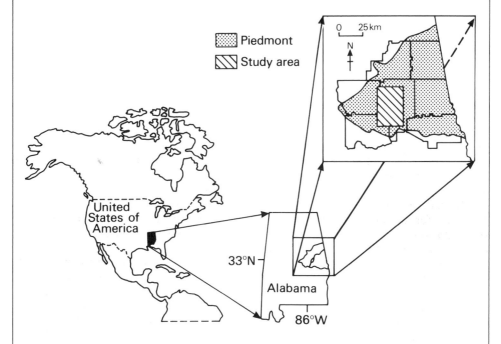

Figure 1.1   The location of the Alabama Piedmont study in the USA (Golden, 1979) (Modified and redrawn with kind permission of *Ecology*)

The Alabama Piedmont lies between New Jersey and Alabama (Figure 1.1). Apart from some early descriptions as 'oak-pine forest', detailed quantitative descriptions of the forests were lacking. Although the forests are very extensive, almost all of them have been cleared and allowed to regrow at some point in their history. Most flatter plateau areas had been cleared for agriculture, although some had been allowed to regenerate through natural colonisation (old-field succession). Thus human impact has been very important, immediately demonstrating that there are very few parts of the world where the vegetation does not show some effects of human activities.

A representative area of the Piedmont was chosen for study, and data on the trees and the ground flora underneath were collected from 84 different locations which were carefully selected to show differences in both forest type and soil and land management practices. The data on the vegetation were analysed using various methods described later in this book, and 10 major forest vegetation types were identified (see Chapter 6). These 10 types were also shown to be differentiated from each other first on the basis of topography and drainage and second by the different management histories of the sites. These variables represent the primary and secondary environmental gradients.

On the basis of a careful interpretation of the data, various recommendations were made about woodland management practices in the Piedmont region, demonstrating that even a principally academic study invariably provides information which may assist with future land management practice.

Case study 2 — The ecology of the Serengeti grasslands (McNaughton, 1983; 1985)

The tropical grasslands or savannas are one of the major subdivisions of world vegetation, and are known as a **vegetation formation type** and a **world biome**. Such grasslands are predominantly composed of grass and herb species but often have a significant tree component. The factors determining the status of savanna vegetation have long been a matter of debate. Seasonality of climate, fire, grazing, edaphic or soil factors and geomorphology have all been cited as of importance in the origin and maintenance of the vegetation (Huntley and Walker, 1982; Bourlière, 1983; Cole, 1986; 1987).

Plate 1.1 Grazing elephants on the savannas of the Serengeti National Park, Tanzania (photo: M. Kent/C. Breen)

The Serengeti takes its name from the the Masai word for a broad open

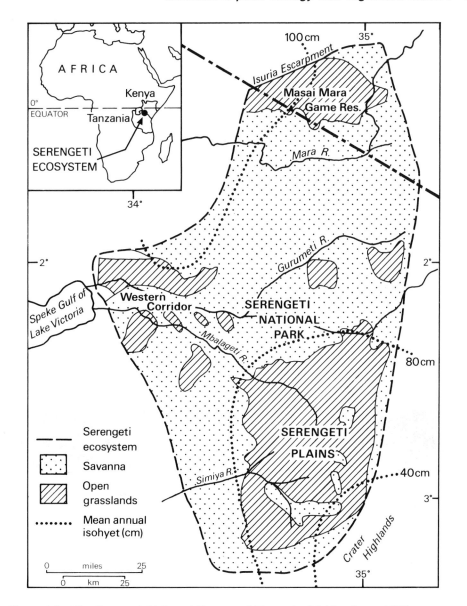

Figure 1.2 The Serengeti Plains of Kenya and Tanzania (McNaughton, 1983)
(Modified and redrawn with kind permission of *Ecological Monographs*)

plain and is situated on the border between Kenya and Tanzania, stretching from Lake Victoria in the west to the volcanic Crater Highlands in the east (Figure 1.2). The Serengeti is famous for its fauna, both the large herds of ungulates such as the wildebeest, topi and zebra (Plate 1.1) which roam the open plains and also its carnivores including lions, cheetahs and jackals. According to McNaughton, three distinct sub-regions can be identified: (a) the open Serengeti Plains of the south-east; (b) the Western Corridor which is used by grazers during seasonal changes and whenever rain falls on the

Serengeti Plains during the wet season; and (c) the north which is used by migrants at the height of the dry season. The greater part of the plains area is included within two national parks that have been established largely to protect the fauna: the Serengeti National Park in Tanzania and the Masai Mara Game Reserve in Kenya.

McNaughton makes the point that prior to his work in the Serengeti, most studies of tropical grassland had been based on descriptive observation and physiognomic classification — description by growth or life form (see Chapter 2). Such studies had often been made in relation to soils and to the distributions of the large herds of ungulates that are characteristic of the biome (Milne, 1935; Pratt *et al.*, 1966; Bell, 1970).

Detailed quantitative studies of plant community structure (species composition and association or groupings of species growing together at the same location) were virtually absent. At the start of his article, McNaughton poses several questions concerning community structure in the vegetation:

(1) Are there repeating combinations of species and community types that occur more or less frequently?
(2) Are there consistent patterns of species abundance and diversity that provide insight into community organisation?
(3) How important is spatial heterogeneity (local variation of plant communities in space) and what is its role in community organisation?
(4) What environmental factors influence species abundance, spatial distribution and community organisation?
(5) If consistent patterns of community properties and environmental relationships exist, why have they developed and what can be inferred from them about the mechanisms and functional consequences of community organisation?

These questions are typical of those asked at the start of a great deal of work in vegetation science; in many parts of the world rigorous description of vegetation has only comparatively recently been completed or has yet to be attempted.

As with the first case study, although this research could be seen as primarily academic in emphasis, there were also a number of applications deriving from the survey. Following control of rinderpest, an acute viral disease of various ungulates and cattle, the population of grazing animals, particularly wildebeest, had increased dramatically from about 220,000 in 1961 to 1.4 million in 1977. The scale and extent of firing had declined following the designation of the greater part of the area as a national park. Elephants, only reintroduced to the Serengeti in 1951, had undergone a population explosion and were causing much damage to vegetation by uprooting trees and laying waste large areas around vital water holes in the dry season. All of these factors were undoubtedly exerting serious and often detrimental effects on the grassland ecosystems and community structure.

For all these reasons, McNaughton selected 105 sites within both national parks to sample the vegetation. Detailed aspects of sampling are described in the paper, and the data were analysed by methods of ordination and classification described in Chapters 5 and 8. In the final analysis, 17 different community types could be recognised, largely characterised by perennial grasses. Examination of environmental controls demonstrated that while rainfall and seasonality were significant in separating different community types, grazing intensity was critical with a secondary gradient being attributable to soil texture.

At the end of the paper, there is an excellent discussion of where vegetation research should proceed following the results of the survey. McNaughton argues that formulation of hypotheses and design of experiments to quantify the exact effects of grazers within each of the vegetation types is necessary. However, there are many other questions which might be asked:

(a) What is the species composition of the grassland and particularly its local spatial variation?
(b) What are the phenological strategies of the grasses (how does each species time the major events of its life — existence as a seed or vegetative rhizome, germination, growth and leafing, flowering, fruiting, death)?
(c) Is there an overlying tree canopy? How dense is it and what is its impact on the underlying vegetation?
(d) What are the grazing species and what are their densities? Do they preferentially graze certain plant species at certain times of the year?
(e) Has the vegetation been burned lately? If so, how often and how extensively?
(f) Is the soil wet or dry? What are the detailed chemical and physical properties of the soils?

All of these questions and many more are immediately suggested and require careful thought, hypothesis generation and experimentation (Figure 1.4a) to be answered. A further problem is that many of these questions overlap: for example are the effects of grazing or burning the same on different soil types?

In summary, McNaughton's work demonstrates how a relatively thorough and complete survey of the vegetation of an area in response to one set of aims and objectives generates even more questions which may require many more years of detailed experimental ecology in order to be answered. Having completed this primary survey of vegetation community types and environmental factors, McNaughton went on to examine primary productivity of vegetation in relation to grazing in a subsequent paper in 1985. This work in turn demonstrates the importance of vegetation as both food supply for higher trophic levels and as a habitat for both herbivores and carnivores.

Case study 3 — Britain's railway vegetation (Sargent, 1984)

There are around 18,000 km of railway line in Britain. While the railway tracks themselves may not be a very suitable place for many plants to grow and may often have been treated with herbicides, the zones adjacent to the track represent an enormous area where natural and semi-natural vegetation is able to flourish. The ecological and conservation potential of these areas was recognised in Britain by the Institute of Terrestrial Ecology in 1977, when a team of researchers were employed to study the plant communities of the railway network (Plate 1.2).

Prior to this date, very little was known about the plant assemblages which were found on British Rail land. At the start of the survey four questions were asked (Sargent, 1984; p. 1):

(1) What kinds of habitats occur and what areas do they cover? There are

Plate 1.2 A typical example of British railway vegetation near Plymouth, England (photo: M. Kent)

distinct differences between cess (permanent way) and the verge, but is the slope, aspect or kind (cutting or embankment) of engineered formation important in determining the distribution of vegetation (and hence animals)? What are the effects of management and disturbance?

(2) Does the railway provide a refuge? Which species move along or are blocked by this linear environment?

(3) What kinds of vegetation occur? Are these associations unique to the railway or are they essentially continuous with neighbouring forms?

(4) Is the system comparatively stable or are irreversible changes occurring? Is intervention needed to prevent such change or to protect particular areas?

Having decided on the project, one of the biggest problems was that of sampling. Details of this are given in Chapter 2. However, a total of 480 sites were sampled around the whole railway network. In all, 1,632 higher plant species (phanerogams) and 323 mosses and ferns (cryptogams) were recorded. As with the previous studies, a primary aim was to examine plant community structure. Data from 3,502 samples were used in a fairly complex analysis described in Chapter 8, to give 32 major groups of community types known as noda. As a result of this, 185 sites of particular biological interest were identified and have been recommended for conservation protection.

One of the most interesting aspects of the conclusions of the work is the idea that many sites surveyed were very dynamic in nature and undergoing active change through successional processes. One of the problems for future management was shown to be that of understanding and controlling successional change.

Another aspect of this survey was the volume of data generated. The number of vegetation samples collected was so great that a carefully structured series of analyses had to be carried out in order to make sense of the data. This is a feature of quantitative plant ecology — very large amounts of data may be collected in a relatively short space of time. It is for this reason that vegetation scientists have had to resort to mathematics, statistical analysis and computers to assist with making sense of data from even a small survey.

Case study 4 — Methods of selecting lake shorelines as nature reserves (Nilsson, 1986)

This Swedish study is very much an example of applied plant ecology. The aim was to select shoreline ecosystems within one region of Sweden for protection as nature reserves. Shorelines are habitats that are increasingly coming under pressure from development and recreational activities such as fishing and boating. The aim was to conserve the whole range of lakeshore types and in the process to ensure protection of all the plant species. Nilsson was also very concerned that too many ecological assessments of sites for potential nature reserves are based upon the criteria of diversity (species richness — Chapter 3) and rarity of species rather than on a full appreciation of community ecology and the idea of conserving representative examples of complete communities and ecosystems.

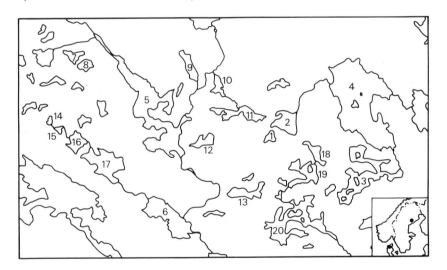

Figure 1.3 Sites included in the Swedish lake shoreline study (Nilsson, 1986) (Redrawn with kind permission of *Biological Conservation*)

The area lies in north-east Sweden about 80 km inland from the Gulf of Bothnia (Figure 1.3) and is described as a plateau with 20 lakes ranging in shore length from 0.77 to 15.73 km, scattered among monadnocks. These shores were first surveyed from aerial photographs to give eight general habitat types within which 418 samples of vegetation were recorded. Sixteen different plant communities were then identified by using classification methods (Chapter 8). On the basis of this identification, the characteristics and mix of shorelines around each lake could be assessed. Various ways of using this information to select sites for conservation along with the more traditional criteria of diversity and species rarity are then discussed. Detailed presentation of Nilsson's methods of analysis will be kept until Chapter 8.

The interesting features of this study are its obvious applied conservation emphasis and the focus upon one particular habitat type within a local region. There are many such examples where one specific habitat type is selected for detailed vegetational survey and analysis of the variability of its plant species composition.

The scientific approach

As with all aspects of science, vegetation description and analysis must be approached in a **logical and systematic fashion**. In addition, all vegetation description and analysis must have a **purpose**. Nevertheless, there are many different approaches to vegetation study. The collection and analysis of vegetation data provides the principal form of ecosystem description and classification. However, simply to describe and observe patterns of variation in vegetation data over space is not the sole aim of vegetation science. A further very important concept is the idea of **explanation**. Explanation is the attempt to answer the question 'why?' For example, 'why is one area of vegetation different from another?' 'Why are certain plant species found in some locations but not in others?' 'Why do certain vegetation types appear to be undergoing change — due either to natural processes of succession or to human-induced effects?' 'Why is the vegetation of a particular area under stress and showing signs of damage or disease?'

To attempt answers to such questions and provide explanations, we have to have some view or **theory** of the manner in which the world functions. For any part of science, there is always an existing body of knowledge and theory which is generally accepted at the time. On the basis of this existing knowledge and theory, **a scientific approach** will usually involve the generation and testing of **hypotheses** concerning observed variations in vegetation cover and their causes. This is usually known as the **deductive** approach (Figure 1.4a). The sequence of activities shown in Figure 1.4a is commonly described as **scientific method**. It follows a rational and logical sequence of thought processes and actions. Most scientists would probably subscribe to that view and are thus known as **rationalists**. An important point is that hypotheses are generated from the existing body of theory and data collection, and analyses are then based on the notion of accepting or rejecting the hypothesis. The basic principles of deductive thinking are explained in much greater detail in many texts, for example Harvey (1969), Moss (1977) and Haines-Young and Petch (1986).

The testing of hypotheses and accepting them as true (verification) or rejecting them as false (falsification) is central to the process of deduction and is discussed at length by Popper (1972a,b; 1976) and Haines-Young and Petch (1986). The emphasis in much scientific publication on trying to prove hypotheses acceptable (verification) rather than rejecting them (falsification), has led to the notion of **logical positivism**. Other views are those of Kuhn (1962; 1970) and Feyerabend (1975; 1978). They are known as **relativists**. Kuhn argues that the history of science demonstrates that most scientific work is not concerned with trying to overthrow existing theories and develop new ones; rather, scientists tend to accept existing established theories and ideas and use them to solve 'puzzles' within different parts of their subject. (Kuhn deliberately chose the word 'puzzles'.) Well-established theories and concepts which are generally believed to be tried and tested are known as **paradigms**. The ecosystem concept is a good example of such a paradigm. Large

a)

b)

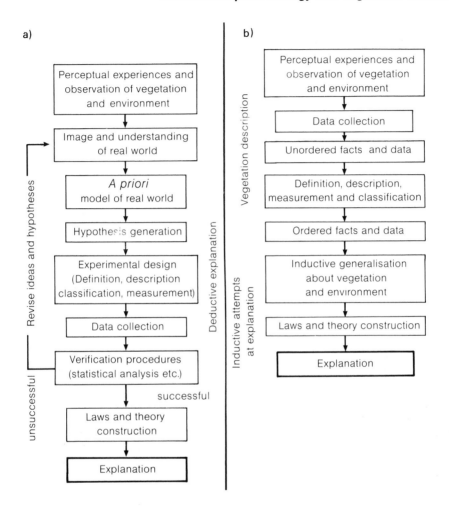

Figure 1.4 (a) Deductive and (b) inductive approaches to scientific enquiry

numbers of scientists will normally be working within different areas of a subject, generally accepting their paradigms. This is known as **normal science**. Only occasionally are a large number of results taken together, or a particularly radical piece of research is completed, and existing theories and paradigms drastically revised. The time when this occurs is known as a period of **scientific revolution**. Certain paradigms in plant ecology, such as the concept of the plant community and successional processes (see below), although readily accepted by most ecologists, are still matters of considerable debate, and the nature of these paradigms is almost certain to change over the next century in the way described by Kuhn.

Still more controversial views come from Feyerabend, who has attacked the whole of scientific method (hypothesis generation, verification, falsification) and argues that there are many different routes to scientific results, many of which are far less precise than most scientists would admit. His contention is that there is thus no one scientific method, but a series of alternative approaches which on some occasions may partly overlap and which work with varying degrees of objectivity. He states that the only principle that can be applied to all circumstances is that 'anything 'goes' (Feyerabend, 1978, p. 39). While this in itself may seem an extreme viewpoint, it is of considerable relevance to quantitative plant ecology and vegetation science.

In many parts of the world, it is not possible to devise hypotheses immediately, since basic descriptions of the vegetation are completely absent or at best extremely generalised. Nowhere is this more true than in the tropics. As a result, a large amount of vegetation research is purely **descriptive** in nature and **inductive** in approach (Figure 1.4b). With **induction**, the data are collected without formulation of a prior hypothesis and, if necessary, explanations are then derived from the data collected. Ideas of induction date from Francis Bacon, back in the early-seventeenth century. He argued that one can only start to make valid explanations once all the necessary facts and information relevant to the situation have been collected. Such data were thought to be particularly secure because the collection and ordering of facts for their own sake are not based on any biased or selective guesses or hypotheses. The problems of the inductive approach are many and particularly centre on the inefficient process of collecting a large amount of data which may not be immediately (or even eventually) relevant or useful.

Most research programmes involve elements of both induction and deduction. The vegetation of an area may be described and analysed without any prior hypotheses, resulting in a fuller understanding of the major types, their environmental controls and man's impacts. Study of the results should then lead through the process of hypothesis generation, testing and deduction to more detailed work on specific questions generated by the original survey. A large amount of vegetation work still remains purely descriptive and ends with a list of the major vegetation types and their controlling factors, often with a minimal attempt at explanation. Ideally, however, this should only be a starting point for more detailed research on specific aspects and problems, involving the deductive approach and hypothesis generation and testing. Hairston (1989), Wiegleb (1989) and Eberhardt and Thomas (1991) discuss these issues further and provide an introduction to experimental ecology in relation to hypothesis generation.

To make explanations, many ecologists now use Chamberlin's method of **multiple working hypotheses** (Chamberlin, 1890; 1965). This involves the consideration of as many likely explanations of the distribution of vegetation in terms of environmental controls and the effects of humans and animals as possible, followed by a series of experimental studies designed to reject the hypotheses until the most likely ones remain as reasonable explanations. Often it may be possible to assess the likelihood of particular explanations being correct. Scientists adopting that approach have often been termed **probabilists**. A related idea is **Occam's Razor**, named after William of Occam (*c.* 1300–49), who argued that the simplest and most straightforward explanation should be sufficient until there is good reason to believe that a more complex one is required.

A final stage of scientific method is **prediction**. Once hypotheses have been generated, tested and accepted, it becomes possible to use the results to predict future outcomes. In quantitative plant ecology, this can be particularly important for understanding in environmental management and biological conservation.

Understanding the relationships between theory and practice in vegetation science is extremely important. Valuable further discussion is provided in Podani (1984), Austin (1986), Noy-Meir and van der Maarel (1987), Roberts (1987), Grubb (1989), Pickett and Kolasa (1989), Roughgarden *et al.* (1989) and van der Maarel (1989).

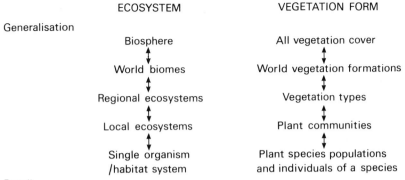

Figure 1.5 The various scales of ecosystem and vegetation study (after Rowe, 1961)

Scales of study

Ecosystems may be defined at a range of scales from an individual leaf up to the level of the whole **biosphere**. The various levels of ecosystem recognition are summarised in Figure 1.5. The concept of **nesting of systems** is important. All levels of ecosystems may be seen as **sub-systems** of the biosphere (Rowe, 1961). Each sub-system occupies a progressively smaller area in space. As vegetation is the most obvious external feature on the basis of which ecosystems are defined and classified, it follows that smaller and smaller units of vegetation can be recognised, from the biosphere through vegetation formations down to the individual plant and leaf (Figure 1.5).

Much early work in both plant ecology and biogeography concentrated on the **world biomes and vegetation formations**, which represent the first major subdivision of the biosphere. Inevitably, most of these studies were extremely generalised. In this century, studies have become focused at the lower scale of **the plant community**, first because this is the scale at which populations and individuals of a plant species can be identified and grouped together to characterise the vegetation of an area of a few square metres to several square kilometres. Second, the community scale is important because it is at this scale that humans can make best sense of the nature and variation of the vegetation cover of the earth. Third, it is at this scale that human activity in changing vegetation cover takes place and thus conservation and environmental management practices and policies may be applied.

This problem of scale in ecology has received close attention in recent years and a number of papers have been published on the subject (Allen and Wileyto, 1983; Giller and Gee, 1987; May, 1989; O'Neill, 1989; Allen and Hoekstra, 1990; Rahel, 1990). One of the biggest problems is the diverse and conflicting use of terminology.

The concept of the plant community

When an ecologist stands on a hilltop and surveys a landscape dominated by natural or semi-natural vegetation in any part of the world, the main differences in pattern visible in the landscape will be those of plant communities. Major distinctions will be made on the basis of **physiognomy** or **the growth form** of the vegetation, for example woodland as opposed to scrub or grassland. These units will also represent the major subdivisions of the landscape in functional terms as ecosystems. More subtle changes in the landscape will also be evident in variations in colour between different areas of vegetation with perhaps the same physiognomy. These colour variations will be a reflection of differences in **plant species composition** and stage of development. A considerable part of plant ecology and vegetation science is concerned with methods for actually characterising and defining these areas as different plant communities. Thus the concept of the plant community is absolutely fundamental to the whole discipline.

The plant community can be defined as the collection of plant species growing together in a particular location that show a definite **association** or **affinity** with each other. The idea of association is very important and implies that certain species are found growing together in certain locations and environments more frequently than would be expected by chance. Other groups of species will grow together in other environments. Most environments of the world support certain associated species which can therefore be characterised as a plant community.

Environmental limiting factors

The reason that certain species grow together in a particular environment will usually be because they have similar requirements for existence in terms of environmental factors such as light, temperature, water, drainage and soil nutrients. They may also share the ability to tolerate the activities of animals and humans such as grazing, burning, cutting or trampling.

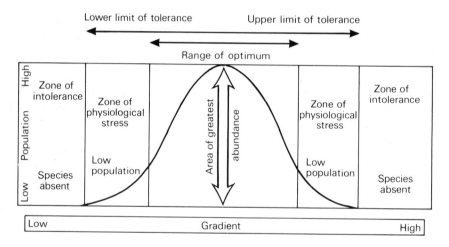

Figure 1.6   The Gaussian or normal curve of plant species response to a single environmental factor and zones of tolerance (Redrawn from Cox and Moore, 1985; with kind permission of Blackwell Scientific Publishers Ltd.)

If one environmental factor is taken, for example soil moisture, and the abundance of a species is plotted across its range of variation, the result may approximate to a **normal or Gaussian curve** (Figure 1.6). This variation of species abundance in response to an environmental factor is known as an **environmental gradient**. If several species are associated in a community, it is often assumed that their abundance curves in relation to environmental factors will be broadly similar. However, studies of species' response to environmental gradients suggest that in practice, species curves vary enormously (Figure 1.7). The width and height of the curve for each species will be very different, indicating differences of **tolerance range**. Also

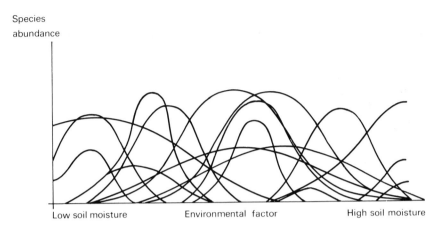

Figure 1.7 Overlapping Gaussian curves of species response to an environmental factor

the form of the curve will rarely be of the idealised perfect bell shape of Figure 1.6. Instead, curves will be skewed, bimodal or have a 'plateau' shape (Austin and Austin, 1980; Austin *et al.*, 1984; Austin, 1987; Austin and Smith, 1989).

A further complication is that a species found growing at a point on the earth's surface will usually be responding to more than one environmental factor. Thus each species will have a different environmental response curve for every environmental factor and each curve will differ in form. The ultimate favourableness or unfavourableness of a site for growth of a certain species will be represented by the collective positions of the site on each of the environmental response curves. For some factors, the site will be near the central, most abundant part of the curve and there the environmental conditions will be near optimal. For other factors, it may be at or near the extremes. If one of the points is at or beyond the limits of the curve, the conditions will be too unfavourable for the species to grow. The factor for which this occurs is known as the **master limiting factor**. However, since several environmental factors determine the growth of a plant, unfavourable conditions for a species in terms of one factor may sometimes be compensated for by another. As an example, a deficiency of one soil nutrient may be compensated for by the abundance of another. This is known as **factor compensation.**

A related concept is that of the **niche**. This is defined as 'the limits, for all important environmental features, within which individuals of a species can survive, grow and reproduce' (Begon *et al.*, 1990). The concept of the niche has been widely debated and discussed, for example in Whittaker *et al.* (1973) and Schoener (1989).

An appreciation of these ideas is important in the understanding and conceptualisation of the plant community. If species do grow in association with each other, many of them should have similar although rarely identical response curves and hence tolerance to the prevailing environmental conditions.

The debate on the existence of plant communities

The idea of the plant community was hotly debated by early plant ecologists. Two American ecologists, F. E. Clements and H. A. Gleason, expressed the most extreme viewpoints.

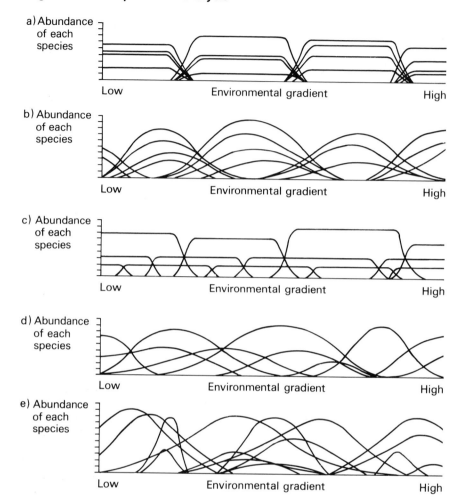

Figure 1.8 Five examples of different hypothetical patterns of species abundances along an environmental gradient, assuming that the environmental factor changes at a gradual, uniform rate along the gradient. (a) sharp community boundaries with dominant species evenly spaced along the gradient and subdominant species strongly correlated with dominants. (b) boundaries between communities more gradual due to Gaussian abundance curves, with subdominant species strongly correlated with dominants. (c) dominant species evenly spaced with sharp boundaries but subdominants not correlated with dominants. (d) Gaussian abundance curves for both dominant and subdominant species with curves and tolerance ranges tending to vary. (e) highly variable tolerance curves with species appearing and disappearing at random along the gradient. (Redrawn and adapted from Whittaker, 1975; with kind permission of Macmillan Publishing)

Clements' view of the plant community

Clements (1916; 1928) saw plant communities as clearly recognisable and definable entities which repeated themselves with great regularity over a given region of the earth's surface. Clements' view of the plant community is known as the **organismic**

concept, in which the various species comprising the vegetation at a point on the earth's surface were likened to the organs and parts of the body of an animal or human. Putting all the parts together made a kind of super-organism which was thus the plant community and the organism (the plant community) could not function without all its organs present.

Within North America, Clements defined three major groups of vegetation which he called **climaxes**: forests, scrub and grasslands. Each of these three climax types was then divided into a number of formations: for example the forest climaxes were woodland (*Pinus* and *Juniperus*), montane forest (*Pinus* and *Pseudotsuga*), coast forest (*Thuja* and *Tsuga*), sub-alpine forest (*Abies* and *Picea*), boreal forest (*Picea* and *Larix*), lake forest (*Pinus* and *Tsuga*), deciduous forest (*Quercus* and *Fagus*), isthmian forest and insular forest. Each of these associations was further divided into 'associations', which were recognised on the basis of one or more characteristic dominant species. Clements also saw each of these associations as a 'climax community' which given sufficient time and relative long-term stability would come into equilibrium with climate. For this reason his theory was known as the **climatic climax** or **monoclimax theory**.

If Clements' ideas are applied to the concept of an environmental gradient, then the distribution of species along that gradient to form different communities would be similar to that in Figures 1.8a or b. Species growing together have similar tolerance ranges and subordinate species are closely correlated with dominant species.

Gleason's view of the plant community

Gleason (1917; 1926; 1939) saw all plant species distributed as a **continuum**. He argued that plant species respond individually to variation in environmental factors and those factors vary continuously in both space and time. As a result, the combination of plant species found at any given point on the earth's surface was unique. Every species has a different distribution or tolerance range and abundance over that range — a different size and shape of curve as in Figure 1.7. The assemblage of plants growing in an area is not only the result of environmental conditions but also species migration. Any area is continuously receiving propagules of species. The success of these species depends on the combination of environmental factors at that site and the tolerance ranges of the invading species. Gleason argued that the range of permutations of combinations of environmental factors, together with the different tolerance ranges of the species, would always give a different combination and abundance of species. In diagrammatic terms, his views would accord with graphs d and e of Figure 1.8. Sampling along those gradients would always produce a different mix of species composition and abundance, and samples could thus never be generalised into clearly defined plant communities.

For these reasons, Gleason's view is known as the **individualistic concept** of the plant community. Taken to its extreme, Gleason's view was that plant communities, although they exist in the sense of a group of species at one point in space, cannot be identified as combinations of associated species repeating over space. He was thus fundamentally at odds with Clements.

Interesting discussion of these early viewpoints on the nature of the plant community are found in Tansley (1920; 1935), Cain (1934a) and Whittaker (1951). Sobolev and Utekhin (1978) and Robotnov (1979) introduce the work of L.G. Ramenskii, a Russian ecologist, who has largely been ignored in the Western literature. They make the point that Ramenskii published ideas on the continuum concept and the plant community in the early decades of this century and that independently, he evolved theories that were very similar to those of Gleason in America.

Present-day viewpoints on the plant community

Even today, ecologists still differ in their conceptualisation of plant communities. However, a majority of ecologists would agree about the existence of plant communities that repeat themselves over space. Their viewpoint lies somewhere between the two extremes of Clements and Gleason.

Most quantitative plant ecologists who use classification methods (Chapters 7 and 8) would tend towards the views of Clements because by definition classification assumes that samples of vegetation composition (species and their abundances) can be grouped into types. Other workers, however, particularly in America, would argue that classification of vegetation samples into groups is impossible. Instead, vegetation samples can only be arranged along environmental gradients as continua, using ordination techniques as described in Chapters 5 and 6. They thus tend to agree more with the views of Gleason. Whittaker (1951), Goodall (1966) and McIntosh (1967b; 1986) include further useful discussion on this topic.

The most realistic present-day view of plant communities is probably that of the **community-unit** theory and the idea of the vegetation of a particular region being distributed as a **mosaic**. These ideas derive from the work of Whittaker (1953) and Whittaker and Levin (1977) and what they described as **climax pattern**. Whittaker argued that in any region, such as the Great Smokey Mountains of Tennessee and North Carolina, broadly similar conditions in terms of environmental factors and biotic pressures will occur over considerable areas. Where these combinations repeat themselves, the vegetation is also repeated, like similar fragments within a mosaic (for example ROC and OCF in Figure 1.9). However, not all areas could be placed within one or other of these forest types, nor were the boundaries as clear as shown in the idealised diagram of Figure 1.9. Often, one forest type would grade into another across an **ecotone** and while perhaps 60–80 per cent of the vegetation could be put into one definite forest type, 20–40 per cent could not because they were **transitional** or **ecotone areas** between types. Such ecotone areas have been largely neglected in plant ecology and most research has concentrated on plant communities themselves. However, as van der Maarel (1990) emphasises, ecotones are of great interest ecologically and deserve more attention in research.

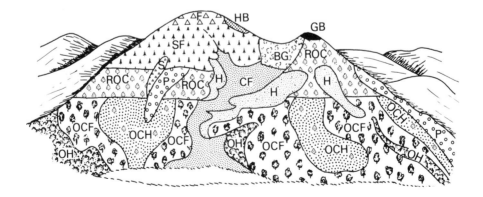

Figure 1.9 Topographic distributions of vegetation types on an idealised west-facing mountain and valley in the Great Smokey Mountains, USA. Vegetation types: BG, beech gap; CF, cove forest; F, Fraser fir forest; GB, grassy bald; H, hemlock forest; HB, heath bald; OCF, chestnut oak–chestnut forest; OCH, chestnut oak–chestnut heath; OH, oak–hickory forest; P, pine forest and pine heath; ROC, red oak–chestnut forest; S, spruce forest; SF, spruce-fir forest. (Redrawn from Whittaker, 1956; with kind permission of *Ecological Monographs*)

Recent literature on the nature of the plant community

A very interesting recent discussion of the nature of the plant community is provided by Shipley and Keddy (1987), who combine the differing views on the nature of the plant community with hypothesis generation and the idea of multiple working hypotheses. Austin (1980; 1987) and Austin and Smith (1989) have proposed a new model of the continuum concept which challenges all existing assumptions of the response of species to their environment but Austin and Smith admit that their ideas require further validation and testing. The main conclusion is that an adequate general model of species response to environment does not yet exist and present models are too simplified. Since species/environment response models still require verification, then there must still be many aspects of the nature of the plant community which have yet to be clarified. Other recent reviews on the nature of communities are presented in McIntosh (1980; 1986), Greig-Smith (1986), Southwood (1987), Roughgarden (1989) and Cody (1989).

Human activity, plant communities and land use

Over large parts of the globe, human populations have modified plant communities extensively. Where natural and semi-natural vegetation remains, one of the most obvious effects has been to sharpen community boundaries so that species and community distributions are often much more like those in Figure 1.8a.

Where human activity has completely removed the former vegetation cover, it has normally been replaced with some form of land use such as agriculture, industry or urbanisation. With some land uses, vegetation becomes almost completely absent. However, in others, highly modified vegetation types (for example parks and gardens) may thrive. Often, between different land use types, there are small or even quite extensive areas where the previous natural or semi-natural vegetation may thrive, albeit often in highly modified form.

Several ecologists have recently put forward a 'new' approach to the ecology of highly modified areas under the heading of 'landscape ecology' (Naveh and Lieberman, 1984; Forman and Godron, 1986). The plant communities of highly modified landscapes are often more clearly defined and demarcated. For that reason, they are often simpler and easier to study. Ecologists and biogeographers have traditionally tended to work only within natural and semi-natural vegetation-dominated ecosystems. However, there is plenty of scope for research into the vegetation of these areas where vegetation is just one component of a land-use type. Sprugel (1991) presents an interesting discussion on human disturbance to the environment and the concept of 'natural' and 'modified' plant communities.

The intrinsic properties of plants themselves

Plant species strategies

The species that grow together to form a community have usually also proved that they can coexist with each other. Much early plant ecology was highly deterministic in nature and tended to assume that the combination of environmental factors occurring at a particular point in space were the only controls of the plant species growing there. Now it is generally agreed that the properties of the plants themselves are also very important. These properties are related to plant physiognomy and physiology and include the idea of **plant species strategies** and **plant population biology** (Grime, 1979; van der Maarel, 1984b; Harper, 1977; Silvertown, 1987; Tilman, 1988). Through **evolution**, each species has evolved a set of characteristics of physiognomy and physiology which improve its chances of

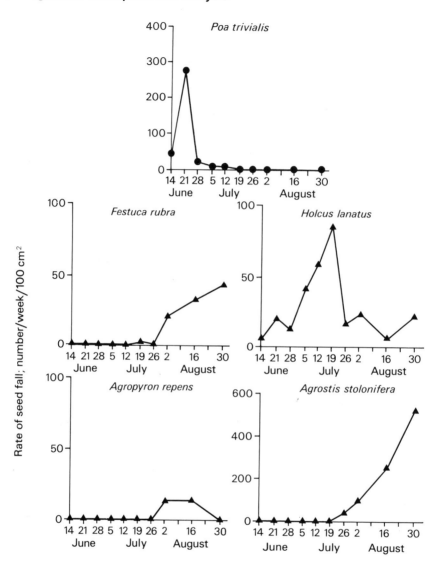

Figure 1.10 The time count of release of seeds of five grass species in a meadow (Redrawn from Mortimer, 1974)

survival in certain environments. Numerous components of the life history of a plant vary to produce the different strategies. Reproductive mechanisms (seed versus clonal growth or vegetative reproduction), growth form and rate, timing of germination, growth, flowering and fruiting, reproduction and death (phenology) are all examples of aspects of physiognomy and physiology that are different for each species. The timing of the single function of seed production shows how varied species strategies may be (Figure 1.10). The grasses have varied strategies so that in each case their seed has an optimal chance of germination success. No two species have evolved identical strategies for timing of seed dispersal. In other examples, however, similar strategies may enable some species to coexist in a certain location.

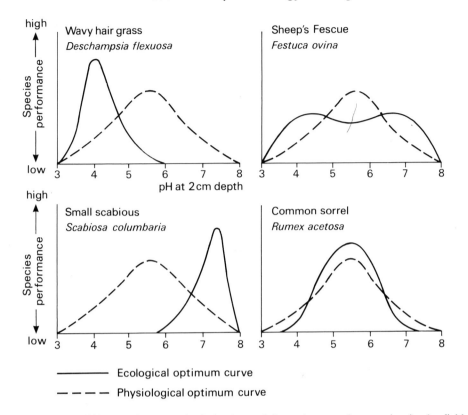

Figure 1.11   The difference between the behaviour of four plant species growing in the field (ecological optimum curve) and under non-competitive conditions in controlled laboratory conditions (physiological optimum curve). (From Rorison, 1974; with kind permission of Blackwell Scientific Publications)

Alternatively, they may cause them to come into direct competition with each other. It is important to be aware that some plant ecologists do not like the use of the species strategy concept and that it represents a controversial idea.

Competition and coexistence

Given the combination of environmental controls at a certain location, plant species may **compete** to occupy a position and **coexist** with other species which may already be there. Mechanisms of competition are many and varied. Growth form and physiognomy, growth rate, shading effects, deposition of litter, releasing of toxic substances from roots and in litter (allelopathy) and differences in reproductive strategy and output are all examples of such mechanisms. Competition can also occur between individuals of the same species (intra-specific competition) or between individuals of different species (inter-specific competition). Often species may survive within a community just by being there first and pre-empting space. Walker *et al.* (1989) developed this idea with their 'ecological field theory'. Good introductions and reviews of competition and coexistence are provided in Fitter (1987), Keddy (1989) and Law and Watkinson (1989).

The importance of **competition, interaction and coexistence** is shown in Figure 1.11. Rorison (1974) carried out experiments on the environmental tolerance of four plant species grown under both field and laboratory conditions along an environmental gradient of pH. The resulting abundance curves for the four species are very different both one from another and between field and laboratory. The abundance distribution along the gradient for the field situation (the ecological curve) is in all cases different and more limited than for the laboratory situation (the physiological curve). Rorison argued that one of the reasons for this difference was the removal of the interspecific competition found in the real world (field) from the laboratory experiment. In the absence of competitors, species were able to grow across a much wider range of pH.

The time factor in the study of vegetation

Concepts of succession and climax

Plant communities are dynamic in nature. In all parts of the world with seasonality of climate, the community changes with the seasons in terms of both presence of species and in their relative abundance. Over longer periods of time, community composition will often change according to the principles of **succession and climax**. Succession involves the immigration and extinction of species together with changes in their relative abundances. **Primary succession** occurs on bare ground where vegetation has not previously been found. Such sites are relatively small in extent on a world scale and are represented by landslips, new surfaces created by human disturbance or new volcanic islands emerging out of the ocean, as in the case of Krakatoa in the Western Pacific in 1883 and Surtsey in the mid-Atlantic, south of Iceland in 1963. Several successive groups of species may invade as **seral stages**. **Secondary successions** occur when an established vegetation cover is removed or modified to an earlier seral stage. The end product of succession is the **climax community**.

Whereas succession implies changes in species composition through time, the climax concept is based on the idea of relative stability. The nature of climax communities has always been controversial. Clements (1916) was responsible for the original ideas of succession and climax and it was he who first applied the terms 'sere', 'seral stage' and 'climax' to successional processes and plant communities. He also introduced the concept of the **climatic climax**, arguing that, ultimately, all communities in a given region come into equilibrium with the regional climate. For this reason, his ideas were known as the **monoclimax theory**. Tansley (1935) argued that observation of communities showed that, although in theory most vegetation might reach an equilibrium with climate, this was unrealistic because the required time periods were so long. Also factors other than climate such as soil (edaphic), topographic (geomorphic) and biotic (human and animal) factors could hold a community relatively stable for considerable periods of time. He thus proposed the **polyclimax theory**, whereby within a given region communities could be in relatively stable equilibrium with any one or a combination of the above factors.

The notion of climatic climax has also been questioned in the very large amount of research completed over the past 20 years into **Quaternary ecology and palaeoecology** (Birks and Birks, 1980). World climates are shown to have been increasingly dynamic over the Quaternary period; consequently the idea of stable climax communities existing for long periods of time must be called into question. Wholesale movements of world vegetation formation types are known to have occurred throughout the Quaternary at all latitudes with many smaller but still significant fluctuations superimposed on them.

Although concepts of long-term stable climaxes are now questioned, there is no

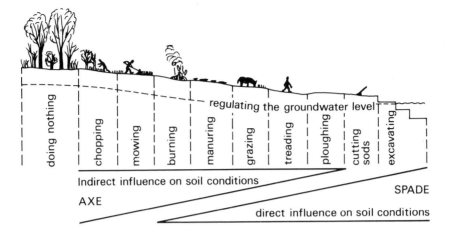

Figure 1.12 The spectrum of biotic controls on plant communities (Redrawn from Bakker (1979) with kind permission of Kluwer Academic Publishers, The Hague)

doubt that prior to the arrival of humans, extensive regional vegetation types existed over large tracts of the globe. The problem for present-day community ecologists is that only a comparatively small proportion of this former natural vegetation remains. Thus in the majority of situations, study of plant communities involves the examination of communities which are subject to some form of successional change or which are held as **plagio- or biotic climaxes** by human activities (burning, cutting, mowing, treading) or animals (grazing, treading, dunging). The nature of these controls are summarised in Figure 1.12, where the spectrum of biotic controls of vegetation is illustrated. Maintenance of many communities requires the continuance of those land management activities that created the communities in the first place. This is one of the most important reasons why active management of vegetation and communities for conservation is necessary. However, the application of management practice usually requires detailed research to provide information to form the basis of management planning. The importance of methods for vegetation description and analysis in this process has already been stressed. Re-examination of the contrasted case studies at the start of this chapter emphasises this point.

A number of valuable summaries and criticisms of successional and climax theory have been published in recent years: Drury and Nisbet (1973), Connell and Slatyer (1977), Noble and Slatyer (1980), Crawley (1986), Gibson and Brown (1986), Gray *et al.* (1987), Huston and Smith (1987), van der Maarel (1988a), Smith and Hutson (1989), Burrows (1990) and Luken (1990). The major debate at present centres on whether it is still valid to see succession as a community process where seral stages replace each other, or whether it is more realistic to concentrate on the level of the individual plant and plant species in terms of immigration, interaction and extinction.

Problems in quantitative plant ecology

There are a number of difficulties which face the student taking a course or carrying out research involving the use of vegetation description and analysis and quantitative plant ecology.

Temperate and tropical ecology

Most ecological theory and method has evolved through the activities of academics and researchers working in the temperate regions of the world (Northern Europe and North America). While many ideas and techniques transfer readily from these areas to other biomes and vegetation formation types, some do not. This problem is most acute in the tropical regions of the world and in particular in the tropical rain forest. Research over the past decade has shown that concepts of the plant community, succession and climax, diversity and species richness all require careful application and often drastic revision when dealing with tropical forest environments. It is also increasingly argued that many of the conceptual and practical problems in the ecology of non-tropical areas of the world may be solved by research carried out in the tropics. Describing tropical ecology, Deshmukh (1986) states (p. iv): 'Many notions that seemed well-founded a few years ago are actually based upon flimsy evidence (sometimes none at all) and many recent ideas contradict rather than confirm what seemed like established ideas.' This view is reinforced by various authors, notably Golley (1983), Myers (1984) and Proctor (1985; 1990). The same problems apply to methods of vegetation description and analysis. Standard methods such as the use of quadrats (see Chapter 2) and methods of classification and ordination (Chapters 5–8) may be totally inappropriate or require considerable modification in order to work in rain forest areas. Further introductory discussion of these issues is found in Whitmore (1990) and Mabberley (1991), while information on the rate of destruction of tropical forest in Amazonia is given in Eden (1990) and Fearnside (1990).

Species identification

Problems of plant identification present enormous difficulties to students as well as more advanced researchers. For most developed regions, published **floras** exist, although even these require careful training in use. However, if students are just beginning their studies, it is best to keep to relatively simple local situations, where the flora may be well known. Where reliable identification is important, samples may be taken to a local expert or museum for identification. Particularly where rarities are known to occur, plant specimens should never be collected; instead, good quality colour photographs should be taken.

Sampling designs

Sampling is discussed at length in Chapter 2. However, in general, this subject has received less attention from vegetation scientists than it should. For example, important differences exist in approaches to sampling where vegetation is simply being described and classified perhaps for phytosociological purposes (inductive approaches — Figure 1.4b); as opposed to research where plant species' response to environmental factors is of interest and is being examined by experimental methods involving hypothesis generation and testing which may require random sampling (deductive approaches — Figure 1.4a).

The apparent complexity of techniques for analysing vegetation data

Plant community data are **multivariate** in nature. Data are usually collected for samples and species giving a **raw data matrix** (Tables 3.1–3.4). To make sense of these data requires methods for **data reduction** — techniques that create order out of chaos and which will search for and summarise patterns of variation within the data. All the methods of classification and ordination described in Chapters 5–8 are of this type and are based on methods of matrix analysis. In detail, these are very difficult techniques for the average student to understand. The approach taken in this book is that what the methods can show ecologically and in their application are what matters. While it is important eventually to try to understand how the more complex methods work mathematically, for the student beginning studies in plant ecology the real understanding lies in the demonstration of ecological applications and in the ability to interpret results in the context of the ecological problems for which the data were collected in the first place. That is not to deny that the methods do not have their problems and limitations. Almost without exception, they do. However, once again, these can be appreciated most effectively at the stage of interpretation of results.

The choice of methods for classification and ordination

The mathematical emphasis which underlies the analysis of the site/species matrices which are typical of vegetation data has provided a large number of different methods for analysis, each with particular advantages and disadvantages. For both numerical classification and ordination, there are at least three or four methods in fairly common use and a choice of 15–20 methods overall in each category. This array of methods is bewildering to beginners and even to practiced researchers. Wherever a particular method has been in favour, it has been displaced by another a few years later. By common consensus in Britain and North America at present, TWINSPAN (Two-way indicator species analysis) is the most popular method for community classification, while DECORANA (Detrended correspondence analysis or reciprocal averaging) and CANOCO (Canonical correspondence analysis) are the most widely applied methods of ordination. DECORANA, for example, is already being criticised by many vegetation scientists and the improved approach, CANOCO (Canonical correspondence analysis), is being suggested as superior (ter Braak, 1986a; 1988a). Although comparative evaluations of different techniques have been applied on many occasions, objective tests of different methods on real-world data have never been clearly able to suggest one optimal method. This problem will be discussed again in Chapter 6.

Use of computers

Application of multivariate analysis requires the use of computers. Most of the techniques for analysis described in this book are only made possible by using appropriate computer programs. Students are sometimes apprehensive about using computers in their studies, but computer applications are increasing in all areas of science.

Until recently, most methods for multivariate analysis of vegetation data required a mainframe computer. With the advent of microcomputers virtually all methods are now available as packages for the IBM PC and 'clone' machines. Although still relatively expensive for the individual, most students will have access to such technology in the 1990s.

In Chapter 9, sources of IBM-PC compatible computer programs for the analysis of vegetation and environmental data using BASIC and FORTRAN are presented. Also a suite of programs is available on 3.5 inch disk from the authors. The availability of such user-friendly programs for PCs is very important.

Interpretation of results

Most textbooks on vegetation description and analysis are poor at describing and demonstrating the value of the results of community analyses. Students work extremely hard at project planning, quadrat description, species identification, data preparation and computer analysis only to find that they can make little sense of their results. Yet data collection and analysis are only means to an end. If the methods are taught in the context of a useful and valid ecological problem, then this difficulty over interpretation is usually resolved. For this reason, this book places a heavy emphasis on the case-study approach to demonstrate ecological applications.

The links between descriptive ecology (induction) and experimental ecology (deduction)

The bulk of published work using methods of vegetation description and analysis is descriptive and inductive in approach. The aim is to sample vegetation at numerous points in space and then to generalise them to plant communities through methods of phytosociology or to examine variation of plant species and communities in relation to environmental controls. However, most books on this subject describe methods for vegetation description and analysis as being for hypothesis generation. At the end of the description of plant communities and exploration of environmental relationships, the next logical step is to formulate hypotheses and experiments on selected aspects of plant-environment relationships, coexistence and competition between species or succession and vegetation dynamics. This problem has been highlighted by Austin (1985) and Kent and Ballard (1988) who cite only two clear examples in the literature where the progression from description through hypothesis testing to experimental studies has been carried out (Goldsmith, 1973a,b and Gittins, 1979).

Community versus individualistic plant ecology

Prior to 1975, the emphasis in plant ecology was largely at the community level and although, following Gleason's paper in 1926, the community and gradient analysis school in America did question the whole basis of plant communities, the community approach continued to dominate the whole philosophy of vegetation description and analysis. Since 1975 however, partly as a result of some of the problems described above, a number of ecologists led by Harper (1977), Grime (1979) and Silvertown (1987) have offered a very different approach to plant ecology based on the individual plant and plant species. The individual emphasis is expressed in two related concepts — plant species strategies and plant population biology, and an understanding of these subjects is now a vital part of quantitative plant ecology. This approach has produced many extremely valuable insights into the processes determining the ecology of individual plants and plant species, but the extent to which this work has assisted with broader-scale plant community ecology is still uncertain. Some authors appear to reject the classic approach to plant community ecology completely. In a book entitled 'Plant Ecology', Crawley (1986) dismisses the whole of quantitative plant community ecology in one paragraph. Clearly, this represents an extreme viewpoint, but it is an important one to introduce at the start of a text such as this.

The individualistic approach is sometimes described as an example of **reductionism**, where scientists assume that answers can be found by studying phenomena at lower and lower and more detailed levels (in the case of plant ecology to the individual species level and even below — Figure 1.5). The reductionists view is the opposite of the **holistic** view of community ecologists. **Holism** recognises that understanding can also be achieved by looking at plant communities and ecosystems as complete entities, where all species are examined together. Both approaches are valid. The critical point is that neither is more important than the other.

Whether traditional holistic quantitative plant ecologists will be able to find common ground with individualistic plant population biologists and species strategists is a vital question for the next decade. This theme is discussed again at the end of the book.

2

The description of vegetation in the field

Initial considerations

The methods of vegetation description employed in a particular project will depend on a number of factors:

(a) **the purpose of the survey** — the features and characteristics of the vegetation to be described will vary according to the overall aims and objectives.
(b) **the scale of the study** — very different description methods will be required for a survey covering many thousands of square kilometres compared to very detailed studies of a small area of perhaps a few hundred square metres.
(c) **the overall habitat type** — different techniques are necessary for different habitat types and growth forms. Thus a method suited to forests in the south-eastern United States will be totally inappropriate for a tropical grassland or savanna in West Africa or tropical rain forest in Brazil.
(d) **resources available** — finance, equipment, manpower and time. Vegetation description of any sizeable area will require sufficient resources of all four to be available.

Several other questions will also be important:

(1) Is it necessary or relevant to identify all of the plant species present?
(2) If identification is required, do published floras already exist?
(3) Are environmental data to be collected and if so, which variables should be measured and are the appropriate equipment and resources available to do so?
(4) What methods of vegetation data analysis will be used? Certain types of data analysis impose conditions and constraints on the form of data collected in the field.
(5) Has the dynamic nature of the vegetation been taken into account? The time of year when the recording is to take place is critical, and information on the successional and climax status of the vegetation may be extremely valuable.
(6) Where more than one recorder is being used, have the problems of ensuring consistency between different workers been addressed?
(7) The balance between speed and accuracy of the description method should be assessed. Often the time available for survey and the amount of data which can be collected will be limited by seasonality of climate or weather conditions and the resources available for survey. The maxim that the best and most suitable quality of data must be collected in the minimum or available time is a useful one to apply.

Finally, it is important to see field data collection as only part of the whole project. The collection of data should never be isolated from aims and objectives and methods of analysis (Jongman *et al.*, 1987).

Physiognomic and floristic data

Methods of vegetation description fall into two categories:

(a) **Physiognomic or structural** — where description is based on external morphology, life-form, stratification and size of the species present.
(b) **Floristic** — where the species present in the study are identified and their presence/absence or abundance is recorded.

Physiognomic and structural methods have been used primarily for the classification of vegetation at small scales (over large areas) such as world vegetation formations. Floristic analyses have usually been applied at the large scale (over small areas) particularly at the level of the plant community. However, physiognomic and structural methods have their uses at the community scale, for example in habitat classification (see below). Floristic data can only be collected in very generalised form at scales higher than the community level.

Techniques of vegetation description based on physiognomy and structure

Raunkaier's life-form classification

One of the most famous physiognomic techniques of vegetation description is the life-form method of Raunkaier (1934; 1937). He devised a **biological spectrum** based on the height above ground of the perennating buds of each species which are the parts of the plant from which growth commences in the next favourable growing season. His life-form classification is based on the assumption that species morphology is closely related to climatic controls, the humid tropics representing the most favourable conditions for species in terms of solar radiation, temperature and precipitation, while species in other environments, deficient in moisture, solar radiation or temperature show varying degrees of adaptation and response in the positioning of their perennating buds. The main groups of his classification are summarised in Table 2.1. Five main categories are recognised – each of which may be sub-divided further. The phanerophytes, for example, are divided on the basis of height into Nanophanerophytes (less than 2m in height), Microphanerophytes (2–8m), Mesophanerophytes (8–30m) and Megaphanerophytes (over 30m). Crypto-phytes are subdivided into geophytes (plants with rhizomes, bulbs or tubers), helophytes (plants with perennating organs in soil or mud under water and with aerial shoots above water) and hydrophytes (plants with perennating buds under water and with floating or submerged leaves). More detailed categories can be recognised based on the degree of protection of the buds and whether or not the species are deciduous. A full outline of the classification with some minor modifications is given in Mueller-Dombois and Ellenberg (1974).

Application of life-form classification involves categorising all the plants in a study area into Raunkaier groups. The results are usually then presented as a bar graph showing the percentages of species in each. A good example of the use of life-form spectra comes from the work of Hopkins (1965) and Richards (1952). Hopkins drew a spectrum for savanna vegetation in Nigeria (Figure 2.1), while Richards constructed one for tropical rain forest in British Guiana (Figure 2.2). The structure and life-forms of the two different environments are clearly shown with the phanerophytes dominant in the rain forest, where the climate is at an optimum for growth, whereas a much greater diversity of types and adaptations is found in Hopkins' savanna example, where there is a marked dry season. Hemicryptophytes,

Table 2.1   The major categories of the Raunkaier life-form classification system (1937)

Group 1 PHANEROPHYTES

Species with perennating buds emerging from aerial parts of the plant
   (a)   evergreen phanerophytes without bud scales
   (b)   evergreen phanerophytes with bud scales
   (c)   deciduous phanerophytes with bud scales

Each of these types may be classified according to height
Megaphanerophytes ($>$30m)
Mesophanerophytes (8–30m)
Microphanerophytes (2–8m)
Nanophanerophytes ($<$2m)

Group 2 CHAMAEPHYTES

Species with perennating buds borne on aerial parts close to the ground (below 2m)
They may be woody or herbaceous
   (a)   Suffruticose chamaephytes — after the main growth period, upper shoots die so that
        only the lower parts of the plant remain in the unfavourable period
   (b)   Passive chamaephytes — at the onset of adverse conditions, shoots weaken and fall
        to ground level, becoming procumbent. They thus get some protection from
        environmental stress
   (c)   Active chamaephytes — shoots are only produced along the ground and remain so in
        the unfavourable season
   (d)   Cushion chamaephytes — a modification of passive types where shoots are arranged
        so close together that they cannot fall over and the close packing of all shoots forms
        a cushion

Group 3 HEMICRYPTOPHYTES

All above-ground parts of the plant die back in unfavourable conditions and buds are borne
at ground level
   (a)   Protohemicryptophytes — leaves become better developed up the stem of the plant.
        Poorly developed leaves protect the bud in early stages of growth
   (b)   Partial rosette plants — the developed leaves form a rosette at the base of the plant
        in the first year of growth. The following year an elongated aerial shoot may form
   (c)   Rosette plants — leaves are restricted to a basal rosette with an elongated aerial
        shoot, which is exclusively flower-bearing

Group 4 CRYPTOPHYTES

Plant species with buds or shoot apices which survive the unfavourable period below ground
or under water
   (a)   Geophytes — plants with subterranean organs such as bulbs, rhizomes and tubers
        from which shoots emerge in the next growing season
   (b)   Helophytes — plants with their perennating buds in soil or mud below water and
        which produce shoots reaching above water
   (c)   Hydrophytes — species with buds which lie underwater and survive the unfavourable
        season by budding from rhizomes under water or from detached vegetative buds
        which sink to the bottom

Group 5 THEROPHYTES

Plants which survive the unfavourable period as seeds. Species are thus annuals and
complete their life cycle from seed to seed in the favourable summer months.

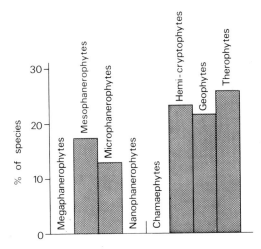

Figure 2.1 Savanna Raunkaier life-form spectrum, Olokomeji Forest Reserve, Nigeria (Hopkins, 1965; redrawn with kind permission of Heinemann Publishing)

geophytes and therophytes constitute around 70 per cent of the species present in the savanna. A more recent example demonstrating close relationships between Raunkaier life-form spectra and climatic gradients in Israel is provided in Danin and Orshan (1990).

Although life-form spectra have traditionally been used to describe world vegetation-formation types, they can profitably be used in certain situations at the community level. If locally severe environments are to be compared, life-form data can provide more information than floristics. An example is in the study of colonisation of derelict land and spoil heaps in industrial and urban areas which have been highly modified by human activity. Here the severity of the micro-environment for plant growth is reflected in the wide range of different life-form types which occur (see case studies).

The structural-physiognomic classification schemes of Dansereau, Küchler and Fosberg

These three systems are grouped together because of similarities in their concept. However, they each have differences in approach.

Dansereau's method (1951, 1957)

Dansereau devised a description method based on six sets of criteria: (a) plant life-form; (b) plant size; (c) cover; (d) plant function (deciduous or evergreen); (e) leaf size and shape and (f) leaf texture (Figure 2.3). No detailed information on the species present other than the dominants is required, allowing the technique to be used rapidly by inexperienced workers over large areas. Each of the dominant plants found within an area is described using first the symbols, shapes and shading shown in Figure 2.3 to produce a symbolic profile diagram and second the lettering system associated with the subdivisions in each category. The Dansereau method, although

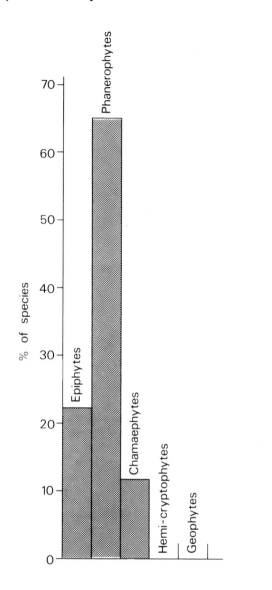

Figure 2.2 Raunkaier life-form spectrum for tropical rain forest, British Guiana (Richards, 1952; with kind permission of Cambridge University Press)

logical and easy to use, has not been widely applied, despite its world-wide scope and potential. Part of the reason for this may be the rather abstract symbols of the profile diagrams plus evidence that when used the method often requires modification to suit particular circumstances, as demonstrated in the case study at the end of this chapter.

Küchler's method (1967)

Küchler's method is hierarchical in nature, beginning with a subdivision of vegetation into two broad categories — woody and herbaceous. Within the first category,

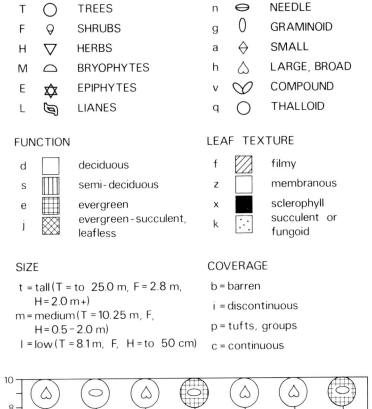

LIFE FORM

T	TREES
F	SHRUBS
H	HERBS
M	BRYOPHYTES
E	EPIPHYTES
L	LIANES

LEAF SHAPE AND SIZE

n	NEEDLE
g	GRAMINOID
a	SMALL
h	LARGE, BROAD
v	COMPOUND
q	THALLOID

FUNCTION

d	deciduous
s	semi-deciduous
e	evergreen
j	evergreen-succulent, leafless

LEAF TEXTURE

f	filmy
z	membranous
x	sclerophyll
k	succulent or fungoid

SIZE

t = tall (T = to 25.0 m, F = 2.8 m, H = 2.0 m+)
m = medium (T = 10.25 m, F, H = 0.5 – 2.0 m)
l = low (T = 8.1 m, F, H = to 50 cm)

COVERAGE

b = barren
i = discontinuous
p = tufts, groups
c = continuous

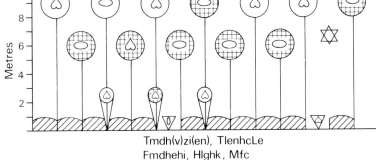

Tmdh(v)zi(en), TlenhcLe
Fmdhehi, Hlghk, Mfc

Figure 2.3 The symbols used in the construction of profile diagrams to denote the six major classes of the Dansereau structural vegetation description method (Dansereau, 1951) and an example of a Dansereau profile diagram from the Killarney *Quercus-Taxus* woodlands at Muckross (Shimwell, 1971; redrawn with kind permission of Sidgewick and Jackson)

Table 2.2 A summary of Küchler's method for structural description of vegetation (after Küchler, 1967)

LIFE-FORM CATEGORIES

BASIC LIFE FORMS		SPECIAL LIFE FORMS	
Woody Plants		Climbers (lianas)	C
Broadleaf evergreen	B	Stem succulents	K
Broadleaf deciduous	D	Tuft plants	T
Needleleaf evergreen	E	Bamboos	V
Needleleaf deciduous	N	Epiphytes	X
Aphyllous	O		
Semideciduous (B + D)	S	LEAF CHARACTERISTICS	
Mixed (D + E)	M	hard (sclerophyll)	h
Herbaceous plants		soft	w
Graminoids	G	succulent	k
Forbs	H	large ($>400cm^2$)	l
Lichens, mosses	L	small ($<4cm^2$)	s

STRUCTURAL CATEGORIES

Height (Stratification)	Coverage
8 = >35.0 metres	c = continuous (>75%)
7 = 20.0–35.0 metres	i = interrupted (50–75%)
6 = 10.0–20.0 metres	p = parklike, in patches (25–50%)
5 = 5.0–10.0 metres	r = rare (6–25%)
4 = 2.0–5.0 metres	b = barely present, sporadic (1–5%)
3 = 0.5–2.0 metres	a = almost absent, extremely scarce
2 = 0.1–0.5 metres	(<1%)
1 = <0.1 metres	

seven woody types are distinguished, while in the second, three herb classes are recognised (Table 2.2). Within these 10 physiognomic groups, further differentiation of vegetation may be achieved by the recording of whether or not plants show certain specialised life-forms and leaf characteristics and on the basis of height and cover. Results are presented as formulae, using letters and numbers. The system is also intended as a basis for vegetation mapping.

Fosberg's method (1961)

Fosberg's technique is important in that he presented his method as a means of describing vegetation within the International Biological Program (IBP), which was established during the 1960s with the aim of quantifying the primary production and energy budgets of all major world ecosystem types (Newbould, 1965; Peterken, 1967). The purpose of the Fosberg approach is to provide a classification of vegetation at the world scale. The criteria used had to be structural because any floristic data would have been far too detailed and difficult to obtain. Also comparisons between differing parts of the world based on floristics would have been impossible and of little relevance.

The method is summarised in Figure 2.4. The classification is hierarchical and starts with a subdivision of vegetation into categories of open, closed or sparse. Within each of these, a series of 31 formation classes is recognised at a second level

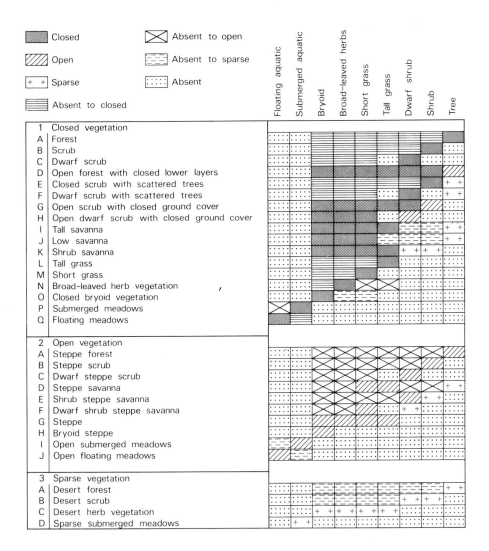

Figure 2.4 A summary of the Fosberg structural system (Peterken, 1967; redrawn with kind permission of Blackwell Scientific Publications)

(Figure 2.4). Height and continuity of the vegetation are the main criteria used. Further subdivision within each of the 31 classes at a third level is possible using plant function (whether the dominant foliage is evergreen or deciduous). Beyond that, a fourth level of division is made using the properties of leaf texture, leaf size and growth form of the dominant species. At this fourth level, mapping of units in the field can be carried out with each category being called a formation group.

Sampling for physiognomic and structural description

The basis of **sampling** for physiognomic and structural description is rarely discussed and this is a major criticism of the whole approach. A physiognomic or structural description of a vegetation type is usually made in an area which the researcher considers to be **typical or representative**. While this is probably the only realistic way in which these methods may be applied, it raises some interesting questions about what is representative. Choosing a representative area implies a degree of familiarity with the vegetation of the region under study and students, in particular, may not possess sufficient breadth of knowledge or experience. The element of bias and subjectivity of sample selection is thus always present. Nevertheless, all physiognomic and structural description methods, serve as a practical means of collecting and organising vegetation data for general purposes at the regional, continental or world scale.

As an example, Goldsmith (1974) carried out an assessment of the Fosberg method as part of a rapid survey of vegetation for conservation purposes on Majorca in the Mediterranean. He concluded that Fosberg's method offered a quick and repeatable approach to vegetation description, providing a useful framework for an ecological inventory of the island. However, he also found that attempts at vegetation mapping using Fosberg's classification groups resulted in some difficulties, largely because mapping on an extensive scale demanded the use of aerial photography or a suitable range of hill-top vantage points in order to show accurately the extent of the Fosberg vegetation types.

Structural and physiognomic methods have been shown to be of considerable value in the tropics and particularly in tropical forest environments where floristic data are both difficult to obtain and to analyse. The value of structural data in Australian rain forest environments has been shown by Webb *et al.* (1970); Webb (1978) and Webb *et al.* (1976); while Werger and Sprangers (1982) have compared structural and floristic descriptions in dry tropical communities in India.

Habitat classification (Elton and Miller, 1954; Elton, 1966)

The structural approach to vegetation description is also well illustrated by the habitat description and classification system devised by Elton and Miller (1954). The method was originally proposed as a rapid method of habitat survey for zoologists but has subsequently been used for more general surveys of ecosystems and habitats for rural planning purposes and as part of techniques for ecological evaluation. The main assumption of the method is that structural complexity in the vegetation, as represented by the degree of layering or stratification present, can be equated with habitat diversity, which will in turn encourage animal and to a lesser extent plant species diversity. Three major **habitat systems** were originally recognised: terrestrial, aquatic and the aquatic-terrestrial transition (Figure 2.5). The **terrestrial habitat system** is of greatest interest and is divided into four categories: open ground, field layer, scrub and woodland on the basis of the height of the dominant species.

When combined with the measurement of the area of each habitat type recognised, the technique provides a rapid method of assessing ecosystem structure and diversity. Where all four layers are present, as for example at the edge of a temperate deciduous woodland or in open woodland, habitat and species diversity will be high. Areas of open grassland will have lower habitat diversity and thus less animal and sometimes plant species diversity.

A further refinement of the method involves counting the number of plant species within a sample quadrat of each habitat layer recognised in an area. The information on habitat structure, area of each type and plant species diversity may then be combined in various ways to enable comparison of differing vegetation types. Examples of the application of this approach are demonstrated in the case study at the end of this chapter.

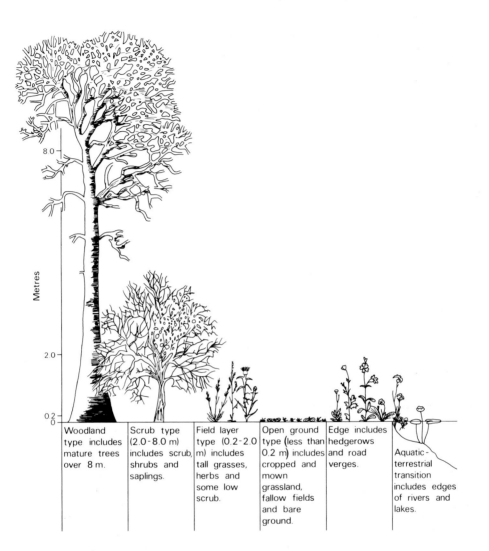

Figure 2.5 The habitat description approach of Elton and Miller (1954)

Linear features, such as hedgerows and road verges, may be counted as 'edge' types and their length between blocks of vegetation or their total length within a convenient size of grid can be measured. Habitats adjacent to water bodies are assigned to the separate aquatic-terrestrial transition category.

Habitat classification has the advantage of being very simple to apply and can be very rapid in its application. Despite its simplicity, it is a very efficient means of collecting information on ecosystem structure and function using vegetation as a general index. The technique is also easily adapted and although originally devised for use in temperate forest regions is capable of modification and refinement for use in many different world vegetation types, depending on the purpose for which it is used and the environment under study.

Methods of vegetation description based on floristics

The nature and problems of floristic data

The description of vegetation on the basis of **floristics** means that individual species within the community being studied must be identified. Floristic description immediately raises three problems. The first is the identification of plant species. The second is whether or not to collect data on the abundance of each species and if so, how to measure abundance; the third problem is the problem of sampling.

Species identification

Most species have both **Latin** and **English**, local or common names. Often the common names are easiest to remember, particularly if they relate to some obvious characteristic of the plant. Unfortunately, however, common names vary from region to region and some plants have several such names. For example, *Arum maculatum*, a frequent plant of hedges and woodlands in England, is known as both 'lords and ladies' and 'cuckoo-pint'. If vegetation description is to be truly scientific, then the use of the **Latin binomial** is essential. In the case of *Arum maculatum*, *Arum* refers to the genus and *maculatum* to the species. Where plants are unknown to the recorder, the correct approach is to use a flora. For Britain, the recommended floras are Clapham *et al.* (1987), which is also produced in a smaller excursion edition (Clapham *et al.*, 1981) and Stace (1991). For Europe, the *Flora Europaea* extends to 5 volumes. Many smaller and more manageable floras exist for all parts of the world and sources of these are summarised in Table 2.3.

Table 2.3 Bibliographies of floras for Britain and the rest of the world

---

Britain and Europe

Huckin, D. (1981)	*Flora and Fauna of Localities* — a bibliography of books and articles dealing with the plant and animal life of specific countries and regions of the world: Vol. 1. Great Britain and Europe. Biggleswade Press, Biggleswade.
Kerrich, G.J. (1978)	*Key Works to the Fauna and Flora of the British Isles and Northwestern Europe.* The Systematics Association, Special Volume 9, 169–75. Academic Press, London.

Tutin, T.G., Heywood, V.H., Burges, N.A., Moore, D.N., Valentine, D.H., Walters, S.M. and Webb, D.A. (1964–80) *Flora Europaea.*

Vol. 1. Lycopodiaceae to Plantanaceae
Vol. 2. Rosaceae to Umbelliferae
Vol. 3. Diapensiaceae to Myoporaceae
Vol. 4. Plantaginaceae to Compositae (and Rubiaceae)
Vol. 5. Alismataceae to Orchidaceae
Cambridge University Press, Cambridge.

Rest of the World

Frodin, D.G. (1985)	*Guide to standard floras of the world.* Cambridge University Press, Cambridge.
Huckin, D. (1981)	*Flora and Fauna of Localities* — a bibliography of books and articles dealing with the plant and animal life of specific countries and regions of the world: Vol. 2. Africa, America, Asia, Australia and the USSR. Biggleswade Press, Biggleswade.

Categories

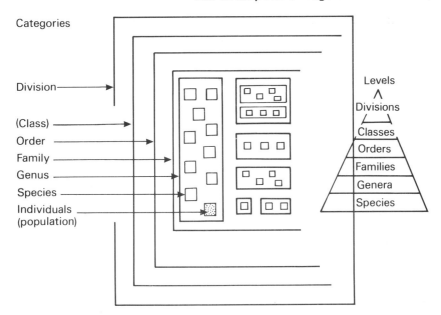

Figure 2.6 The hierarchy of botanical nomenclature (after Tivy, 1982; with kind permission of Longman Publishing and Professor V.H. Heywood)

In many other parts of the world, no systematic description of plant species may be published or only limited local floras may be available for selected areas. In such situations, it may become necessary to compile a flora prior to or during work on vegetation description. Local help is usually invaluable in such circumstances. Sometimes, however, the time-consuming nature of such work may be sufficient to cause a researcher to abandon a project based on floristics or to consider seriously the use of structural and physiognomic methods, identifying only dominant species. As always, the decisions depend on the aims and objectives of the project and the resources and expertise available.

Floras

The classification systems used in botanical floras employ a **hierarchical dichotomous key** for plant identification and use of a flora usually takes much practice to perfect. Use of the key is based on the presence and absence of distinctive morphological and physiological properties of plants and the hierarchy of taxonomic nomenclature is shown in Figure 2.6. There is an **International Code for Botanical Nomenclature** which applies the following rules for plant identification (Gilbertson *et al.*, 1985):

(1) One taxon — a single taxonomic unit of any scale (for example a variety, a species, a genus) may have only one valid name.
(2) Two or more taxa may not have the same name.
(3) In cases not in agreement with the first two points, the decision on which name is to be used is based upon priority of publication.
(4) The identity of the plant to which the name has been given is according to the type specimen designated by the author as representative.

In practice, the application of taxonomic rules is complex and is best left to the experts. Names are, for example, changed from time to time. This occurred for a number of grass species in Britain when the revised *Excursion Flora* was published in 1981 (Clapham *et al*.). One widely known example of such a change is the common bent, a grass which was formerly known as *Agrostis tenuis* but which is now called *Agrostis capillaris* for reasons which probably lie in point 3 above. The name *capillaris* was probably given to the species elsewhere at an earlier date, but the relationship between the two had only recently been recognised. These problems of nomenclature are shown elsewhere in this book with the species known as deer sedge. Some authors have used the Latin name *Trichophorum cespitosum*, while others use *Scirpus caespitosus*. Both Clapham *et al*. (1987) and Stace (1991) indicate that *Trichophorum cespitosum* is correct at the present time.

Separate floras are often compiled for different groups of plants, for example grasses and sedges. In many countries, **pictorial floras** now exist, which make identification very much easier for the student. Such floras are usually structured by families with their own keys, often based on flower structure and colour. Beginners, however, often use them with a random search procedure at first and this quickly enables them to build up an understanding of families and genera and their relationships, so that the working of the key becomes clear.

Quadrats — the sampling frame for recording plant species

Quadrats

The usual means of sampling vegetation for floristic description is the **quadrat**. Traditionally, quadrats are square, although rectangular and even circular quadrats have been used. The purpose of a quadrat is to establish a standard area for examining the vegetation. **Quadrat size** is very important and will vary from one type of vegetation to another. Methods have been devised to estimate the optimum size of quadrat for a particular community type and are based on the concepts of **minimal area** and **species-area curves** (Cain, 1938). The most frequently used approach is derived from the methodology of the Braun–Blanquet school of vegetation classification and phytosociology. It involves starting with the area which is considered to be the smallest feasible quadrat size — usually just containing one or two species — then doubling the size of the quadrat; counting the number of species present again; doubling the size of quadrat again; counting the number of species and so on until no new species are recorded at a doubling of quadrat size. The resulting graph of species numbers against quadrat size is known as the minimal area or species-area curve (Figure 2.7). If a homogeneous area of vegetation is taken, then the curve of species numbers levels off and the point at which this occurs is taken as the minimal area for sampling that community. The recommended quadrat size should then be a little larger than the minimal area.

In practice, minimal area curves are often less easy to define and much confusion surrounds their use. The reason is that the method is really only suitable as part of the overall Braun–Blanquet approach to subjective vegetation classification (Chapter 7), where a vegetation sample or **relevé**, as it is known, is deliberately chosen as being a uniform and representative sample of the plant community being described. The method only works well if the vegetation being sampled is truly homogeneous and is not an edge or ecotone between two vegetation types. If non-uniform areas are taken, then the minimal area curve may level off within one locally homogeneous area but then start to rise again as the doubling of quadrat size starts to sample a different community type or ecotone between community types. The selection of homogeneous areas also presupposes a certain amount of

a) Progressive doubling of quadrat size

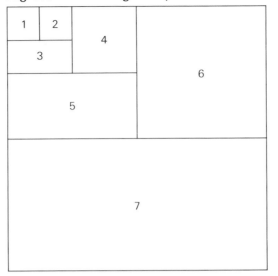

b) The resulting species – area curve

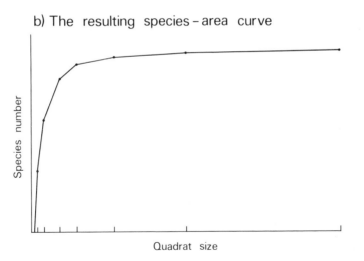

Figure 2.7 Progressive doubling of quadrat size for minimal area and derivation of species-area curves

knowledge about the vegetation being studied and students, in particular, are unlikely to have this knowledge. It also assumes that community types can be clearly defined and does not make it clear how to treat transitional zones between communities. Despite these comments, minimal area curves can still help in the establishment of quadrat size, provided that the points above are realised.

Table 2.4 Suggested quadrat sizes for certain vegetation types

Vegetation type	Quadrat size
Bryophyte and lichen communities	0.5m × 0.5m
Grasslands, dwarf heaths	1m × 1m–2m × 2m
Shrubby heaths, tall herbs and grassland communities	2m × 2m–4m × 4m
Scrub, woodland shrubs	10m × 10m
Woodland canopies	20m × 20m–50m × 50m (or use plotless sampling)

As an alternative to minimal area, general guidelines can be laid down for the optimum size of quadrat for selected vegetation types. These are presented in Table 2.4, but they should not be taken as universally appropriate for every situation and be treated with particular caution in tropical environments.

A conventional square quadrat usually has a wooden or metal frame and is often subdivided with lengths of wire or string (Figure 2.8). This subdivision increases the accuracy of recording when using certain measures of abundance, since each sub-unit of the quadrat can be examined separately. The most common subdivisions are into tenths, fifths or quarters of each side of the quadrat, giving 100, 25 or 16 sub-units respectively (Plate 2.1).

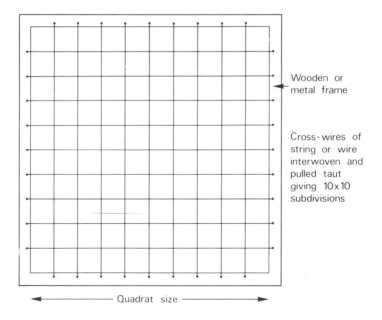

Figure 2.8 label: Wooden or metal frame

Cross-wires of string or wire interwoven and pulled taut giving 10×10 subdivisions

Quadrat size

Figure 2.8 A conventional square quadrat with subdivisions

Vegetation pattern and quadrat size

The concept of **pattern** in vegetation refers to the manner in which the individuals of a given species are distributed within a plant community. It is also dependent on the size of the species in relation to the size of the quadrat. Species can exhibit

Plate 2.1   Using a sub-divided quadrat (5 × 5 subdivisions) (photo: P. Coker)

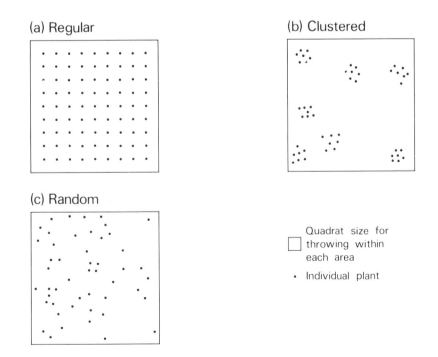

Figure 2.9   Pattern in vegetation: (a) regular (b) clustered (c) random

clustered, random or regular pattern (Figure 2.9). Depending on quadrat size and whether the species is distributed in a clustered, random or regular fashion, different results will be obtained by throwing the small quadrat in Figure 2.9 into each of the larger square areas, even though those square areas all have the same numbers of individual plants. If the quadrat size is changed to the size of the larger square areas, there would, however, be no observable differences in the numbers of individuals and the effects of pattern would be masked. Using the small quadrat in Figure 2.9, the regularly distributed species (a) has a much greater chance of being sampled than the clustered species (b), while the randomly distributed species (c) would be somewhere inbetween. Kershaw and Looney (1985) and Causton (1988) make the point that more plant species tend towards a clustered rather than a random or regular pattern.

A useful laboratory exercise to demonstrate the effects of different quadrat sizes in relation to variations in plant size and the effects of pattern has been devised by Williams *et al.* (1979). Although the problem of pattern is well known to ecologists, the recognition and solution of the problem that every species of plant in a community has a different pattern and size is virtually impossible, since, theoretically one would need a different size of quadrat for every species. The problem is reduced, however, if the largest size of quadrat to suit all species in a particular community or survey is chosen. The other alternative is to use **nested quadrats**, with different sizes of quadrats for the different-sized components of the vegetation. For example, in a wood, $1m^2$ for ground flora, $5m^2$ for shrubs, saplings and scrub and $20m^2$ for trees. If such different sized quadrats are used, the data will have to be analysed separately for each quadrat size and component of the vegetation.

Recently, interest has been shown in **fractal geometry** as a means of examining problems of quadrat size and vegetation pattern (Palmer, 1988). Preliminary results indicate that fractal analysis will have an important role in the future.

Permanent quadrats

Studies of vegetation dynamics, succession or the effects of management practices on vegetation may require the establishment of **permanent quadrats**, where the species may be recorded at intervals over a long period of time. Corners of quadrats are usually marked out with coloured pegs and accurate survey, including plane table methods (Gilbertson *et al.*, 1985) may be useful if individuals of a species are being studied. Individuals may also be tagged with a marker within the permanent quadrat if, for example, competition within and between species or species population dynamics are being studied. A very useful review of permanent quadrats and the philosophy behind their use is provided in Austin (1981).

Measurement of species abundance

Presence/absence or qualitative data

In the measurement of abundance, a very important distinction is made between **presence/absence** and **abundance data**. As the name suggests, with presence/absence data, only the occurrence of a species within a quadrat is noted, and there is no measurement of the amount of each species. Hence the data are qualitative. The decision on whether to record abundance values or not depends on the aim of the project and the time and resources available for recording. Presence/absence methods are extremely rapid to use, and the results represent the simplest form of vegetation data. All the following measures of abundance implicitly include the assessment of presence/absence and can be reduced to this form if necessary. The abundance data contain additional information.

Abundance measures or quantitative data

Abundance measures can be categorised into two types:

(1) Subjective — these are estimated by eye and thus values will vary from one recorder to another.
(2) Objective — where more accurate and precise measures are taken which should not vary from one recorder to another.

Subjective measures

FREQUENCY SYMBOLS

The most subjective and descriptive method of vegetation description is that of **frequency symbols**. The following DAFOR scale is used to characterise species: dominant (d), abundant (a), frequent (f), occasional (o) and rare (r), with the prefixes 'locally' and 'very' where appropriate. This scale is used most often where an area of grassland or a small wood is being studied as a whole site rather than by using a series of quadrats. Another example is the ACFOR scale: abundant, common, frequent, occasional and rare.

COVER ESTIMATED BY EYE

Cover is defined as the area of ground within a quadrat which is occupied by the above-ground parts of each species when viewed from above. Cover is usually estimated visually as a percentage, but stratification or multiple layering of vegetation will often result in total cover values of well over 100 per cent. A number of recording scales are available. Some workers use values in the range 0–100 per cent at 5 per cent or 10 per cent intervals giving even-sized classes. Others use the **Domin** scale and the **Braun-Blanquet** scale, where the range 0–100 per cent is partitioned into five or 10 classes with smaller graduations nearer to the bottom of the scale (Table 2.5). Since estimation is done by eye, there is certain to be a degree of error

Table 2.5 The Braun-Blanquet and Domin cover scales

Value	Braun-Blanquet	Domin
+	Less than 1% cover	A single individual. No measurable cover
1	1–5% cover	1–2 individuals. No measurable cover. Individuals with normal vigour
2	6–25% cover	Several individuals but less than 1% cover
3	26–50% cover	1–4% cover
4	51–75% cover	4–10% cover
5	76–100% cover	11–25% cover
6		26–33% cover
7		34–50% cover
8		51–75% cover
9		76–90% cover
10		91–100% cover

Plate 2.2 Estimation of percentage cover by eye using a 0.5 × 0.5m
quadrat (photo: M. Kent)

in recording. A recorder will tend to over-estimate species which are in flower,
attractive and conspicuous and which he knows and to under-estimate others.
Nevertheless, the method is rapid to use, and the problems of subjectivity may have
been over-emphasised (see below). Use of cover estimates is essential when describing grass species where individuals of a species cannot be identified (Plate 2.2).

Objective measures

DENSITY

Density is a count of the numbers of individuals of a species within the quadrat. Subdivided quadrats are useful to increase accuracy in counting and the method presupposes that individuals of a species can be recognised. Most **monocarpic** species (largely annuals and biennials) grow as identifiable individuals, while herbaceous **polycarpic** species (perennials) possess a more complex growth form, the best examples being many species of grass. Species with rhizomes can cause similar problems since they will often be connected underground. Density is most frequently measured in studies of herb species or tree saplings and is rarely used in the description of whole communities. Good examples are population studies of rare orchids in chalk grassland nature reserves or censuses of tree saplings in woodlands. The method is time-consuming and is entirely dependent on quadrat size, which must be kept constant because the number of individuals counted is entirely a function of the size of area examined. Density is also affected by pattern of species in relation to quadrat size, as discussed above. Last, comparison of densities for different-sized species using the same quadrat size is nonsensical because many more individuals of a small species can grow within the same area compared to a large species.

FREQUENCY

Frequency is defined as the probability or chance of finding a species in a given sample area or quadrat. Recording frequency involves either throwing a series of quadrats within a local area and recording the presence and absence of species in each (local frequency) or by using sub-units of a quadrat and recording presence/absence of a species in each sub-unit. If a relatively small area of vegetation is taken and 100 quadrats are thrown within it, then the proportion of quadrats which contain a species is the local frequency. Thus if a species occurs in 63 of those quadrats it has a frequency of 63 per cent.

The most effective use of frequency is, however, where a quadrat is subdivided into 10×10 units, as in Figure 2.8, and presence/absence is recorded in each of the 100 contiguous sub-units to give a percentage score. This technique probably provides the most accurate data of the various measures discussed so far but is time-consuming to use. Frequency is also dependent on quadrat size, plant size and patterning in the vegetation.

COVER ESTIMATION USING A COVER PIN FRAME

Objective measurement of cover can be achieved by using a **cover pin frame** (Figure 2.10), which consists of a row of 10 pins in a wooden frame, with the length of the frame equal to one side of the quadrat. The pins are lowered vertically on to the ground and the species which they touch are recorded. This procedure is repeated 10 times at 10 equal intervals across the quadrat to give a total of 100 pin samples (Plate 2.3). The pin diameter should be as small as possible, since tip size has been shown to be correlated with exaggeration of cover values. If more than one species is touched as the pin is lowered, more than one 'hit' is recorded which can result once again in a total cover value for a quadrat of over 100 per cent. The method is lengthy to use and is affected by patterning in the vegetation. It is also difficult to use in tall-growing or shrubby vegetation and is best suited to grassland.

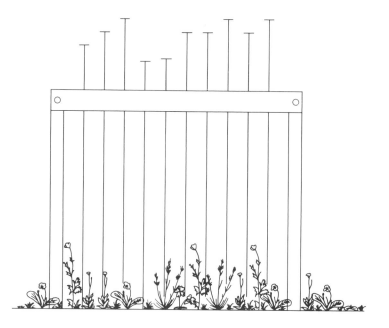

Figure 2.10 A cover pin frame

Plate 2.3 Measuring cover with a pin frame (photo: P. Coker)

BIOMASS, YIELD AND PERFORMANCE

Experiments to determine primary productivity and studies of competition between plant species often require information on the quantities of plant tissue present in a quadrat. **Biomass or standing crop** is the amount of plant material present in a quadrat at a given time. The figure is obtained by clipping the above-ground vegetation using shears and secateurs and sorting the resulting crop into species. The dry weight of each species is then found by drying them for 24 hours at 105°C and weighing the dried plant material. The result is expressed in g/unit area.

If samples of vegetation are taken from adjacent plots at different times, any increase in biomass between the first and second harvest is known as the **yield** or **productivity** of the species and the community. Care must be taken in the sampling of such adjacent plots where the assumption is made that they are representative of the same community type. With both yield and biomass estimates, the point of clipping at the ground surface must be carefully chosen and standardised. Ideally **root biomass** and yield should also be measured, but this can only be done with great difficulty. No completely satisfactory method of assessing root biomass exists, although methods such as coring followed by estimation of the volume of roots in the core and techniques involving the washing of roots have been tried. All methods are extremely time-consuming.

Biomass data are only occasionally used directly as a measure of abundance because of the time and effort involved in gathering the data. Also, in order to eliminate the possible over-riding effect of between-quadrat biomass differences, it may be necessary to transform data by expressing the weight of each species as a percentage of the total biomass in the quadrat. This measure is known as **percentage biomass**.

Assessment of **performance** involves the measurement of some relevant part of a plant to provide an index of growth rate or vigour. Typical measurements are leaf size, length and shape, plant height and flower or fruit characteristics, the latter providing information about reproductive allocation.

Root and shoot frequency and presence/absence data

In any survey, it must be clearly stated whether species are recorded only if they are rooted in the quadrat, or if any of their above-ground parts extend into the quadrat, even though they may be rooted outside. This problem has traditionally been described for frequency measures but clearly is equally applicable to most other measures and particularly presence/absence data.

Unbounded, bounded and partially bounded data

Smartt *et al.* (1974; 1976) made a useful division of abundance measures into:

(a) unbounded measures — those with no fixed upper limit — biomass, density.
(b) partially bounded measures — where there is a limit in that the quadrat area is equated with 100 per cent, but multiple layers can cause abundance values to exceed this figure — percentage cover.
(c) bounded measures — where the upper limit is fixed — frequency recorded in a subdivided quadrat.

Comparisons of the effectiveness of different abundance measures

Smartt *et al.* (1974; 1976) investigated the properties of different measures of species abundance by recording the same vegetation using a range of different methods. They concluded that although the measure chosen for a given project should depend primarily on the purpose for which the data are being collected, measures of cover,

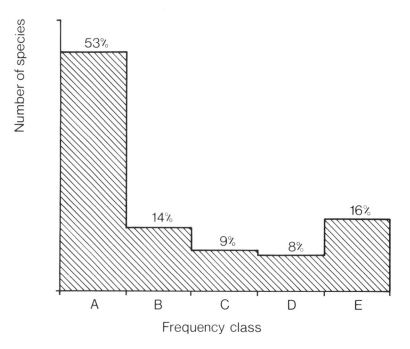

Figure 2.11 The Raunkaier J curve (Redrawn from Kershaw and Looney, 1985; with kind permission of Edward Arnold Publishers/Hodder & Stoughton)

even assessed subjectively, were a good approximation to superficially more accurate measures such as frequency and biomass. Another reason why subject cover scales often provide rapid, yet relatively accurate description lies in the evidence of **Raunkaier's 'J' curve** (Figure 2.11) and the Law of Frequencies (Raunkaier, 1928). This law states that the numbers of species of a community in the five percentage frequency classes 0–20, 21–40, 41–60, 61–80, 81–100 are distributed:

$$A > B > C \gtreqless D < E$$

The graph shows that, as might be expected, the greatest number of species occur with low frequency (A (0–20 per cent); = 53 per cent) but the next highest percentage of species is for species with high frequency (E (81–100 per cent); = 16 per cent). These results have been interpreted in many ways and have to be treated with caution because of the usual factors of quadrat size, plant size and pattern. However, what they also suggest is that a majority of species occur with either very low or relatively high frequency. The same principle probably applies to cover in that many (but not all) communities contain several dominant species with high frequency and cover, together with a number of associated species with low frequency and cover. Thus, usually, minor variations in recording of percentage frequency and cover do not greatly affect results of subsequent community analyses.

Line-intercept methods

In environments where the vegetation is sparse, such as in some Mediterranean,

Plate 2.4 The line-intercept method being used to describe semi-arid Mediterranean vegetation in Murcia, south-west Spain (photo: M. Kent)

semi-arid, and hot and cold desert environments, the normal square quadrat may not be suitable for sampling. Instead **line-intercept methods** may be used. Typically a 10m-length of tape is laid out and all species intercepting or touching the line are recorded (Plate 2.4). A form of percentage-cover data can be collected by measuring the length of line touching each species present or over bare ground or rock. The method can also be extended, so that all species within a certain width on either side of the line are included which is analagous to a very small belt transect (see below). This approach can also be valuable in woodland and scrubland, where shrubs are often difficult to measure quantitatively, and in studies of plant colonisation of derelict and degraded land, where plants will often be sparsely distributed.

Sampling design for vegetation description and analysis

The **sampling** of vegetation for description and analysis is an area which has received insufficient attention in the literature. Even in published work, sampling designs are often insufficiently well explained or sometimes ignored altogether, yet choosing the right approach is a vital part of any vegetation survey work. The primary determinant of the sampling design should be the aims and objectives of the project. However, factors such as the time and resources available for study, the type of habitat and proposed methods of data analysis and presentation must also be considered. Kenkel *et al.* (1989) present a good review of sampling problems in community ecology. **Spatial autocorrelation** is also a problem which has attracted recent attention (Fortin *et al.*, 1989). This topic is covered at length in Chapter 4.

Stratified sampling

A large amount of work in plant community ecology, either deliberately or implicitly, employs **stratified sampling**. The principle of **stratification** is that the vegetation of the area under study is divided up before samples are chosen on the basis of major and usually very obvious variations within it. The samples or quadrats are then allocated to these different areas using the various types of sampling strategy discussed below. Stratification is normally carried out by an initial reconnaissance of the survey area and/or examination of any aerial photography which may be available. The major divisions are made primarily on the basis of differences in growth form, physiognomy and structure of the vegetation (for example woodland, scrub, tall herb, short-cropped grassland, open ground) with secondary divisions made on variations in dominant species. Other criteria for stratification are obviously available, for example, areas of vegetation subject to different management regimes, important environmental differences such as aspect, geology or slope form or areas which are known to have been undisturbed for differing time periods and have undergone different degrees of successional change.

   Some form of stratified sampling design has much to recommend it because major sources of variation in the vegetation are recognised before sampling commences. Smartt (1978) has taken this idea even further in proposing his **'flexible' systematic model** for sampling vegetation. The sampling design is based on the assumption that the greater the diversity and the rate of change of the vegetation cover over a given distance, the more intensively it should be sampled. The approach involves laying down a network or skeleton of primary sampling points over the whole of the study area to define a framework. Analysis of the degree of variation in floristic composition between each of these primary points then enables secondary sampling points to be allocated to areas where a large change in species composition has been found between two primary points. Where there is little change, then further samples are not required. The result of this is that more samples are taken where variation in floristic composition is high and less where it is low. Smartt argues that since most studies of vegetation are looking for variations within and between plant communities, this is a very efficient approach and time is not wasted on repeated sampling of essentially similar vegetation. This method may be viewed as a special form of stratified sampling, since samples are allocated on the basis of some predetermined criterion — in this case the extent to which floristic diversity varies locally over the survey area.

Random sampling

Strict application of random sampling means that every point within the survey area should have an equal chance of being chosen on each sampling occasion. To take a random sample, a grid of coordinates is set up over the survey area and pairs of random numbers are taken to locate each quadrat. An extended discussion of the many difficulties involved in doing this is presented in Causton (1988). Tables of random numbers are available in most elementary statistical textbooks (for example Gilbertson *et al.*, 1985) and can be generated on some hand calculators. Other sources are playing cards, diaries or telephone directories. Where aerial photographs are available, random points can be placed on the photographs, although exact location of those points subsequently in the field can prove difficult on rough terrain or woodland. Where a grid is difficult to set up, for example in scrub or woodland, an alternative approach is to use **random walk procedures**, whereby a sample point is located taking a random number between 0° and 360° to give a compass bearing, followed by another random number for a number of paces. Several of these can be strung together to increase the degree of randomisation between points. However, since the location of each next point is still to some extent dependent on the previous one, random walks are not strictly random.

Most **descriptive** vegetational work, which is inductive in approach (see Chapter 1), does not require **random** sampling. Indeed, much descriptive vegetation survey is biased in its sampling and uses some form of stratified sampling design as discussed in the previous section. Nevertheless, random sampling still remains a possibility for use in descriptive studies, perhaps combined with the stratified approach.

However, if the deductive approach using **hypothesis generation and testing** is being applied, including the application of inferential parametric statistics (statistical methods for hypothesis testing based on probability), then random sampling is essential because most statistical methods for hypothesis testing based on probability assume that all observations or samples are independent of each other. A common situation where this may be necessary is in the exploration and testing of relationships between plant species and environmental factors, using correlation and regression techniques (Chapter 4).

Systematic sampling

Systematic sampling involves the location of sampling points at regular or systematic intervals. The size of the sampling interval is extremely important and is usually a fixed interval, such as 100m or a regular number of paces. If quadrats are taken at fixed intervals, then care must be taken that the sampling interval does not coincide with any **pattern** in the vegetation owing either to the properties of plants themselves or to some regularly distributed environmental control. For example, if the floristics of an old meadow in England are being examined, former

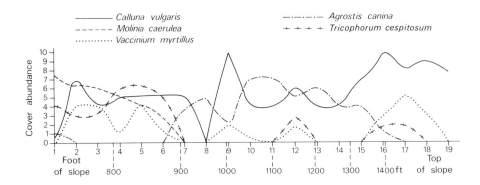

Figure 2.12 Results of a continuous belt transect up a hillslope near Creetown, Kirkcudbrightshire, Scotland, showing variation in cover-abundance of selected species (Tivy, 1982; redrawn with kind permission of Longman Publishing)

agricultural practice may have left some form of ridge and furrow system with damper conditions in the furrows, which supports one set of species, and drier conditions on the ridges supporting another. If the spacing of the ridge and furrow coincides with the sampling interval, only ridges or only furrows could be selected.

Transects

The **transect approach** is very popular in vegetation work. A transect is a line along which samples of vegetation are taken. Transects are usually set up deliberately across areas where there are rapid changes in vegetation and marked **environmental gradients**. Most transects are thus biased in their location, although it is possible to locate the start and end of a transect at random and then take samples along the **line** connecting the two points. Classic examples of laying out of transects across gradients are up hillsides, where slope angle, drainage and altitude combine; across major changes in geology; or through **ecotones** such as terrestrial/aquatic/marine transition zones, for example river and lake edges, salt marshes and sand dunes. The main purpose in using transects in these situations where the change in vegetation is clearly directional is to describe maximum variation over the shortest distance in the minimum time.

Where quadrats are laid out next to each other or contiguously along a transect line, the result is a **belt transect**. Figure 2.12 shows an example where the changes in species composition up a Scottish hillside have been described. Another good example comes from the study of recreation ecology and footpaths, where short belt transects are laid out across paths to illustrate the effects of walker pressure on vegetation (Figure 2.13). If abundance data have been collected, **histograms** may be drawn and environmental data plotted on the same diagram to enable visual comparison of correlations between species distribution and environmental factors to be made.

A sophisticated form of transect sampling across large areas is described by Austin and Heyligers (1989). Using the climatic, topographic and lithological characteristics of a 20,000 km² forested area in New South Wales, Australia, a series of **gradsects** (transects incorporating significant environmental gradients) were selected as representative of the environmental variability in the area. The method

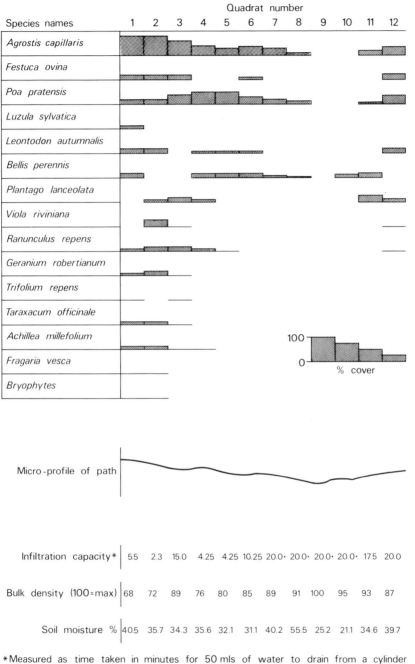

*Measured as time taken in minutes for 50 mls of water to drain from a cylinder pushed 1 cm into the ground.

Figure 2.13 A belt transect across a heavily used footpath at Aysgarth Falls, Wensleydale, Yorkshire, England

uses computerised databases for the environmental data and numerical classification (Chapter 8) to define sampling classes. The aim is to obtain a data set that is representative of the major environmental gradients.

Grids and isonomes

For certain specialised and detailed studies, the transect approach can be broadened out into a **grid**, where a large number of quadrats are placed adjacent to each other in the grid and species abundances and environmental factors are recorded in each quadrat. Plotting of values for abundances or environmental factors for each quadrat or cell of the grid allows a contour diagram to be constructed which is known as an **isonome**. Separate contour maps can be produced for each species or environmental factor and, when overlaid, visual correlation of species distributions and their relationships with environmental factors can be explored. A good example of this approach is provided in Kershaw and Looney (1985), where the distribution of two species of *Carex* is shown to be clearly correlated with micro-topography. Isonomes are very tedious to produce.

Plotless sampling

In forest and woodland ecosystems and where vegetation is **sparsely** distributed, as in certain *maquis* and *garigue* Mediterranean vegetation types, semi-desert or high alpine communities, the use of conventional quadrat analysis and sampling may be limited. In the case of woodlands, sampling of the tree cover requires very large quadrats, and often the ground flora may be totally impenetrable. In sparse communities, the large amount of bare ground causes problems in recording; it is the presence and absence of individual plants and the distances between them that are important. Several sampling techniques have been devised to overcome these problems and are known collectively as **plotless sampling**. The simplest of these is the **nearest individual method**, which involves the location of random sampling points throughout the area. In woodlands, the distance to the nearest n individuals of each tree species is recorded as shown in Figure 2.14 for species A, B and C. Successive distance measurements are taken and the whole procedure is then repeated for a series of random points. The density of each species is then derived from the formulae:

Mean area = (mean distance to nearest n individuals of a species)^2

$$\text{Tree density for a species} = \sqrt{\frac{\text{Mean area}}{2}}$$

A complete review of plotless sampling methods is given in Mueller-Dombois and Ellenberg (1974).

Checksheets

Speed and accuracy in recording will always be helped by the drawing up of a standardised **checksheet** prior to a survey. Usually the checksheet is designed so that all the information relating to one quadrat or vegetation sample is entered on a single

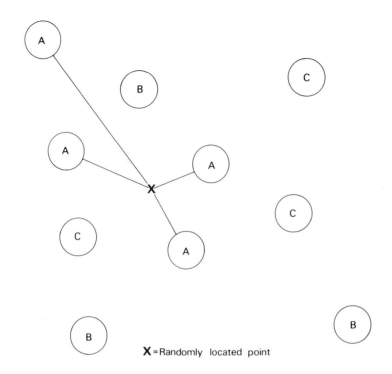

Figure 2.14 The nearest-individual method of plotless sampling

A4 sheet (Figure 2.15). Three sets of information are generally collected:

(a) information on general site conditions — location, grid reference, transect number, quadrat number, surrounding terrain and land use, weather conditions and any other similar information.
(b) a species checklist or space for a species list to be written in by hand in the field. Ideally, these should use the Latin binomial.
(c) environmental data concerning the quadrat and the vegetation being described. The nature of these data will depend on the aims and objectives of the survey. Typically, in an initial descriptive survey, data enabling the primary gradients of variation in vegetation should be collected, for example soil moisture, aspect, soil depth, light conditions, soil nutrient status, information on grazing, burning and management. Many methods for the study of soil variables are given in Smith and Atkinson (1975), Briggs (1977), White (1979), Ball (1986) and Allen *et al.* (1986), while techniques for geomorphic and climatic factors are presented in Hanwell and Newson (1973) and Briggs and Smithson (1985). Further useful advice on procedures for quadrat survey and checksheet design are given in Bunce and Shaw (1973).

In addition to ensuring standardisation of recording, use of a checksheet makes loss of data during and following field survey less likely, particularly if several recorders are working together (Plate 2.2). The phenomenon of the **'data drain'**, whereby part or sometimes all of the data disappear or suddenly appear unintelligible is greatly reduced.

Department of Geographical Sciences	Department of Archaeology & Prehistory
University of Plymouth	University of Sheffield

Vegetation Survey
South Uist - Outer Hebrides

RECORDERS : *MK/BB*

SITE CODE : *F1*	QUADRAT NUMBER : *15*
GRID REF : *729282*	ALTITUDE : *5m*
SLOPE ANGLE : *1°*	ASPECT : *290°*
SOIL TIN NUMBER : *725*	pH : *7.4*
% SOIL MOISTURE : *23.4*	ORGANIC MATTER : *8.4%*

PEAT/SOIL DEPTHS : *25/25/25/25/25*

Species :

	% cover
Plantago lanceolata	*50*
Trifolium repens	*10*
Festuca rubra	*60*
Anacamptis pyramidalis	*1*
Rhinanthus minor	*5*
Sinapis arvensis	*5*
Euphrasia officinalis (agg.)	*3*
Cerastium glomeratum	*1*
Equisetum arvense	*25*
Luzula multiflora	*2*

Site description and comments :

Edge and back slope of machair sand dunes towards Upper Loch Kildonan.

Figure 2.15 A checksheet for a vegetation survey across South Uist, Outer Hebrides, Scotland

Case studies

Use of the Raunkaier life-form spectrum to describe vegetation communities on derelict colliery spoil heaps (Down, 1973)

Most spoil heaps that result from mining activities are extremely harsh environments for plant colonisation and establishment. Spoil produced by deep coal mining activity presents some of the most severe problems of all. Colliery spoil is a mixture of shales, mudstones and sandstones, within which the coal seams were interbedded underground. By volume, the shales are dominant in the heaps, and it is their physical and chemical characteristics which determine the severity of the surface of the heap for plant growth. (Kent, 1982; 1987).

The shales are predominantly grey-black in colour, with a low albedo, resulting in very high summer temperatures (up to 60 °C) and a high diurnal variation in temperature. Water availability is low in summer, yet waterlogging may occur in winter. The spoil is often tipped at very steep angles, so that slope wash and erosion are serious problems, and the shales weather down to fine clays that are easily transported. Many essential plant nutrients, particularly nitrogen and phosphorus, are deficient or in chemical forms unavailable to plants. Toxicity problems occur owing to the evolution of forms of sulphuric acid from the mineral pyrites (FeS_2) during weathering of shales and to high concentrations of soluble salts. Soil organism populations are low or non-existent.

Reclamation is desirable for both landscape and safety reasons, but because of cost this usually means only the establishment of some form of grass sward. However, prior to reclamation of a site, information on problems of vegetation establishment can be obtained from a survey of the existing natural colonising vegetation, particularly if there appear to be large variations from one part of a tip to another. The life-form of the species present is also of interest, since those species that can colonise such a harsh environment are presumably well adapted to the prevailing conditions.

Vegetation description using floristic methods is often of limited value, since these sites are at early successional stages and contain many **ruderal** species. Ruderals are colonisers of bare and disturbed ground and take advantage of the lack of competition from other species that enter later in succession. Also on most sites, the source of colonising species is the surrounding semi-natural vegetation which can be highly variable from one site to another. The result is that there is a wide variety of early natural colonisers.

As a result of these problems, researchers have turned to life-form methods to describe the vegetation of spoil heaps. Down (1973) worked on five disused colliery spoil heaps in Somerset, England, where the age of each heap was known, ranging from 12 to 98 years. He recorded both life-form and floristic data with a view to assessing the severity of the micro-climate and environment for plant growth. The results of the floristic data showed that 73 vascular plant species were found, but only 10 of these showed a definite trend with age and site condition. The life-form spectrum for each tip was also recorded and the results are shown in Table 2.6.

The life-form data demonstrate two important points. First, the percentages of rosette species (Hr) declined from 31.8 to 11.8 as the tips aged. This was accompanied by an increase in protohemicryptophytes (species with partially

Table 2.6   Life-form spectra of plant species on colliery spoil heaps of five differing ages in Somerset (Down, 1973). Results as percentages of total number of species on each tip (Reproduced with kind permission of Academic Press and *Environmental Pollution*)

Life form	Spoil age (years)				
	12	15	21	55	98
MM	0	7.1	0	7.9	0
M	4.5	0	7.1	10.5	2.9
Chh	4.5	7.2	3.6	5.3	5.9
H	9.2	7.2	2.6	2.6	5.9
Hp	4.5	10.7	3.6	21.1	32.4
Hs	22.7	32.1	35.7	21.1	29.4
Hr	31.8	25.0	33.1	23.6	11.8
Total H	68.2	75.0	75.0	68.4	79.5
Gr	0	3.6	3.6	0	0
Grh	0	0	0	0	2.9
Th	22.8	7.1	10.7	7.9	8.8

Key to life forms

MM — Phanerophytes, woody plants with buds more than 25cm above soil level, plants higher than 8m

M — Phanerophytes, plants 2–8m high

Chh — Herbaceous chamaephytes with buds above soil level but below 25cm

H — Hemicryptophytes, herbs with buds at soil level

    H — undivided hemicryptophytes

    Hp — protohemicryptophytes with basal leaves smaller than stem leaves

    Hs — partial rosette hemicryptophytes (semi-rosette), basal leaves larger than stem leaves

    Hr — rosette hemicryptophytes, leafless flowering stems with basal leaf rosette

Gr — Geophytes with buds or roots

Grh — Geophytes with rhizomes

Th — Therophytes, passing the unfavourable season as seeds

formed basal leaves which protect the perennating buds and with better-developed leaves higher up the stem) with age. Second, therophytes (species which spend the unfavourable period of the year as seeds) represented 22.7 per cent of the original flora but declined to 9.8 per cent over 98 years.

The colonising flora were thus dominated by hemicryptophytes and therophytes. The importance of the rosette hemicryptophytes as colonising species had not been noticed before but appears to be a function of their life-form. They have long tap roots which ensure that they are firmly anchored in unstable soils and which enable them to extract water from a depth when the soil is subject to drought in the summer months. The rosette form of the basal leaves also serves to reduce temperature variability around the sensitive perennating buds of the plant and cut down evapotranspiration around the stem/spoil interface. The decline in therophytes is due to the dominance of ruderal therophyte species at early stages of colonisation and their inability to compete with perennial grasses and herbs during later stages of succession.

Down concludes with the comment that in most schemes for the reclamation of colliery waste, perennial grasses and legumes are planted. The life-forms of these species are unlike most early natural colonisers of spoil heaps, and many examples exist where heaps which have been reclaimed using grass species often suffer serious die-back and regression. He suggests that a partial solution may lie in the sowing of more rosette species in reclamation seed mixes.

This case study shows how life-form description methods, originally devised for use at the world scale to show plant response to gross differences in world climate, may be applied equally successfully at the much more local and detailed scale of a colliery spoil heap to assist with reclamation studies. However, at both scales, it is plant response to limitations of climate and habitat that makes the use of Raunkaier's methods possible, together with the limitations of floristic techniques at this scale.

Application of Elton and Miller's habitat classification system within methods of ecological evaluation and habitat assessment (Kent, 1972; Kent and Smart, 1981; Goldsmith, 1975)

The conservation and management of semi-natural vegetation in Britain involves not just the establishment of a national network of nature reserves and protected sites but also concern for all remnant areas of semi-natural vegetation, especially within the highly modified agricultural landscape of lowland Britain. For a conservation policy to be established and implemented for the remaining vegetation, some form of inventory and assessment of such sites is necessary and furthermore, some estimation of their 'ecological value' may be attempted.

In recent years, a wide range of techniques have been proposed for assessing 'islands' of semi-natural vegetation that remain within highly developed landscapes (Usher, 1986). In Britain, a substantial impetus came in the 1970s from the requirement to provide information for county structure plans. Methods devised by Kent (1972), Goldsmith (1975) and Kent and Smart (1981) have included the use of Elton and Millers's habitat classification system, although with some modification in each case.

In the method of Kent (1972) and Kent and Smart (1981), all areas of remnant semi-natural vegetation within a small study region of 10×10 km of predominantly arable agriculture were regarded as individual sites (Figure 2.16). Each site was first divided into Elton and Miller habitat-formation types: open ground, field layer, scrub, woodland, edge and aquatic-terrestrial transition (Figure 2.5). Numbers of plant species were then counted for all layers in each habitat type recognised. One problem was the minimum scale at which a single habitat type could be identified. Some areas were a mosaic of two or even three habitat types, for example a tall grass meadow invaded by intermittent patches of scrub. The identification of each patch of scrub as a separate formation type was impossible. Instead, a combination of habitat types was recognised — field layer/scrub. Some woodlands thus had three or four layers or habitat types present.

The area of each site was measured from maps using a planimeter, the edge length of each site was recorded, distances were measured to the nearest other sites of semi-natural vegetation and various aspects of the existing management

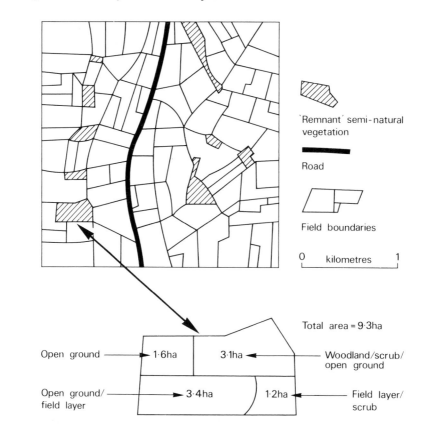

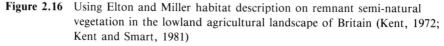

Figure 2.16   Using Elton and Miller habitat description on remnant semi-natural
vegetation in the lowland agricultural landscape of Britain (Kent, 1972;
Kent and Smart, 1981)

and use of sites were noted. Sites were then grouped together, using the
ordination and classification methods described later in this book, and a series
of groups of different sites emerged, representing the main types of habitats
within each 10km² area. The number of individual sites within each group
and their characteristics put a relative scarcity value on the amount of that
habitat within the local region. Although numerical methods were used, in a
small area, the sites could easily be grouped subjectively on the basis of the
habitat data collected.

This classification of habitats could then be used as an index for planning
purposes. If development or modification of a site was proposed, its position
in the classification could be looked up and the planner or decision-maker
could then assess the habitat characterisation of that site in an informed
manner.

In the second study (Goldsmith, 1975), Elton and Miller's habitat classification was used in a different manner. Rather than simply making an inventory
of sites of semi-natural vegetation, the aim was to survey the whole
agricultural landscape, including arable and grazing land, and an attempt was
made to assess 'ecological value'. The basis of 'ecological value' is discussed
at some length in Goldsmith's article, but the basic concept is that

some habitats are commonly agreed to be more 'interesting' and 'valuable' than others by professional and academic ecologists and planners.

Following a transect along the Wye Valley to the Black Mountains on the English–Welsh border, 1 km^2 blocks were taken and each block or square was allocated to one of three **land system types**:

I.   Unenclosed upland, mostly moorland over 300m;
II.   Enclosed cultivated land, mostly permanent pasture;
III.   Enclosed flat land, most arable, in the valley bottoms.

Within each of these land systems, habitats were recorded as:

1. Arable and ley
2. Permanent pasture
3. Rough grazing
4. Woodland
    (a)   deciduous/mixed
    (b)   coniferous
    (c)   scrub
    (d)   orchard
5. Hedges and hedgerows
6. Streams

For each habitat type within each grid square, four variables were measured:

(1) Extent (E) — the area of habitat types in 1–4 in hectares and the lengths of linear habitats 5 and 6.
(2) Rarity (R) — recorded for each habitat type in each land system and calculated from R = 100 per cent — the per cent area of each habitat type within each land system.
(3) Plant species diversity (S) — 20 × 20m quadrats were located in each habitat type within each land system.
(4) Animal species diversity (V) — the stratification of the vegetation using Elton and Miller's ideas, where the number of vertical layers was recorded from 1 for open ground to 4 for well-developed woodland.

Thus the habitats were recorded first and then the structural nature of the vegetation within each habitat type was taken as an index of animal-species diversity. The approach is clearly derived from the original ideas of Elton and Miller.

The index of ecological value for each grid square was finally derived by multiplying together the four variables for each habitat type and then adding the results for each of the six habitat types present within a grid square. The results were then rescaled in the range 1–20 to enable them to be mapped in convenient classes.

The main point to stress is the assumption that animal and to a lesser extent plant species diversity is correlated with stratification and vertical structure of vegetation. One problem, however, is that both methods were applied to relatively small areas in the first instance and would require modification if used more widely or if an attempt was made to formulate a standardised approach that could be used by all planning authorities across the country.

Use of the Dansereau universal structural system for recording and mapping vegetation in South-East Queensland, Australia (Dale, 1979)

The aim of this survey was too make an inventory of the vegetation of South-East Queensland in Australia prior to the production of a vegetation map. As such the aim was primarily academic but it could also be seen as laying the foundation for later applied work. The categorisation of vegetation types and the map would then provide a framework for more detailed research in the area. Structural and physiognomic approaches to vegetation description were considered to be the most suitable because of the large area to be surveyed.

The problems of sampling for physiognomic survey are highlighted, in that five major vegetation types were recognised subjectively prior to applying the Dansereau method for description. Diagrams constructed from a transect through each type are shown in Figure 2.17. No mention is made of how the transects were located, although it would be logical to assume that they were placed across what were believed to be the most typical or representative areas.

Problems were also encountered in the application of the standard Dansereau method to the range of semi-open Eucalypt woodlands. The main difficulties came in the representation of all forms of lianes, determining seasonality for all species and in the over-representation of certain species, such as the bottle tree (*Brachychiton sp.*), on the diagrams. As a result, Dale produced his own set of modified symbols to suit his particular circumstances (Figure 2.18), which were partly adapted from the work of Dansereau *et al.* (1966). They described modifications of the standard types with respect to crown shape, leaf size, stem diameter, angle of branching and height of principal branching. In Figure 2.18, crown shapes are drawn to scale, and the line on the trunk section of each tree corresponds to the height of principal branching. The diamond shape below each trunk refers to the scrub nature of the Eucalyptus species.

Thus while the Dansereau method was shown to be valuable in distinguishing between major vegetation types, for more detailed surveys of sub-groups of Eucalyptus types in Queensland, modifications and revisions were required. Such adaptations of existing methods to suit varying circumstances and environments is commonplace in vegetation description. This study also shows how physiognomic methods have their place in certain world vegetation-formation types where floristic description can only be carried out with difficulty.

Domin, percentage cover and frequency scales and the study of floristic changes in grassland on the Isle of Rhum, Scotland, following the reduction or exclusion of grazing (Ball, 1974)

The island of Rhum (Rum) in the Inner Hebrides of the west coast of Scotland is Britain's largest nature reserve, with an area of 10,620 ha. The greater part of the island consists of upland grassland communities dominated by the grasses *Agrostis capillaris* (common bent) and *Festuca rubra* (red fescue) and wet heathland and bog areas with *Calluna vulgaris* (ling or Scots heather), *Molinia caerulea* (purple moor grass) and *Scirpus caespitosus* (deer sedge). The vegetation has been subjected to intensive grazing and burning practices, the main grazers being red deer, sheep, cattle and to a lesser extent, ponies and goats.

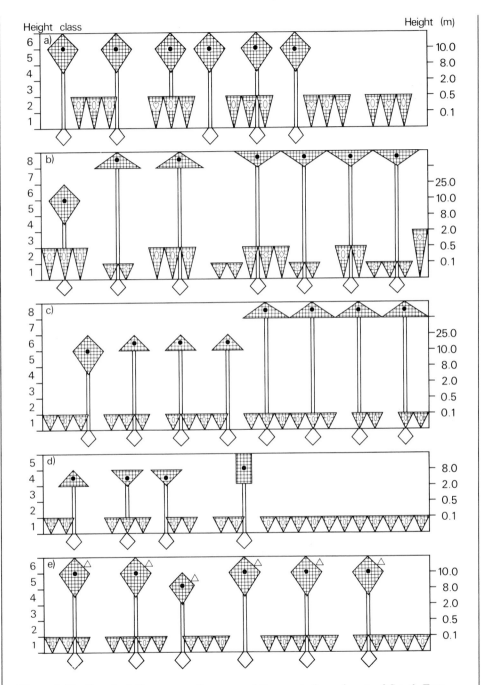

Figure 2.17 Standard Dansereau diagrams of the vegetation of part of South-East Queensland, Australia (Dale, 1979; redrawn with kind permission of Kluwer Academic Publishers)

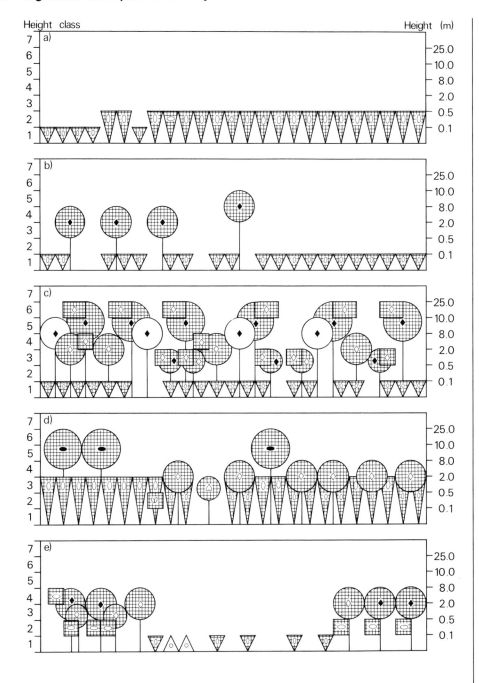

Figure 2.18 Modified Dansereau diagrams for *Eucalypt* woodlands of South-East Queensland, Australia (a) *E. maculata* (b) *E. moluccana* (c) *E. tereticornis* (d) *E. melanophloia* (e) *E. crebra* (Dale, 1979; redrawn with kind permission of Kluwer Academic Publishers)

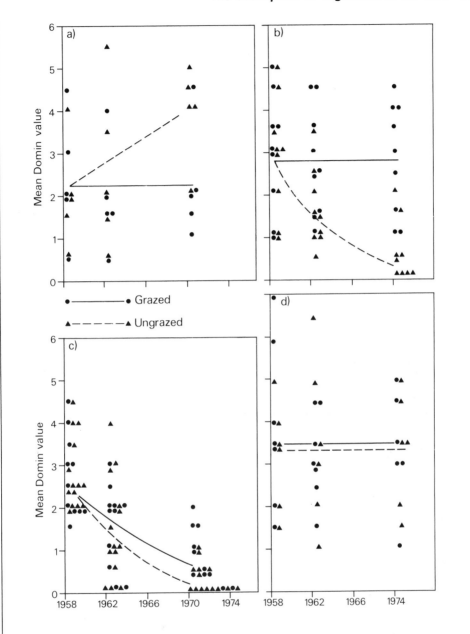

Figure 2.19 Changes in mean Domin value of four groups of species in grazed and ungrazed plots on species-poor *Agrostis-Festuca* grassland on the Isle of Rhum, Scotland, 1958–70. Means are taken from three replicate plots (Ball, 1974; redrawn with kind permission of Academic Press and *Journal of Environmental Management*)

In 1957, when the Nature Conservancy took over the whole island as a nature reserve, the sheep were removed and the deer population was reduced. Burning practices were also controlled and carried out on a more regularised basis. In order to examine changes in the vegetation resulting from the new pattern of management, thirteen 5 × 5m experimental exclosures or permanent plots were established on various community types in 1958. The vegetation of each area was described by a 10-point Domin scale using randomly located 4m^2 quadrats inside and outside each exclosure in 1958, 1962 and 1970.

Figure 2.19 shows the changes observed within one community type, the species-poor *Agrostis-Festuca* grasslands. The Domin values for four groups of species (a–d) in the grassland sward have been plotted for the three sampling intervals over the 12-year period. The trend in the mean Domin values of the enclosed plots across the three sampling intervals is shown by a dashed line, while that in the grazed plots is shown by a solid line.

The ungrazed plots show a marked increase in the cover and dominance of group (a) species, notably *Poa pratensis* (meadow grass) and *Festuca rubra* over the experimental period and a marked fall in the Domin cover values of other grasses and herbs in groups (b) and (c), notably the grasses *Anthoxanthum odoratum* (sweet vernal grass) and *Cynosurus cristatus* (crested dog's tail) and the herbs *Plantago lanceolata* (ribwort plantain), *Polygala serpyllifolia* (heath milkwort), *Euphrasia officinalis* (eyebright), *Achillea millefolium* (yarrow) and *Veronica officinalis* (common speedwell). Species in group (d) have remained constant after the exclusion of grazing and include *Agrostis capillaris*, *Holcus lanatus* (Yorkshire fog), *Koeleria gracilis* (crested hair grass), *Ranunculus acris* (common meadow buttercup), *Potentilla erecta* (tormentil) and *Rumex acetosella* (sheep's sorrel).

In contrast, the grazed plots show that species composition and abundance has not changed at all in groups (a), (b) and (d), whereas there is a significant fall in group (c) species, although to a lesser extent than in the ungrazed plots. Thus the effect of a relaxation in grazing pressure is to reduce the abundance of many species quite dramatically to the advantage of the two main grass species, which, owing to their greater height and tussock structure, can gain competitive advantage over the herb species, that are of lower stature.

One problem of this study was that the experimental recorder changed between 1962 and 1970. Therefore in 1969, a check was made on the results by carrying out a percentage cover and frequency analysis along a transect located by random numbers which traversed the boundary of each exclosure. Three replicate transects of 16 quadrats were thrown, eight inside each exclosure and eight outside. A ½m^2 quadrat, sub-divided by cross-wires into 25 × 10cm squares was used, and the cross-wire intersections were used to provide a 16-point sampling frame in each quadrat for cover analysis.

The principle which underlay this experimental check was that the original study using Domin values was carried out **through time** by recording change at the **same point in space**. However, with the change of researcher, a different approach was adopted whereby **at the same time**, quadrats at **different locations in space** were examined both inside and outside the exclosures. Thus in the second analysis, variation in space was substituted for variation through time, and it was assumed that the grazed grasslands around the enclosures in 1969–70 were the same as those **inside** the exclosures when they were originally established in 1958. The other interesting aspect of

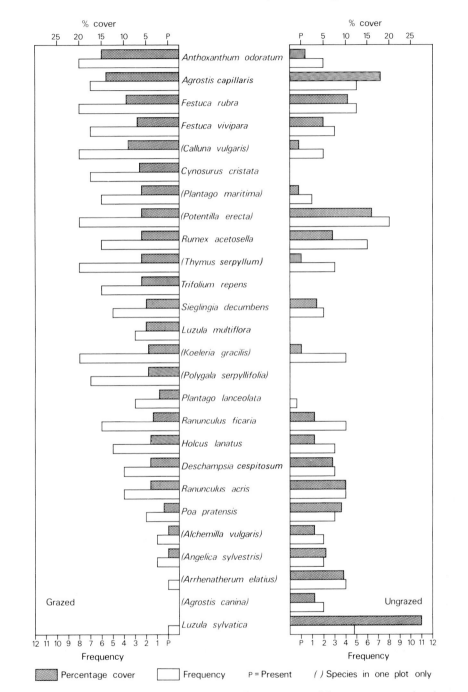

Figure 2.20 Floristic differences as shown by percentage cover and frequency on grazed and ungrazed plots in species-poor *Agrostis-Festuca* grasslands on the Isle of Rhum, Scotland in 1969. P — present, species in parentheses in one plot only. Solid columns show percentage cover, open columns frequency. Percentage cover and frequency data are averaged from 24 plots in each case (three replicates × 8 quadrats) (Ball, 1974; redrawn with kind permission of Academic Press and *Journal of Environmental Management*)

this survey was that in the 1969 survey percentage cover and frequency were recorded, as opposed to domin cover values alone.

The results of this second survey for the species-poor *Agrostis-Festuca* grasslands are shown in Figure 2.20. Two points emerge:

(1) The two measures of abundance, frequency and percentage cover, show significant differences in results. Most important is that three species — *Arrenatherum elatius* (false oat-grass), *Luzula sylvatica* (great wood-rush) and *Plantago lanceolata* (ribwort plantain) — were recorded only by frequency and were missed in the point cover estimates. Also the degree of difference between the two measures appears to be much greater in the grazed than in the ungrazed plots.

(2) When compared to the first survey, some predictable similarities occur but differences also exist. Following exclusion of grazing, *Poa pratensis* and *Festuca rubra* increase in cover, although *Festuca rubra* does not increase in frequency and the species of groups (b) and (c) in the domin analysis are shown to be much reduced or eliminated. However, *Agrostis capillaris*, *Potentilla erecta* and *Ranunculus acris* are much increased in cover but not in frequency, and certain species emerge as being important in the enclosed plots, notably *Luzula sylvatica*, *Agrostis canina* (velvet bent) and *Arrenatherum elatius*. Thus although overall trends are the same, possibilities exist for differences in detailed interpretation as a result of variations in the abundance measures used.

The implications of these results for the future management of the *Agrostis-Festuca* communities are interesting. The changes in the dominance of the grasses in the sward following the exclusion of grazers are probably not of great significance in terms of the food supply for the grazing animals. The general level of grazing on the island had been reduced and from the point of view of range management, the only problem was that a reduction in grazing leads to an increase in plant litter and taller vegetation, causing an increased level of flammability of the vegetation. However, one of the main reasons for declaring the Isle of Rhum a nature reserve, and a major goal of the management plan, was to maintain high floristic diversity in all the vegetation types present. Thus, although the observed changes may not be important for upland agriculture, species diversity and the aims of nature conservation were affected. The answer appeared to lie in increasing the grazing pressure on the swards again in order to keep the more aggressive pasture grasses in check. Towards this end, the annual cull of the red deer on Rhum was severely reduced.

Sampling design for a vegetation survey of the Faröe Islands (Tomlinson, 1981; Milner, 1978)

The flora of the Faröe Islands, which lie midway between Shetland and Iceland in the North Atlantic, is relatively well documented. However, the nature of the plant communities and their similarity to the vegetation on other Scottish islands and mainland Britain had not been studied in such depth. Thus in June–July 1978, a team of ecologists visited the two islands of Vagar and Mykines with the aim of describing the vegetation and plant communities.

They were confronted with a typical problem in terms of vegetation sampling which was to describe the maximum variation of the plant communities of the area but in a relatively short space of time. Tomlinson devised a method of stratified sampling based on information published from a similar survey of the Shetland Islands completed by the Institute of Terrestrial Ecology in 1975 (Milner, 1978). In the Shetland survey, 150 physical attributes on the 1-inch Ordnance Survey map were recorded for each 1km grid square. These attributes included factors such as distance from the coast and aspect and could be grouped into three categories:

(a) factors related to the coast and coastal features;
(b) factors related to high altitude;
(c) factors related to the size and type of waterbodies.

The measurements for all grid squares on Shetland were then analysed using numerical classification (Chapter 8) and ordination (Chapters 5 and 6) from which 16 groups emerged. The result was effectively a broad-scale classification of habitats based on physiographic characteristics. Since the Faröes are similar to Shetland in their general physiography, Tomlinson decided to apply the 16-group classification for Shetland to the Faröes in order to provide a basis for stratification for vegetation sampling. Nine of the 16 Shetland land types were represented on the two Faröe islands and these are summarised in Table 2.7.

Table 2.7 The nine physiographic categories from the ecological survey of Shetland (Milner, 1978) used as the basis for stratified sampling in the vegetation of the Faröe Islands (Tomlinson, 1981; with kind permission of *Biological Conservation*)

Land Classes	General physical characteristics	Number of squares	Proportion surveyed (%)
A	West-facing exposed coasts, typically with high cliffs	82	3.6
B	East-facing exposed coasts	39	7.6
C	West-facing sheltered coasts, typically with gently sloping shores	19	15.8
D	East-facing sheltered coasts	48	6.2
E	Uplands (altitude exceeding 250m) with prominent freshwater features	14	21.4
F	Upland plateaux and summits lacking freshwater features	295	1.0
G	Lowlands with prominent freshwater features	71	4.2
H	Uncultivated lowlands lacking freshwater features	70	4.3
I	Villages and associated cultivated lands (always by the sea)	38	7.8

Total number of 0.25km squares containing land = 676
Total land area of the two islands = 169km^2
% includes lake shores, pools, rivers and multiple stream confluences

In order to carry out the stratification, a 0.25km^2 grid was superimposed on the 1:25,000 base maps of the islands. Each square was assigned to one of the nine classes in Table 2.7 and three squares were then taken at random for each land class with the condition that one of the three should be from the smaller island of Mykines if the class was present there. The number of squares and the proportion of each surveyed (3 ÷ number of squares in each class × 100 per cent) are shown on the right of Table 2.7.

For each square selected in this manner, four sampling points were located at random within each quarter. This involved taking each quarter, gridding it and locating the point by random numbers. These points were then visited in the field, and data were collected in a 200m^2 area around each point using an approach similar to that recommended by Bunce and Shaw (1973):

(a)   all vascular plants together with bryophytes and macro-lichens were listed.
(b)   estimation of the abundance of each species was made using the Domin scale (Table 2.5).
(c)   a description of the soil profile at each site was completed with pH deter-mination of samples from 100mm depth.
(d)   a record of the habitats within each plot was compiled.

The random point was then used as the centre of each 200m^2 plot and lines were pegged out to each of the four corners of the square. The diagonals were marked at certain intervals to give quadrats of 4, 25, 50, 100 and 200 m^2. The smallest quadrat was examined first, and then by working outwards, complete coverage of the plot was ensured. In all, 104 such plots were sampled and 227 plant species recorded.

This study demonstrates a number of important principles relating to vegeta-tion sampling. First, a logical system of sampling was established which approximated to a stratified random sample. This approach minimised the chances of omitting important vegetation types, while maintaining a high level of objectivity. Thus sampling was specifically designed to ensure that as much variation in the vegetation of the Faröes was included. Second, the stratifica-tion process was based on a previous survey of similar environments in Shetland. Often, existing published information may be used to provide a basis for sampling and stratification. Third, the expedition had only a relatively short period within which to work. The exact location of field plots was decided in advance of departure using the stratified random approach. Once in the field, this made data collection and planning of routes and camps much easier. Tomlinson states that because of this forward planning, they were able to complete between 8–12 field plots in one day.

Sampling and the survey of Britain's railway vegetation (Sargent, 1984)

This case study was introduced in Chapter 1. The overall aim of the survey was to describe the nature and variation of the railway vegetation of Britain and to attempt to explain some of the observed variations. Of particular interest was the sampling method employed. The 18,000 km of railway were subject to stratification for sampling, with the rural railway network being divided into 893 measured, 10-mile units. Certain geographic attributes of each 10-km section of track were described: for example whether the railway was single or multiple tracked, went through a cutting or was on an embank-ment, the adjacent geology and soil types, selected general climatic variables and neighbouring habitat types and land uses.

These characteristics for the 893 track sectors were then classified using

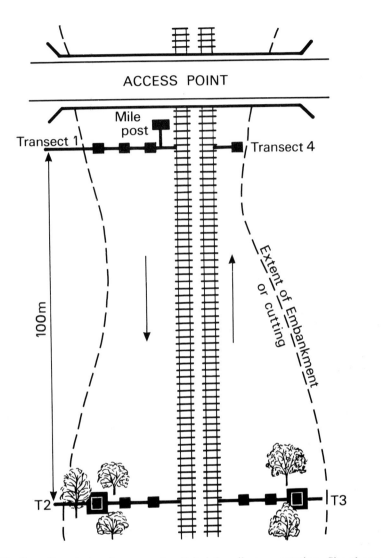

Figure 2.21 Sampling design for surveying Britain's railway vegetation. Site chosen by stratified random sampling, 100m long. Arrows indicate the direction in which the survey team walked, facing oncoming traffic for safety (Sargent, 1984; redrawn with kind permission of Caroline Sargent and the Institute of Terrestrial Ecology)

indicator species analysis, which is described in Chapter 8, to give 26 different classes or types of railway track, each corresponding to a geographic region. The main characteristics of each of the 26 classes are shown in Table 2.8. These 26 classes were then used as the strata for sampling of the vegetation.

A total of 480 sampling sites was then distributed proportionately across the 26 track classes. Samples were placed randomly within each track class at British Rail mile posts and at each post, a 100m length of track was examined

Table 2.8 Characterising attributes of the 26 railway track classes defined by classification of 83 geographic attributes of 893 10km sections of railway track (Sargent, 1984; with kind permission of the Institute of Terrestrial Ecology)

	Southern Chalk Uplands	South Eastern	Chilterns	South Coastal	Weald	South Midlands	Midlands & East Anglia	Northern Sandstones	Fens	Eastern Lowlands	Central Southern	South Western	West Coastal	Lancashire Plain	Pennine Coal Measures	Scottish Lowlands	Midland Hills	Pennines	Western Coal Measures	North West Coastal	North Coast Carboniferous	Welsh Uplands	Igneous Coastal	Central Highlands	Highland Coastal	West Highlands
<7.0C January	x	x																								
Well-drained calc soil	x																									
>6.0 hrs sun July	x	x	x	x	x			x	x		x	x														
Chalk and oolites	x			x																						
<10 days snow cover		x	x							x	x	x	x	x												
Electrified		x	x														x									
<400' ASL	x				x	x									x				x							
<25' ASL				x					x																	
Alluvium		x			x								x													
Drift			x	x	x					x	x	x	x						x			x				
Stagnogleys			x	x	x	x	x								x	x		x	x		x					
<6.0 hrs sun July						x	x		x	x			x	x			x									
<20 days snow cover						x	x		x	x	x		x	x	x	x					x		x			
<100' ASL								x					x							x						
Salt marsh								x																		
Bunter														x		x										
Coal measures															x			x	x		x					
<200' ASL												x			x	x			x	x						
<30 days snow cover																		x							x	

Table 2.8 contd

	40	41	32	6	30	70	70	42	33	28	28	40	29	15	51	56	29	51	36	16	28	18	16	38	26	24
<6.0C January	x	x								x																
Non-calc. brown earths			x	x	x			x													x					
<5.5 hrs sun July				x	x	x	x	x		x																
<6.5C January						x	x	x	x	x																
Carboniferous & magnesian								x	x	x		x														
Igneous & intrusive								x		x	x															
>400' ASL							x		x				x													
Boulder clay							x	x	x		x		x			x	x									
Lowland podzols													x						x	x						
Heath/rough pasture										x	x		x	x					x		x		x			
Single track														x	x			x				x		x		
Metamorphic																									x	
Upland gleys																										x
Number of 10km sections	40	41	32	6	30	70	70	42	33	28	28	40	29	15	51	56	29	51	36	16	28	18	16	38	26	24

Total 893

as shown in Figure 2.21. Four transects were then located at right angles to the track which usually reflected the line of greatest variation. A varying number of 4m^2 (nested 4 and 25m^2 in woodland) quadrats were recorded, depending on the width of the verge. Percentage cover was measured to the nearest 5 per cent with separate categories of 1 and 2 per cent together with vegetation height, pH, slope and certain other environmental measurements. A total of 3,502 quadrats were recorded, containing some 667 species.

A last interesting point is that these stratified random samples were supplemented by a further 241 sites which were purposely located at sites of known ecological interest. These were thus chosen in an entirely subjective and biased manner. This demonstrates a point made previously, that very often, sampling for vegetation survey is deliberately biased in order to sample known or easily recognised areas of high floristic and ecological variation and interest.

Taken together, these case studies show the remarkably wide range of aims and objectives of vegetation description and demonstrate how methods almost always have to be adapted to suit particular scales and circumstances. **Adaptability** and **flexibility** are two key words when carrying out fieldwork in plant ecology and biogeography.

3

The nature and properties of vegetation data

Tabulation and checking of data

Virtually all methods for vegetation data analysis are applied to data based on **floristics**. Most physiognomic and structural data are presented in tabular or graphic form and are rarely suitable for any form of statistical manipulation.

The raw-data matrix

When a series of vegetation samples or quadrats have been thrown and floristic data collected, the results are usually drawn up in a table known as the **raw-data matrix** (Table 3.1). Typically, quadrats are displayed in columns and species in rows in the matrix, but it is equally valid to write the matrix in **transposed** form, with the quadrats in rows and species in columns (Table 3.2).

Table 3.1 New Jersey salt marsh data (Fresco in Cottam *et al.*, 1978)

	Quadrats											
Species	1	2	3	4	5	6	7	8	9	10	11	12
1. *Atriplex patula var hastata*	1	10	2	1	1	2	5		1		5	2
2. *Distichlis spicata*		15	80	2	10	15	30	1	10	10	20	
3. *Iva frutescens*							5	1	2	1	20	10
4. *Juncus gerardii*			1			40	1					
5. *Phragmites communis*								1	10	20	5	30
6. *Salicornia europaea*	5	10	2	1	1		2			2		
7. *Salicornia virginica*				5	10							
8. *Scirpus olneyi*						5	20				1	
9. *Solidago sempervirens*									1	5	1	2
10. *Spartina alterniflora*	75	30	5	20	5	1		10	1	2		
11. *Spartina patens*							20	10	50		2	5
12. *Suaeda maritima*				20	10							

The ordering of species in a matrix depends on the aims of the project and the type of analysis. Most data are listed in rows, with the species names written horizontally at the start and end of each line. The Latin binomial should always be used and species are usually listed alphabetically, as in Table 3.1, although sometimes they may be shown in groupings such as trees, shrubs, grasses, herbs, lichens and bryophytes. Species can also be listed by **constancy**, which is defined as the number of samples or quadrats in which each species occurs. Species are then ranked downwards from high to low constancy and the **commonness or rarity** of species within the data set is clearly displayed.

Within the matrix, species scores for either presence/absence or some form of

Table 3.2 The New Jersey salt marsh data transposed and in presence/absence form

| | Species | | | | | | | | | | | |
| Quadrats | 1 | 2 | 3 | 4 | 5 | 6 | 7 | 8 | 9 | 10 | 11 | 12 |
	Atriplex patula var. hastata	*Distichlis spicata*	*Iva frutescens*	*Juncus gerardii*	*Phragmites communis*	*Salicornia europaea*	*Salicornia virginica*	*Scirpus olneyi*	*Solidago sempervirens*	*Spartina alterniflora*	*Spartina patens*	*Suaeda maritima*
1	1	0	0	0	0	1	0	0	0	1	0	0
2	1	1	0	0	0	1	0	0	0	1	0	0
3	1	1	0	1	0	1	0	0	0	1	0	0
4	1	1	0	0	0	1	1	0	0	1	0	1
5	1	1	0	0	0	1	1	0	0	1	0	1
6	1	1	0	1	0	0	0	1	0	1	0	0
7	1	1	1	1	0	1	0	1	0	0	1	0
8	0	1	1	0	1	0	0	0	0	1	1	0
9	1	1	1	0	1	0	0	0	1	1	1	0
10	0	1	1	0	1	1	0	0	1	1	0	0
11	1	1	1	0	1	0	0	1	1	0	1	0
12	1	0	1	0	1	0	0	0	1	0	1	0

abundance data may be entered. Presence/absence data are usually shown with a 1 to denote presence and 0 or a blank to indicate absence (Table 3.2). Where abundance data are recorded the actual species scores within each quadrat are displayed with blanks or zeros where a species is absent (Table 3.1). An important characteristic of most vegetation data matrices is that they are **sparse**, meaning that there are a large number of zero or blank entries indicating species absences. In most raw data matrices, the number of blank entries far exceeds the number of filled entries or presences.

Great care must be taken in the preparation of a raw data table. The transfer of data from check sheets or field notebook is often extremely error-prone. Mistakes are most common with species that are found in only a few quadrats. Once completed, the table should always be **double-checked** against the original data.

Example data matrices

Three different sets of vegetation data are used throughout this book to explain methods of data analysis. All three sets are small in size and can thus be used to demonstrate hand-worked examples of various techniques for data analysis. Nevertheless they contain interesting internal variation.

Data set 1 — The New Jersey salt marsh data (Fresco in Cottam et al., 1978) (Tables 3.1 and 3.2)

This small data set has been chosen for two reasons. First, it has been presented and used quite widely in the literature on vegetation analysis in order to demonstrate and test different analytical methods. It can thus be used to test and calibrate both the methods and the computer programs available with this book. Second, it is an American data set in contrast with the other two which are

Table 3.3 Garigue and maquis data from Garraf, north-east Spain

Species	1	2	3	4	5	6	7	8	9	10	11	12	13	14	15
							Quadrats								
1. Bare rock/earth	3	3	6	4	3	4	6	6	6	3	6	3	6	5	6
2. *Brachypodium ramosum*	3				4										
3. *Ceratonia siliqua*			2				2		1						1
4. *Chamaerops humilis*	1			1										2	
5. *Cortaderia selloana*		1				2				2		4		2	
6. *Erica multiflora*	4			3	3					3					
7. *Euphrasia sp.*	2	1		2						1		2		2	
8. *Lavandula augustifolia*	1			1	1										
9. *Olea europaea*			3				1	4	2		4		3		3
10. *Phillyrea augustifolia*	2	3		3	2	3				3		1		4	
11. *Pinus halepensis*	3	2		3	3	2				2		3		3	
12. *Pistacea lentiscus*	3			2	3				4						
13. *Quercus coccifera*	5	4	1	6	4	5	2		1	4	1	4	1	5	1
14. *Rosmarinus officinalis*	3	1	2	2	3		2	3	3	1	2	2	1	1	2
15. *Salvia sp.*	1			3	3										
16. *Sedum sp.*		1				1			2			1		1	
17. *Smilax asper*			1				2	3		1			2	2	2
Aspect	N	N	S	N	N	N	S	S	S	N	S	N	S	N	S

Domin cover scale: 1 = present–1%; 2 = 2–5%; 3 = 6–10%; 4 = 11–20%; 5 = 21–50%; 6 = >50%

European. The data are in the form of percentage cover values and are part of a much larger data set.

Data set 2 — Garigue and maquis vegetation from the Garraf Massif, north-east Spain (Table 3.3)

The Garraf massif lies 20km south of Barcelona in north-east Spain and comprises a large area of limestone overlain by degraded Mediterranean vegetation, known as **maquis and garigue** (Figure 3.1). The former climax forest was probably either Kermes Oak Forest (*Quercus coccifera*) and/or Aleppo pine (*Pinus halepensis*). The vegetation has been subject to clearance for viticulture in many places over the historical period, but the terraces and vineyards have subsequently been abandoned. Regular burning has taken place and the exposed *terra rossa* soil cover has largely been eroded and washed into the valley bottoms, where it is still cultivated. The vegetation represents a post-fire successional stage which is still subject to intermittent burning (Plate 3.1). Aspect is also a very significant control of floristic variation in the Mediterranean region. The matrix comprises a small sub-set of quadrats collected from a transect across an east–west orientated hill ridge, with half the quadrats from the north-facing slope and half from the south-facing slope. A quadrat size of 5m × 5m was used to allow for the shrubby nature of the vegetation and the abundance data were collected using a six-point Domin cover scale.

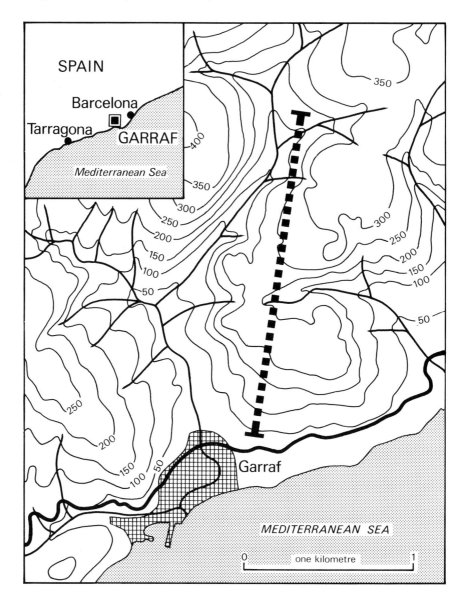

Figure 3.1 The location of the vegetation transect at Garraf, south of Barcelona, north-east Spain

Data set 3 — Data on upland grassland and heathland at Gutter Tor, Dartmoor, south-west England (Table 3.4)

These data comprise 25 quadrats, from a student exercise, collected along a transect up a tor on south-west Dartmoor, Devon, England (Plate 3.2). The vegetation is typical of the uplands of western and northern Britain with grazed, short-cropped grassland, in some places invaded by bracken (*Pteridium aquilinum*), patches of heath, represented by species of heather (*Calluna vulgaris*, *Erica cinerea*) and bilberry (*Vaccinium myrtillus*). At the bottom of the hill is a small valley bog,

Plate 3.1 Garigue vegetation near Garraf, south of Barcelona, north-east Spain (photo: M. Kent)

within which are found species such as the round-leaved sundew (*Drosera rotun-difolia*) and bog asphodel (*Narthecium ossifragum*). The area is heavily grazed by sheep, cattle and horses and has been subject to burning at infrequent intervals. A quadrat size of 1m² was used with subjective estimates of percentage cover.

Multivariate data and multivariate analysis

Floristic data recorded in the two-way (quadrat by species) marix are **multivariate data**. Each species or quadrat added to the data set represents a potential source of variation and each extra variable or sample constitutes an extra dimension. The data are said to be **multidimensional**, and the methods for analysing such data are thus techniques for **multivariate analysis**.

In order to understand how multivariate methods of vegetation analysis work, it is necessary to explain the concept of **species** and **sample space**. Table 3.5 presents data for six species in three quadrats or samples. Figure 3.2a shows the species in the one-dimensional sample space of quadrat X, Figure 3.2b shows the species within the two-dimensional sample space of quadrats X and Y, while Figure 3.2c shows them within the three-dimensional space of quadrats X, Y and Z. The position of each species is thus determined by the coordinates of its scores in quadrats X, Y and Z. Any further quadrats added to these would increase the dimensions by one for each additional quadrat. However, although this is simple to describe mathematically, it cannot be shown geometrically beyond three dimensions.

It is clearly possible to construct this geometrical diagram the other way around, with quadrats defined in terms of species space. However, again this could only be

Table 3.4   Vegetation and environmental data from Gutter Tor, Dartmoor, S.W. England

Species	Quadrats (% cover)																								
	1	2	3	4	5	6	7	8	9	10	11	12	13	14	15	16	17	18	19	20	21	22	23	24	25
1. *Agrostis capillaris*			80						1				70	90				90		75				65	
2. *Agrostis curtisii*				80						25										15					
3. *Bryophytes*				15		70	15															5			
4. *Calluna vulgaris*				10	35						25										90	5			2
5. *Carex nigra*	15	10							5	25	10					100									10
6. *Cerastium glomeratum*													1					2							
7. *Cladonia portentosa*									10	10	5					15					20	10			
8. *Danthonia decumbens*													5	15											
9. *Drosera rotundifolia*	2											2													
10. *Erica cinerea*										5						20									
11. *Erica tetralix*				5							20	25										25			
12. *Festuca ovina*			10		40		40	40					30	20	30	5	20	20	10	25	1		50	50	
13. *Galium saxatile*			10		1	5	35	40	5	10		5	10		20	5	5	10	25				50	25	
14. *Juncus effusus*	5	15							5			5													20
15. *Juncus squarrosus*																			5	3					
16. *Luzula sylvatica*																			3						
17. *Molinia caerulea*	35	50		40		20				25	75						90					95			10
18. *Narthecium ossifragum*	20											10													
19. *Plantago lanceolata*																								2	
20. *Potentilla erecta*			20	3		5		10	2	10	5		10	1	5	10	10	5	2					15	
21. *Pteridium aquilinum*								90	75						60				100				95		
22. *Sphagnum sp.*	80	80									55														90
23. *Taraxacum officinale*																		1							
24. *Trichophorum cespitosum*		5							2			10													5
25. *Trifolium repens*			3															3							
26. *Vaccinium myrtillus*					50		20			20	10			20							20	5		5	
27. *Viola riviniana*													2			5			5					5	

Table 3.4 contd.

Total cover	157	160	123	154	127	102	110	190	105	130	150	119	128	147	115	127	130	136	147	128	131	145	195	162	144
Number of species	6	5	5	7	5	5	5	5	8	8	7	8	7	6	4	6	5	8	6	6	4	6	3	6	8
Environmental data																									
Soil/peat depth (cm)	54	42	12	15	16	11	9	15	40	38	31	25	17	10	14	15	23	18	20	10	8	16	27	13	55
Slope angle°	3	1	20	6	18	10	9	20	4	3	4	4	15	12	25	9	10	15	17	5	7	2	12	13	1
pH	4.8	4.6	3.7	3.9	4.0	3.7	4.0	3.7	4.4	4.2	3.9	4.7	3.8	4.0	4.0	3.7	4.2	3.9	4.2	3.9	3.6	4.0	4.1	4.2	5.2
% soil moisture	95	110	34	75	55	52	41	31	95	76	67	105	35	43	15	50	72	36	21	43	50	72	33	21	112
Grazing (Number of faecal units)																									
Cattle	—	—	2	1	1	1	—	1	—	2	2	2	2	1	—	2	2	3	1	2	—	1	—	1	—
Sheep	—	—	4	4	3	2	3	1	—	1	6	5	6	—	1	—	6	—	2	4	—	—	1	1	—
Horses	—	—	1	—	1	—	—	1	—	1	—	1	1	—	1	—	1	—	2	2	1	—	1	2	—
Rabbits	—	—	1	1	—	—	1	1	—	—	1	1	—	1	—	1	1	1	1	1	—	1	—	—	—
Total	0	0	7	5	5	3	3	4	0	3	5	8	9	1	1	2	11	3	9	3	1	2	2	4	0

Plate 3.2 Vegetation at Gutter Tor, Dartmoor, south-west England (photo: M. Kent)

Table 3.5 A matrix of 3 quadrat samples × 6 species to illustrate the geometrical models of species in sample space and samples in species space

	Samples/quadrats		
Species	X	Y	Z
a	2	7	2
b	6	0	9
c	8	1	5
d	1	9	6
e	5	3	5
f	9	4	4

done graphically for three species, although mathematically there is no limit — every further species included adds another dimension. Figure 3.2d shows the three quadrats X, Y and Z defined in terms of only the first three species: a, b and c — d, e and f, the fourth, fifth and sixth cannot be used in the geometrical diagram but can be used mathematically.

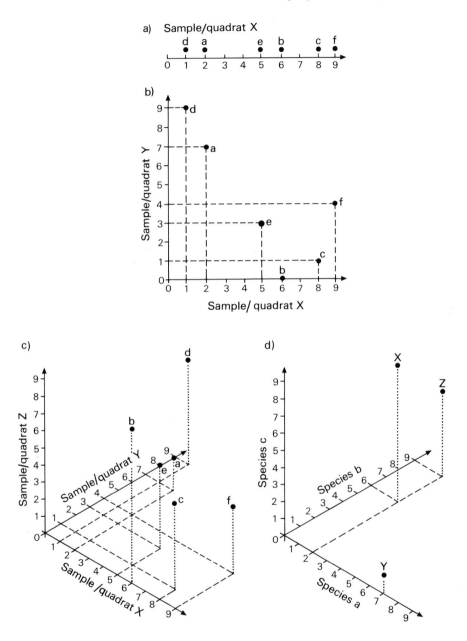

Figure 3.2   The geometrical model of species in sample space: (a) one dimension (sample X); (b) two dimensions (samples X and Y); (c) three dimensions (samples X, Y and Z); (d) samples X Y and Z in three species' space (species a, b and c)

Redundancy and noise

A two-way species by samples matrix is often described as having a moderate to large amount of **redundancy** within it, in that many samples or quadrats are very similar to others and thus duplicate variation in the data. Equally, many species are like other species in their distribution across a series of samples or quadrats, and again the variation is effectively duplicated. A related concept is that most methods for analysing vegetation data are concerned with **data reduction**, whereby the original data, containing a lot of redundancy, are reduced down to the major sources of variation within them, eliminating the duplication or redundancy in the process.

Another property of vegetation data matrices is **noise**. Noise is best explained in terms of plant response to environment. If two samples are taken from the same local plant community, although they will be very similar to each other, they will rarely be identical. Thus a dominant species in one sample could be recorded as having 90 per cent cover, while in another sample from the same community, only 70 per cent cover. Similar variations will occur for all other species recorded in the two samples. The observed differences between the two samples can partly be explained by very local differences in environmental factors, the chance occurrence of individuals, local variations in biotic pressures such as burning and grazing or errors or variations in the recording of abundance. However, these minor sources of variation are usually of little interest compared to the much greater differences among the samples from this community type and those from other contrasting community types. At this broader scale, the data from the original community are 'noisy' because nearly identical samples show variation which is of no great ecological significance at the broader scale. All data sets thus contain two kinds of variability; one which is important and interesting and assists with explanation and understanding of plant-environment relationships, and other, more detailed variation, which is of very little interest and can interfere with and blur the overall picture.

While this description of noise is somewhat simplified, it is an important concept in vegetation analysis. Further discussion of the subject can be found in Pignatti (1980) and Gauch (1982a,b).

Types of data

The values held within raw data matrices vary in type, and this has important implications for the various statistical methods described in subsequent chapters. Of the range of data types which can be identified, four main categories are described:

(a) **Nominal** — this type of measurement involves categorisation without numerical values or ranks. Presence/absence data coded as ± or 1/0 is an example of this type. However, the data may also be in a series of divisions by name. Colours are a good example, or trees, shrubs, grasses, herbs and bryophytes. Such data cannot be manipulated according to the basic rules of arithmetic. One colour is not greater or less than another, for example. Nor can a colour be divided!

(b) **Ordinal** — such data can be placed in rank order along a continuum. Some of the simpler abundance scales for recording vegetation data are of this type, for example dominant, abundant, frequent, occasional and rare. All four basic arithmetical operations can be performed on ranked data. The Braun–Blanquet scale (Chapter 2) is another good example.

(c) **Interval** — interval data, have a constant unit of measurement and thus differences between values can be compared. Temperature is the most widely quoted example. A difference of 3 degrees means the same across the whole scale. However, the position of zero in an interval scale is arbitrary. A consequence of this is that a given difference can then be said to be twice as great as another, but not that a certain object is twice as large as the other when the corresponding values are in the ratio 2:1. For example, a temperature of 40°C is not twice a temperature of 20°C.

(d) **Ratio** — these are the most precisely defined type of data which differ from the interval scale in one respect only — they have an absolute zero point. It is thus possible to define ratios. Again to illustrate the difference, if one quadrat is 2m from a tree and another is 4m, then the second is twice as far from the tree as the first. However, if the temperature one day is 10°C and 20°C the next, it is not twice as warm on the second day; those would be interval data.

For the purposes of this book, the differences between the interval and ratio scales are of little consequence, since most vegetation data are ratio data, but the differences between interval/ratio, ordinal and nominal types are very important. Also data collected on the interval/ratio scales usually take more time to collect than data on ordinal and nominal scales. This is because greater precision is required. However, in some situations, nominal data, such as presence/absence, may suffice and save a great deal of time, effort and money.

Measurement of association and similarity between species and samples

General

One of the simplest means of analysing floristic vegetation data is to look at the degree of **association** between species and the **level of similarity** between quadrats or samples. Such ideas are fundamental to most of the more complex multivariate methods covered later in this text.

Chi-square (χ^2) as a measure of association between species

In Chapter 1, the plant community was defined as an assemblage of plant species which show a definite **affinity or association** with each other. There is no doubt that certain species tend to grow together in certain locations, while others never ever coexist. These ideas can be represented by the concepts of **positive** and **negative** association. If two species are found to be positively associated, this means that they are found growing together more often than would be expected by chance or random events. Conversely, negative association means that one species is found growing **without** the other more often than would be expected by chance.

The degree of association between species in a set of samples can be quantified. One of the most widely used methods of measuring association is that of chi-square (χ^2) using 2×2 contingency tables.

The contingency table

In order to calculate χ^2 as a measure of association, the data on presence/absence of the two species from a set of quadrats need to be arranged into a 2×2 **contingency table** of the form shown below:

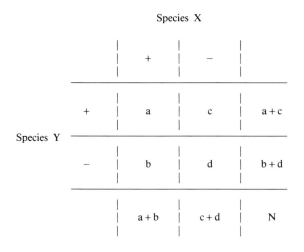

The table is drawn up for two species X and Y whose presence/absence is measured in N quadrats. In each quadrat or sample, there are four possible situations in terms of presence and absence. Both species X and Y can be present ($+X/+Y$ — cell a); X can be present but Y not ($+X/-Y$ — cell b); Y can be present but X not (-X/$+Y$ — cell c); and neither X nor Y may be present ($-X/-Y$ — cell d). In the marginal totals, $a+b$ gives the number of occurrences of X, and $a+c$ the number of occurrences of Y; $c+d$ shows the number of quadrats not containing X and $b+d$ the number of quadrats not containing Y.

Calculation of χ^2 can be explained by an actual example. From a set of 180 quadrats thrown on acidic upland grassland in Britain, the following contingency table was constructed:

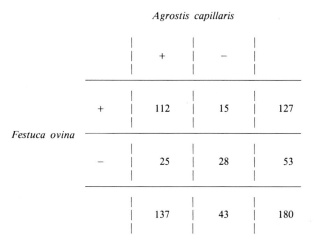

χ^2 is calculated using the contingency table and the following formula:

$$\chi^2 = \frac{(|ad-bc| - 0.5N)^2 \times N}{(a+b)\ (c+d)\ (a+c)\ (b+d)}$$

(3:1)

where a,b,c,d and N are as above.

$|ad - bc|$ = the absolute difference of ad and bc regardless of sign.

This formula includes Yates's correction for small samples, which is the $-0.5N$ on the top line of the formula. For large samples (over 500) this may be omitted. The χ^2 formula is based on the sum of differences between the **observed** and **expected** frequencies of each cell of the contingency table. The method of calculating expected frequencies is explained below. In the above example, the formula gives:

$$\chi^2 = \frac{(|3136 - 375| - 90)^2 \times 180}{137 \times 43 \times 127 \times 53}$$

$$\chi^2 = \frac{1284163380}{39652321}$$

$$\chi^2 = 32.39$$

χ^2 significance tables are available in most statistical textbooks (for example Hammond and McCullagh, 1978; Bishop, 1983; Lee and Lee, 1982; Wardlaw, 1985) and show that with **one degree of freedom, χ^2 has to exceed 3.84 to be significant at the 0.05 level and 6.64 to be significant at the 0.01 level.** Clearly in this example ($\chi^2 = 32.39$), a very high level of significance is achieved which means that the two species *Agrostis capillaris* and *Festuca ovina* are associated with each other more than would be anticipated by chance.

However, although a highly significant association between the two species has been found, the **nature** of the relationship, and in particular whether it is positive or negative, has not been determined. To do this, the **expected frequency** for the cell of the contingency table containing the joint occurrences of species X and Y ($+X/+Y$ — cell a) is calculated and then compared with the value for the **observed frequencies.** The observed value in the above contingency table is 112. The expected value is calculated by taking the probability or likelihood of *Agrostis capillaris* occurring in the data set ($a + b/N$ = 137/180) and the probability of *Festuca ovina* occurring in the data set ($a + c/N$ = 127/180) and multiplying the two probabilities together to get the joint probability or expected frequency.

Expected frequency of cell a (joint occurrences) $= \dfrac{(a + b) \times (a + c)}{N} = \dfrac{137 \times 127}{180} = 96.7$

Thus the observed frequency for joint occurrences of the two species is 112, while the expected frequency is 96.7.

If the **observed** value for joint occurrences **exceeds** the **expected** value, then the two species are **positively** associated. Clearly this is the case in the above example and *Agrostis capillaris* and *Festuca ovina* tend either to be present together or absent together. Only infrequently does one species occur without the other. If, however, the observed value had been **less than** the expected value, then the association would have been **negative** and the species would rarely have been found together: when one species was found, the other would more likely than not be absent.

Exactly why two species should be positively associated is not always as obvious

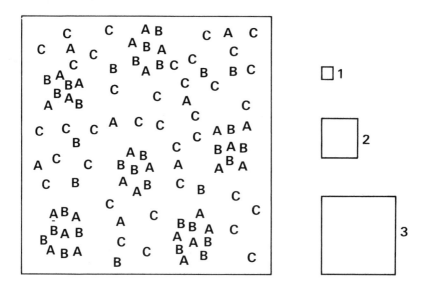

Figure 3.3   The effect of quadrat size on determination of association between plant species (redrawn from Kershaw and Looney, 1985; with kind permission of Edward Arnold/Hodder and Stoughton)

as it might at first seem. Clearly the most common reason why two species regularly appear together is because they prefer the same environmental conditions. However, further investigation often reveals such an explanation to be over-simplistic, and factors such as plant species strategies, competition and interaction also need to be taken into account.

Problems of using χ^2 for the analysis of association

Quadrat size is very important. In the above example, 1m² quadrats were used. If quadrat size is too small and approaches the size of some individual plants, then negative associations will appear which mean nothing more than that two individuals of different species cannot occupy the same location (quadrat size 1, see Figure 3.3). If the quadrat size is too large for the local pattern of the vegetation, then no information on the underlying associations would be obtained (quadrat size 3 — see Figure 3.3 — here every placing of the quadrat would include all three species). If the morphology of some species for which associations are being calculated is very different, then spatial exclusion by the large species may cause positive associations among the smaller species. Thus quadrat size must be chosen in relation to the size of the dominant species, but different sizes may be needed for species of highly differing morphologies and it may not be valid to calculate association between species which differ greatly in morphology. In Figure 3.3, quadrat size 2 is the optimal size and will show up the true associations among the species. This is because the area of quadrat 2 is approximately the same as that of the groups of species A and B, which are highly clustered.

Association between quadrats or samples

Just as it is possible to calculate χ^2 between species to examine association, so it can be calculated between quadrats or samples. To students, this often seems a strange concept at first, but the association in this case is best described as similarity and dissimilarity of floristic composition between samples. Remembering that association is only calculated with presence/absence data, the more species that two quadrats have in common, the more similar they are in terms of species composition and a high χ^2 value coupled with positive association means that the quadrats must contain very similar species. Equally, a high χ^2 value can be obtained with negative association. This means that the quadrats have very few species in common and are very dissimilar or unlike. Also, the χ^2 test has been shown to be unreliable if the expected value in any one cell of the contingency table is less than 5. This is not a rigid rule, although it is worth checking for.

Measurement of similarity and dissimilarity

Many measures exist for the assessment of similarity or dissimilarity between vegetation samples or quadrats. Some are qualitative and based on presence/absence data, while others are quantitative and will work on abundance data. Many will cater for both data types. **Similarity indices** measure the degree to which the species composition of quadrats or sample matches is alike. **Dissimilarity coefficients** assess the degree to which two quadrats or samples differ in composition. It follows that dissimilarity is the complement of similarity.

Of the large choice available, four, in particular, are worth introducing here: they are the Jaccard coefficient and the Sørensen coefficient which are generally applied to qualitative data; and the Czekanowski coefficient and the coefficient of squared Euclidean distance, both of which are suitable for either quantitative or qualitative data.

The Jaccard coefficient

Jaccard (1901; 1912; 1928) developed a very simple mathematical expression of similarity. It was originally used to compare the general floras of larger areas but has subsequently been found suitable for assessing the similarity of quadrats in terms of species composition. The formula is:

$$S_J = \frac{a}{a+b+c} \qquad (3:2)$$

where S_J = Jaccard similarity coefficient
         a = number of species common to both quadrats/samples
         b = number of species in quadrat/sample 1
         c = number of species in quadrat/sample 2

Often the coefficient is multiplied by 100 to give a percentage similarity figure. Dissimilarity is then:

$$D_J = \frac{b+c}{a+b+c} \text{ or } 1.0-S_J \qquad (3:3)$$

The figure can be converted to a percentage if required.

The Sørensen coefficient

This coefficient of similarity (S_S) is defined using the same symbolism, as:

$$S_S = \frac{2a}{2a + b + c} \tag{3:4}$$

Dissimilarity is then:

$$D_S = \frac{b+c}{2a+b+c} \text{ or } 1.0 - S_S \tag{3:5}$$

Again, the figure can be calculated as a percentage if required.

Both coefficients can be used on quantitative as well as qualitative data.

Although Sørensen's coefficient was first published in 1948, it is clearly very similar to that of Jaccard. Generally, Sørensen's coefficient is preferred to the Jaccard because it gives weight to the species that are common to the quadrats or samples rather than to those that only occur in either sample.

As an example, the Sørensen coefficient is calculated for quadrats 11 and 12 of the New Jersey salt marsh data set (Table 3.2). In this case, presence/absence data are being used.

Species	Quadrat 11	Quadrat 12
Atriplex patula	1	1
Distichlis spicata	1	0
Iva frutescens	1	1
Juncus gerardii	0	0
Phragmites communis	1	1
Salicornia europaea	0	0
Salicornia virginica	0	0
Scirpus olneyi	1	0
Solidago sempervirens	1	1
Spartina alterniflora	0	0
Spartina patens	1	1
Suaeda maritima	0	0
Total occurrences	7 (b)	5 (c)

Number of joint occurrences (a) 5

$$S_S = \frac{2a}{2a+b+c} = \frac{2 \times 5}{10+7+5} = 0.45 \ (45\%)$$

$$D_S = \frac{b+c}{2a+b+c} \text{ or } 1.0 - S_S$$

$$D_S = \frac{7+5}{10+7+5} = 0.55 \ (55\%)$$

The Czekanowski coefficient

Czekanowski (1913) devised a further coefficient which is very similar to Jaccard's and which is applicable to either quantitative or qualitative data.

$$S_C = \frac{2 \sum\limits_{i=1}^{m} \min (X_i, Y_i)}{\sum\limits_{i=1}^{m} X_i + \sum\limits_{i=1}^{m} Y_i} \qquad (3:6)$$

where X_i and Y_i = the abundances of species i

$\sum\limits_{i=1}^{m} \min (X_i, Y_i)$ = the sum of the lesser scores of species i where it occurs in both quadrats

        m   = number of species

The coefficient values range from 0 (complete dissimilarity) to 1 (total similarity).

Using the same quadrats (11 and 12) from the New Jersey data but with the quantitative values (Table 3.1), the Czekanowski coefficient is calculated below. The lower values of joint occurrences are underlined and when summed give a total of 20.

Species	Quadrat 11	Quadrat 12
Atriplex patula	5	<u>2</u>
Distichlis spicata	20	<u>0</u>
Iva frutescens	20	<u>10</u>
Juncus gerardii	0	<u>0</u>
Phragmites communis	<u>5</u>	30
Salicornia europaea	<u>0</u>	0
Salicornia virginica	0	0
Scirpus olneyi	1	0
Solidago sempervirens	<u>1</u>	2
Spartina alterniflora	<u>0</u>	0
Spartina patens	<u>2</u>	5
Suaeda maritima	<u>0</u>	0
Total cover	54	49

Sum of the lesser scores of species common to both quadrats (underlined values) = 20

Thus

$$S_C = \frac{2 \times 20}{54 + 49} = 0.39 \ (39\%)$$

A value for dissimilarity can be calculated by subtracting the similarity value away from 1.0 (or 100%). Thus dissimilarity = 0.61 (61%).

This coefficient is probably the most useful of those discussed so far and is also the most widely known because of its use in Bray and Curtis or polar ordination (Bray and Curtis, 1957) — see Chapter 5.

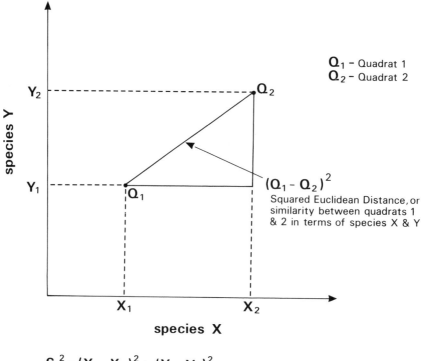

$$S_E^2 = (X_1 - X_2)^2 + (Y_1 - Y_2)^2$$

where S_E^2 = Squared Euclidean Distance between quadrats 1 & 2
        $X_1 X_2$ = Scores for species X in quadrats 1 & 2
        $Y_1 Y_2$ = Scores for species Y in quadrats 1 & 2

Figure 3.4   Squared Euclidean distance between two quadrats (1 and 2) in terms of two species (X and Y)

The coefficient of squared Euclidean distance

This coefficient is based on the Euclidean properties of a right-angled triangle and the fact that the square on the hypotenuse is equal to the sum of the squares on the opposite two sides (Figure 3.4). If two species X and Y occur in two quadrats 1 and 2, the similarity or 'distance' between the two species in geometric space is defined as:

$$S_E = \sqrt{(X_1 - X_2)^2 + (Y_1 - Y_2)^2} \tag{3:7}$$

For more than two species, the generalised formula becomes:

$$D_{ij} = \sqrt{\sum_{k=1}^{m} (X_{ik} - X_{jk})^2} \tag{3:8}$$

where D_{ij} = squared Euclidean distance between quadrats i and j
       m   = number of species
      X_{ik} = the abundance of the kth species in quadrat i
      X_{jk} = the abundance of the kth species in quadrat j

The lower the value of the squared Euclidean distance coefficient between two quadrats, the more similar they are in terms of species composition. If quadrats 11 and 12 of the New Jersey data are again taken to illustrate calculation, then:

Species	Quadrat 11	Quadrat 12	Difference	Difference^2
Atriplex patula	5	2	3	9
Distichlis spicata	20	0	20	400
Iva frutescens	20	10	10	100
Juncus gerardii	0	0	0	0
Phragmites communis	5	30	25	625
Salicornia europaea	0	0	0	0
Salicornia virginica	0	0	0	0
Scirpus olneyi	1	0	1	1
Solidago sempervirens	1	2	1	1
Spartina alterniflora	0	0	0	0
Spartina patens	2	5	3	9
Suaeda maritima	0	0	0	0

$$\sum_{k=1}^{m} = 1,145$$

$$S_E = \sqrt{1145} = 33.84$$

The lower limit of the coefficient is 0, representing complete similarity. However, there is no fixed upper limit for this coefficient. For this reason it is known as a coefficient of dissimilarity.

(Dis)similarity matrices

If χ^2 as a measure of association is calculated between all pairs of species in a data set or the (dis)similarity is calculated between all pairs of quadrats, then a **half-matrix** either of associations (χ^2) or of similarity (similarity coefficients) is formed. As an example, Table 3.6 contains a dissimilarity half-matrix of Czekanowski coefficients for the 12 quadrats in the New Jersey salt marsh data (Table 3.1). These types of matrices form the basis of some methods of classification and ordination discussed in Chapters 5, 6 and 8. Another example is given in the case study at the end of this chapter.

Diversity and species richness

The whole subject of species richness and diversity is fraught with problems and misunderstandings. There is confusion over the meaning of diversity, over methods for measuring and assessing diversity and over the ecological interpretation of different levels of diversity. In particular the relationship between diversity and stability in ecosystems has attracted much research and discussion (May, 1973; 1981; 1984; Goodman, 1975; Jacobs, 1975; Margalef, 1975; Orians, 1975; Pielou, 1975; Pimm, 1984; Hill, 1987; van der Maarel, 1988b; During *et al.*, 1988).

Magurran (1988) has provided a valuable review of concepts of diversity and

Table 3.6   A dissimilarity half-matrix of Czekanowski coefficients calculated between the 12 quadrats of the New Jersey salt marsh data (Table 3.1)

Quadrat number	Quadrat number											
	1	2	3	4	5	6	7	8	9	10	11	12
1	0.00											
2	0.51	0.00										
3	0.91	0.69	0.00									
4	0.66	0.58	0.87	0.00								
5	0.88	0.67	0.73	0.44	0.00							
6	0.97	0.72	0.75	0.93	0.76	0.00						
7	0.96	0.70	0.60	0.94	0.80	0.69	0.00					
8	0.81	0.75	0.89	0.69	0.80	0.95	0.77	0.00				
9	0.97	0.83	0.86	0.94	0.79	0.83	0.58	0.71	0.00			
10	0.93	0.73	0.79	0.89	0.66	0.79	0.79	0.84	0.60	0.00		
11	0.99	0.66	0.69	0.94	0.76	0.69	0.52	0.87	0.67	0.64	0.00	
12	0.99	0.97	0.97	0.98	0.98	0.96	0.82	0.81	0.69	0.48	0.61	0.00

stresses that the word is hard to define. Species richness, meaning a count of the number of plant species in a quadrat, area or community is often equated with diversity. When ecologists talk of high diversity, they often mean a community containing a large number of different species. However, as Magurran states, most methods for measuring diversity actually consist of two components. The first is species richness but the second is the relative abundance (evenness or unevenness) of species within the sample or community. Perfect evenness of five species in a quadrat would mean that in 100 per cent cover, they were distributed 20, 20, 20, 20, 20. Diversity is thus measured by recording the number of species and their relative abundances. The two components of species richness and evenness may then be examined separately or combined into some form of index.

Alpha and beta diversity

Whittaker (1965, 1972a, 1975) made a distinction between two types of diversity — **alpha** and **beta** diversity. **Alpha diversity** is the number of species **within** a chosen area or community such as one type of woodland or grassland. **Beta** diversity is the difference in species diversity **between** areas or communities. Beta diversity is thus sometimes called **habitat diversity** because it represents differences in species composition between very different areas or environments and the rapidity of change of those habitats. It is also related to Whittaker's concept of plant communities as mosaics (Chapter 1). The smaller and more numerous the 'pieces' of the mosaic, the higher the beta diversity. Alpha diversity remains the number of species within each individual piece of the mosaic.

Diversity indices

A large number of indices of diversity have been devised, each of which seeks to express the diversity of a sample or quadrat by a single number. Some indices are also known by more than one name and are presented in different ways (for example Hill, 1973a; Southwood, 1978).

Of the various indices, the most frequently used is the simple totalling of species numbers to give species richness (Magurran, 1988). However, of the indices that combine species richness with relative abundance, probably the most widely used is the Shannon index. The other index occasionally used in plant ecology is McIntosh's diversity index (U).

The Shannon diversity index

This is derived from information theory and the concept that the diversity or information in a sample or community can be measured in a similar way to the information contained within a message or code. The index (H′) is sometimes correctly called the Shannon–Wiener index but elsewhere is mistakenly identified as the Shannon–Weaver index. The index makes the assumption that individuals are randomly sampled from an 'infinitely large' population and also assumes that all the species from a community are included in the sample. This last assumption is not always easy to meet and presupposes that the ecologist knows exactly what the complete species composition of the community is; a very difficult question for most plant ecologists!

The Shannon diversity index is calculated from the formula:

$$\text{Diversity } H' = -\sum_{i=1}^{s} p_i \ln p_i \qquad (3{:}9)$$

where s   = the number of species
$\quad$ p_i = the proportion of individuals or the abundance of the ith
$\qquad\qquad$ species expressed as a proportion of total cover
$\quad$ ln   = log base$_n$

Any base of logarithms may be taken; with $\log_2$ and $\log_{10}$ being the most popular choices. Obviously the choice of log base must be kept constant when comparing diversity between samples. Values of the index usually lie between 1.5 and 3.5, although in exceptional cases, the value can exceed 4.5. If a sample is used, then the true value of p_i is unknown but is estimated as n_i/N, the maximum likelihood estimator (Pielou, 1969).

It is also possible to calculate an equitability or evenness index of the form:

$$\text{Equitability } J = \frac{H'}{H'_{max}} = \frac{\sum_{i=1}^{s} p_i \ln p_i}{\ln s} \qquad (3{:}10)$$
(evenness)

where s   = the number of species
$\quad$ p_i = the proportion of individuals of the ith species or the abundance
$\qquad\qquad$ of the ith species expressed as a proportion of total cover
$\quad$ ln   = log base$_n$

As an example of the use of the index, two sample quadrats from the Gutter Tor, Dartmoor data set (Table 3.4) using $\log_{10}$ are taken.

Quadrat 1 — 6 species present

	% cover	Proportion of total cover (p_i)	$\ln p_i$	$p_i \ln p_i$	$p_i (\ln p_i)^2$
Carex nigra	15	0.096	−2.343	−0.225	0.527
Drosera rotundifolia	2	0.013	−4.363	−0.055	0.242
Juncus effusus	5	0.032	−3.447	−0.110	0.380
Molinia caerulea	35	0.223	−1.501	−0.334	0.502
Narthecium ossifragum	20	0.127	−2.061	−0.262	0.540
Sphagnum sp.	80	0.560	−0.674	−0.378	0.254
Total cover	157			$\Sigma p_i \ln p_i$ (H′) = 1.364	

The diversity of quadrat 1 (H′) = 1.364. The index is best calculated by drawing up a table as above. In cases where a t-test is being used to compare two quadrats, it is convenient to add a further column giving values of $p_i(\ln p_i)^2$. The formula for the Shannon index commences with a negative sign to cancel out the minus created when taking logarithms of the proportions.

Quadrat 25 — 8 species present

	% cover	Proportion of total cover	$\ln p_i$	$p_i \ln p_i$	$p_i (\ln p_i)^2$
Calluna vulgaris	2	0.013	−4.277	−0.056	0.238
Carex nigra	10	0.069	−2.667	−0.184	0.491
Drosera rotundifolia	2	0.013	−4.277	−0.056	0.238
Juncus effusus	20	0.138	−1.974	−0.027	0.538
Molinia caerulea	10	0.069	−2.667	−0.184	0.491
Narthecium ossifragum	5	0.035	−3.360	−0.118	0.395
Sphagnum sp.	90	0.625	−0.470	−0.294	0.138
Trichophorum cespitosum	5	0.035	−3.360	−0.118	0.395
Total cover	144			$\Sigma p_i \ln p_i$ (H′) = 1.037	

The diversity of quadrat 25 (H′) = 1.037. The equitability or evenness of the two quadrats can now be calculated using the formula above:

$$\text{Quadrat 1} \quad J = \frac{H'}{\ln s} = \frac{1.364}{\ln 6} = \frac{1.364}{1.792} = 0.761$$

$$\text{Quadrat 25} \quad J = \frac{H'}{\ln s} = \frac{1.037}{\ln 8} = \frac{1.037}{2.079} = 0.499$$

The higher the value of J, the more even the species are in their distribution within the quadrat. Thus quadrat 1 has a more even distribution than quadrat 25. On the basis of the Shannon diversity index, quadrat 1 is more diverse than quadrat 25, even though quadrat 25 has 8 species as opposed to 6. This demonstrates the manner in which evenness as well as species richness are combined in the index.

Hutcheson (1970) and Magurran (1988) describe a method for calculating the variance of H and a method of calculating 't' to test for significant differences between quadrats or samples using the right-hand column of the above examples.

McIntosh's diversity index (U)

McIntosh (1967a) presented an index (U) with the form:

$$U = \sqrt{\sum_{i=1}^{S} n_i^2} \tag{3:11}$$

where
U = McIntosh diversity index
S = the number of species
n = number of individuals or abundance of the ith species in the quadrat/sample

This index is related to the concept of squared Euclidean distance discussed earlier. The resulting form is dependent on the total abundance of the quadrat sample (N), and it is possible to calculate an index of dominance (D) allowing for N as follows:

$$D = \frac{N - U}{N - \sqrt{N}} \tag{3:12}$$

and a measure of evenness (E) from:

$$E = \frac{N - U}{N - N/\sqrt{s}} \tag{3:13}$$

The calculation of the index for quadrat 1 of the Gutter Tor data is shown below:

Quadrat 1 — 6 species present

	n_i	n_i^2
Carex nigra	15	225
Drosera rotundifolia	2	4
Juncus effusus	5	25
Molinia caerulea	35	1225
Narthecium ossifragum	20	400
Sphagnum sp.	80	6400
Total cover	157	$\sum_{i=1}^{S} n_i^2 = 8279$

$$U = \sqrt{\sum_{i=1}^{S} n_i^2} = \sqrt{8279} = 90.99$$

For quadrat 25 the calculation is:

Quadrat 25 — 8 species present

	n_i	n_i^2
Calluna vulgaris	2	4
Carex nigra	10	100
Drosera rotundifolia	2	4
Juncus effusus	20	400
Molinia caerulea	10	100
Narthecium ossifragum	5	25
Sphagnum sp.	90	8100
Trichophorum cespitosum	5	25
Total cover	144	$\sum\limits_{i=1}^{S} n_i^2 = 8758$

$$U = \sqrt{\sum_{i=1}^{S} n_i^2} = \sqrt{8758} = 93.58$$

Account must be taken of total cover (sample size), and the dominance measure (D) and evenness index (E) are calculated as follows:

For Quadrat 1:

$$D = \frac{N - U}{N - \sqrt{N}} = \frac{157 - 90.99}{157 - \sqrt{157}} = \frac{66.01}{144.47} = 0.456$$

and a measure of evenness (E) from:

$$E = \frac{N - U}{N - N\sqrt{s}} = \frac{157 - 90.99}{157 - 157/\sqrt{6}} = \frac{60.01}{92.91} = 0.711$$

For Quadrat 25:

$$D = \frac{N - U}{N - \sqrt{N}} = \frac{144 - 93.58}{144 - \sqrt{144}} = \frac{50.42}{132.00} = 0.382$$

and a measure of evenness (E) from:

$$E = \frac{N - U}{N - N\sqrt{s}} = \frac{144 - 93.58}{144 - 144/\sqrt{8}} = \frac{50.42}{93.09} = 0.542$$

Thus dominance is greater in quadrat 1 than in quadrat 25.

Dominance-diversity curves (Rank abundance diagrams)

The application of single-figure diversity indices to characterise complex community structure can be criticised because so much of the original species information is lost. In consequence, various workers, notably Whittaker (1965; 1975), have plotted the graph of the proportional abundance of species in a sample or quadrat on a log scale against their rank from most to least abundant (Figure 3.5). The form of the resulting line or curve can be used to describe the evenness of species distribution and relative species dominance within a community. Dominance is the

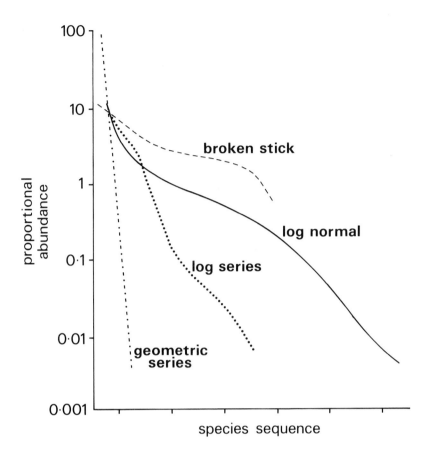

Figure 3.5 Dominance-diversity curves (rank abundance plots) showing the typical form of four species abundance models: (a) geometric series; (b) log series; (c) log normal; (d) broken stick. In these graphs, the abundance of each species is plotted on a logarithmic scale from the most abundant to least abundant species (redrawn from Magurran, 1988; with kind permission of Croom Helm Publishers)

opposite of evenness. Various names such as 'geometric series', 'log series', 'log normal' and 'broken stick' are given to the resulting line (Gray, 1987) (Figure 3.5).

As an example of the use of this method, Hutchings (1983) plotted dominance-diversity curves for species within chalk grassland at a range of downland sites in south-east England (Figure 3.6). His aim was to study aspect and seasonality in relation to dominance and diversity within the plant communities. Figure 3.6 shows the dominance-diversity curves for south- and west-facing slopes. The steepest curves indicating dominance and unevenness were in July for N- and S-facing slopes and January for E- and W-facing slopes. His results are interesting in that in addition to showing how aspect affects dominance, dominance itself also changes seasonally.

Whittaker (1965) shows how these curves may be interpreted through the concept of the **species niche**. The term **niche** was introduced in Chapter 1 and refers to the position of the species within the community, its position in vertical (above ground and below ground) space, horizontal space (internal mosaics and patterns within the

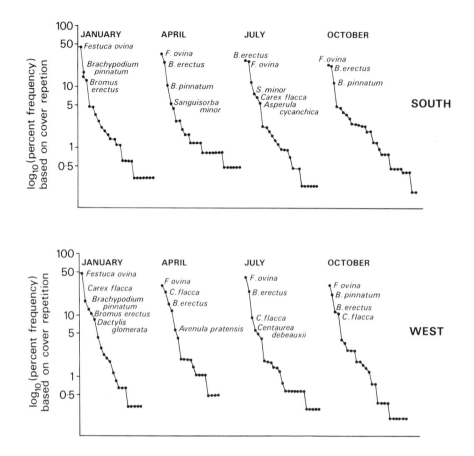

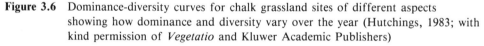

Figure 3.6 Dominance-diversity curves for chalk grassland sites of different aspects showing how dominance and diversity vary over the year (Hutchings, 1983; with kind permission of *Vegetatio* and Kluwer Academic Publishers)

plant community), community functional relationships (such as trophic structure and feeding relationships) and seasonal and daily variations in the species and their interactions with other species. Dominance-diversity curves are thus a means of displaying how the resources of the local area where the plants are growing are partitioned among the species.

Interpretation of diversity indices

Both the Shannon and McIntosh indices are based on both species richness and evenness of species abundances. However, interpretation and comparison of indices from different samples and plant communities is often difficult. Despite arguments to the contrary, most interpretation of diversity indices is still based primarily on species richness rather than both richness and evenness. Separation of the evenness component of the index from the species richness component is not easy, particularly since in the case of the McIntosh index, for any given sample, the resulting index is a function not only of species richness (s) but also the overall abundance (N). For this reason, the Shannon index is often preferred because the

Table 3.7 Frequency of use of criteria for conservation evaluation based on 17 different schemes (Usher, 1986; with kind permission of Chapman and Hall)

Criteria	Frequency of use
Diversity (of habitats and/or species)	16
Rarity (of habitats and/or species)	13
Naturalness	13
Area	11
Threat of human interference	8
Amenity and educational value	7
Scientific value	6
Recorded history	4
Population size	3
Typicalness	3
Ecological fragility, position in an ecological/geographical unit, potential value, uniqueness	2
Archaeological interest, availability, importance for migratory wildfowl, management factors, replaceability, silvicultural gene bank, successional stage, wildlife reservoir potential	1

species abundances are standardised to proportions. This is the reason for the varying figures obtained in the above examples. Many other diversity indices exist of which the most widely used are those of Simpson and Brillouin. These and a number of others are described in Magurran (1988).

Arguments over the ecological significance of high diversity still abound. Most of these arguments assume that diversity is equated solely with species richness and take no account of the relative species abundances and evenness. Most ecologists consider high species richness to be a desirable property of any community or ecosystem and this criterion has dominated most methods for ecological and conservation evaluation techniques (Usher, 1986). Table 3.7 shows species richness to be the most frequently cited criterion in a review of 17 methods of conservation assessment. However, Magurran (1988) showed clearly how the Shannon diversity index was of limited value in assessing 10 woodlands in Northern Ireland for their potential as nature reserves. The two woodlands which actually were nature reserves were bottom of the list in terms of diversity indices calculated on their ground floras. However, the reason for declaring the woods as nature reserves was that the vegetation of these two woods was far more characteristic of the area in general habitat terms and they were also two of the largest sites. Magurran makes the important point that diversity is only one of the various factors listed in Table 3.7 which may be used to select sites for conservation protection.

The long-running argument on the relationship between diversity and stability (Goodman, 1975; Walker, 1989) has probably now been laid to rest. There is no automatic relationship between high diversity or species richness and ecosystem/community stability. Many species-poor communities are extremely stable, for example some heather moorland in Britain, while many species-rich communities show considerable instability, for example chalk grassland communities in south-east England. Pielou (1975) discussed this problem at length and showed that ecosystem and community stability is ultimately dependent on environmental stability. A stable environment will always encourage stability in the biota of a community or ecosystem and over time, depending on factors such as position on

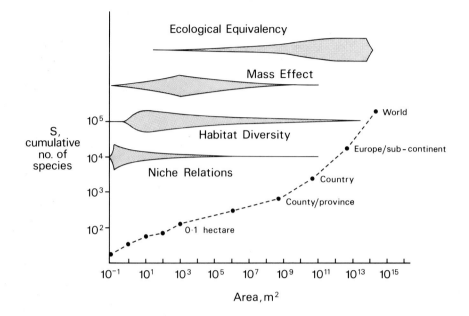

Figure 3.7   Biological mechanisms which determine species diversity (Shmida and Wilson, 1985; redrawn with kind permission of *Journal of Biogeography*)

a world scale and speciation (rate of production of new species through evolution), species numbers and richness *may* increase. Even this argument needs to be treated with care. Tropical rain forest communities have the highest species diversities of all ecosystems, and although this has often been attributed in part to the absence of glaciation in tropical regions during the Quaternary era, research has increasingly demonstrated that the tropics did undergo substantial climatic and environmental perturbation in that period. Prance (1977, 1978), working in the Amazonian Basin, has shown that the old 'heartland' regions to which the forest retreated during the times of extreme climatic change in the Pleistocene are still the most diverse. However, although they were relatively the most 'stable' areas, they still underwent considerable climatic and environmental change, which may also have promoted their diversity (Barrett, 1980; Eden, 1990; Mabberley, 1991). Grubb (1987) has also demonstrated the importance of latitude in determining species richness.

Useful perspectives on complexity, diversity and stability are also presented by Kikkawa (1986), who stresses that the difference between species richness and diversity in one component of an ecosystem, such as the plant community, is a very different concept from, for example, food web diversity, which defines the number of feeding links within and between all trophic levels in the same ecosystem.

Shmida and Wilson (1985) and Auerbach and Shmida (1987) present one of the most interesting recent reviews of the factors affecting species diversity at different scales (Figure 3.7). Most discussion of diversity is at the local community scale described by Shmida and Wilson as **niche relations** (Whittaker's alpha diversity). However, overall diversity is increased much more at the next scale of **habitat diversity** (Whittaker's beta diversity) and is further accentuated by **mass effects**, which represent the flow of propagules of individuals from areas of high diversity (core areas) to unfavourable areas, where they achieve viability. These marginal species from adjacent areas around one particular habitat or community type can greatly increase species diversity. This effect also explains the high species diversities of many transitional zones or ecotones between differing plant communities or habitat types. Finally, at the large scale, diversity is increased by **ecological equivalency**. In one part of the world, a certain collection of plants will have evolved and comprise a particular community. Elsewhere in the world, in a very similar environment, an equivalent but totally different set of species can have evolved to make up the communities in that region. Thus diversity is maintained simply by the spatial separation of the two locations and their very different species compositions. Problems, of course, now occur when species from one of these environments are transported by humans into the other similar environment. A good example of this is the *Rhododendron*: many species have been brought to the British Isles for horticultural purposes but one species in particular, *Rhododendron ponticum*, having 'escaped' from suburban gardens or parkland, is now seen as a pest in natural communities such as woodland and moorland. It was introduced from the Mediterranean in the mid-eighteenth century. Once established, *Rhododendron ponticum* has the ability to grow very rapidly and to kill all species on the ground beneath itself, greatly reducing the diversity of the former community.

The purpose of this chapter has been to introduce the basic nature of vegetation data and the ideas of comparing samples in terms of floristic composition (similarity and dissimilarity indices). These concepts underlie many of the numerical methods for analysing vegetation data presented in Chapters 5, 6 and 8. Ideas of diversity and species richness have also been presented. Again these are important in assisting with interpretation of the results obtained from those more complex methods.

Case studies

Use of χ^2 and 2×2 contingency tables to establish species assemblages in the Narrator Catchment, Dartmoor

A large number of experimental catchments have been set up in Britain and elsewhere to enable detailed studies of the movement of water through the land phase of the hydrological cycle. One such catchment has been operating since 1973 in the Narrator Brook, on the south-west margins of Dartmoor, south-west England. Research has concentrated on the effects of different vegetation types and land-use management practices on both the quantity and quality or chemistry of water moving through the watershed. The features of plants which are significant in determining the rate of movement of water through vegetation are their vertical structure and physiognomy, together with the floristic composition. Thus in the initial stages of the research, a survey of the nature and extent of the plant communities and a detailed knowledge of the species assemblages present were required.

A total of 162 1m² quadrats were collected from within the 4.35 sq. km of the catchment and analysed in different ways to establish the different plant communities and their environmental controls. One of the first stages of analysis involved the use of 2×2 contingency tables and χ^2 to determine significant associations between species. Using a computer program for speed and accuracy, the matrix of 162 quadrats × 82 species was analysed resulting in significant positive associations between 26 species. The results were displayed as a half-matrix (Figure 3.8), where positive associations significant at the 0.01 level are shown by ● and at the 0.05 level by •. All

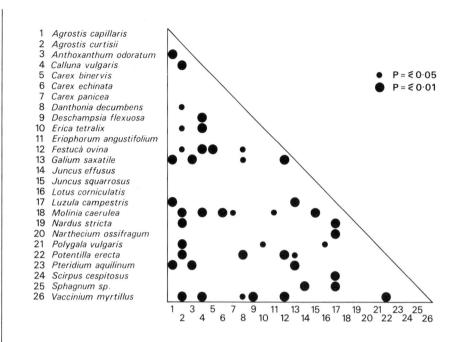

1 *Agrostis capillaris*
2 *Agrostis curtisii*
3 *Anthoxanthum odoratum*
4 *Calluna vulgaris*
5 *Carex binervis*
6 *Carex echinata*
7 *Carex panicea*
8 *Danthonia decumbens*
9 *Deschampsia flexuosa*
10 *Erica tetralix*
11 *Eriophorum angustifolium*
12 *Festucà ovina*
13 *Galium saxatile*
14 *Juncus effusus*
15 *Juncus squarrosus*
16 *Lotus corniculatus*
17 *Luzula campestris*
18 *Molinia caerulea*
19 *Nardus stricta*
20 *Narthecium ossifragum*
21 *Polygala vulgaris*
22 *Potentilla erecta*
23 *Pteridium aquilinum*
24 *Scirpus cespitosus*
25 *Sphagnum sp.*
26 *Vaccinium myrtillus*

● $P = \leqslant 0.05$
● $P = \leqslant 0.01$

Figure 3.8 Chi-square matrix of positive associations between 26 species in the Narrator catchment, Dartmoor

positive associations with expected cell frequencies of less than 5 were rejected.

The information in the half-matrix was then used to construct a constellation diagram, where species are shown as points with lines connecting those species between which significant positive associations exist (Figure 3.9). Three clear groupings of species emerged. The first, on the left side of the diagram, included the pasture species *Agrostis capillaris* (common bent) and *Anthoxanthum odoratum* (sweet vernal grass), plus *Pteridium aquilinum* (bracken), *Galium saxatile* (heath bedstraw) and *Luzula campestris* (field woodrush). These species corresponded with upland pastures which had been improved in the nineteenth century and had become invaded by bracken. The second group were representative of the upland heath communities of south-west Britain and centred on the species *Calluna vulgaris* (ling), *Agrostis curtisii* (bristle bent), *Deschampsia flexuosa* (wavy hair grass), *Potentilla erecta* (tormentil), *Vaccinium myrtillus* (bilberry) and *Festuca ovina* (sheep's fescue). The third group was characterised by species found in the blanket bog communities of Dartmoor and elsewhere in upland Britain, the most important being *Molinia caerulea* (purple moor grass), together with several species of the sedge *Carex* and the rushes *Eriophorum angustifolium* and *Juncus effusus* with several species of the moss genus *Sphagnum*. In Figure 3.9, several species appear to lie between two of the groups, suggesting that they may have affinities with species in more than one group. Good examples are *Galium saxatile* (heath bedstraw), *Erica tetralix* (cross-leaved heath) and *Nardus stricta* (mat grass).

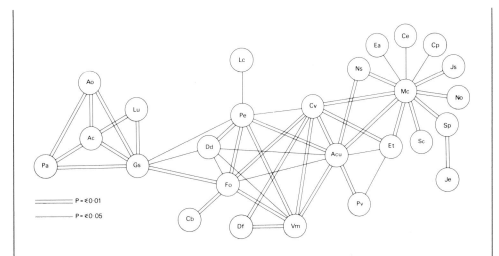

Figure 3.9 Constellation diagram of positive associations between species from the Narrator catchment on Dartmoor, south-west England (derived from Figure 3.8)

Subsequently, more detailed analyses (Kent and Wathern, 1980), confirmed that these three groups of species form the core of all the communities in the catchment. Of great importance was that by cover and area, the communities were by far the most dominant types. On the basis of these results, it was possible (1) to establish an experimental design to study the quantity of precipitation being intercepted by each of the community types using the differences in species composition as a means of stratified sampling; and (2) to determine the extent to which each of these major cover species was capable of modifying the chemical composition of the water as it fell through the vegetation layers. An example of the published results for the bracken-infested grassland is described in Williams *et al.*, 1987. Research suggests that plants differ greatly in their ability to contribute solute ions to intercepted precipitation. Some species, partly because of their morphology, can trap aerosols of certain ions, such as sodium and chloride, by dry deposition. Others produce leaf and stem exudates which when washed off in throughflow and stemflow change the chemistry of the water entering the soil.

The use of measures of association to show relationships between species in this example demonstrates that inventories of vegetation are often required as a starting point for more specific research projects. Although some of the associations between the species in the constellation diagram could be seen by eye, others could not, and the method provides a rigorous way of establishing this. As an example, the key position of *Agrostis curtisii* and its associations with various other species and communities had not been noticed before.

Shannon–Wiener diversity indices and the succession of plant species on Surtsey (Fridriksson, 1975; 1987; 1989)

The volcanic island of Surtsey appeared in the mid-Atlantic, south of Iceland,

Table 3.8 Calculation of diversity indices for colonising vascular plants on the island of Surtsey between 1965 and 1973, together with additional data for 1980 and 1988 (extracted from Fridriksson, 1975; 1989)

Species	65	66	67	68	69	70	71	72	73	80	88
Cakile arctica	23	1	22		2			1	33	1	
Elymus arenarius		4	4	6	5	4	3		66	5	1000
Honckenya peploides			24	103	52	63	52	71	548	50000	∞
Mertensia maritima		1		4				15	25	7	400
Cochlearia officinalis					4	30	21	98	586	75	25
Stellaria media						4	2	2	1		2
Cystopteris fragilis							1	4	3	5	1
Angelica archangelica								2	2		
Carex maritima								1	1	1	2
Puccinellia retroflexa								2	1	7	30
Tripleurospermum maritimum								1	5	1	1
Festuca rubra									1	3	2
Cerastium fontanum										150	30
Rumex acetosella										40	500
Cardaminopsis petraea										8	25
Poa pratensis											25
Sagina saginoides											1000
Armeria vulgaris											1
Poa annua											15
Agrostis stolonifera											1
Unidentified plants				1				4	2	1	
Total number of plants	23	5	51	114	63	101	83	199	1273	50000	∞
Species richness	1	2	4	4	4	4	5	11	13	13	18
Shannon-Wiener diversity index (H')	0.00	−0.500	−0.994	−0.251	−0.644	−0.910	−1.049	−1.255	−1.118	—	—
Evenness (J)	0.00	0.721	0.717	0.181	0.465	0.657	0.586	0.528	0.436	—	—

Calculation of the value for 1973 is as follows:

Species	Number of individuals in 1973	Proportion (p_i)	$\ln p_i$	$p_i \ln p_i$
Cakile arctica	33	0.0259	−3.653	−0.0946
Elymus arenarius	66	0.0518	−2.959	−0.1533
Honckenya peploides	548	0.4305	−0.843	−0.3629
Mertensia maritima	25	0.0196	−3.930	−0.0770
Cochlearia officinalis	586	0.4603	−0.775	−0.3567
Stellaria media	1	0.0008	−7.149	−0.0057
Cystopteris fragilis	3	0.0023	−6.051	−0.0057
Angelica archangelica	2	0.0016	−6.455	−0.0103
Carex maritima	1	0.0008	−7.149	−0.0057
Puccinellia retroflexa	1	0.0008	−7.149	−0.0057
Tripleurospermum maritimum	5	0.0039	−5.540	−0.0216
Festuca rubra	1	0.0008	−7.149	−0.0057
Unidentified plants	1	0.0008	−7.149	−0.0057
Total	1273	1.000		−1.1188

Thus H' = −1.1188

$$\text{Evenness (J)} = \frac{H'}{\ln s} = \frac{-1.1188}{2.5649} = 0.436$$

where s = number of species

ln = $\log_{10}$

following a volcanic eruption in 1965. Between 1965 and 1973, annual records of the numbers of individuals of each plant species were made. The changing pattern of diversity can be examined by calculating the Shannon-Wiener diversity index for each year. Table 3.8 shows the raw data and the calculated Shannon diversity indices over that period. In the early stages (1965–73), although it is clear that species richness increases, particularly at the end of the study period, the indices are variable, although they show a definite increase, again towards the end of the study period. This is presumably closely related to the increase of species richness. The evenness figures (J) are even more variable, with a trend towards greater unevenness as time progresses (the lower the value of J, the more uneven the species distribution). This is partly a response to the greater number of species in later years but also illustrates the problems of interpretation discussed earlier in this chapter. The value of the extra information obtained from the values for the diversity index and the evenness statistic over and above the simple statistics on species richness could be said to be questionable.

By 1980 and 1988, the number of species and individuals has changed dramatically. Further new species have arrived while some early colonisers have gone. The most spectacular change is the explosive growth of *Honckenya peploides* from just 548 plants in 1973 to an estimate over 50,000 in 1980. By 1988 a count or even an estimate of the number of plants was impossible. Since the data are based on the number of individuals, the calculation of diversity indices becomes invalid. This shows their limitations even further.

Another excellent example of this type of work on the Krakatau Islands, Indonesia, using many of the techniques described later in this book, is presented in Whittaker *et al.* (1989).

Dominance-diversity curves and the study of urban grassland (Wathern, 1976)

An increasingly important part of landscape architecture and design involves the sowing of grassland using standard seed mixes. Building projects, such as new housing schemes, industrial estates and road and motorway construction, often require large areas to be reinstated and resown following the total removal of the previous vegetation cover and ecosystem. In Britain, usually a standard seed mixture, dominated by grasses such as *Lolium perenne* (perennial ryegrass), *Agrostis capillaris* (common bent), *Festuca species* (fescues) and *Poa species* (meadow grasses), is sown, perhaps accompanied by a legume, such as clover *Trifolium* species. Such swards are inexpensive to produce, and large areas can be sown rapidly. These reinstated areas offer considerable potential for wildlife conservation. One of the most obvious possibilities is the opportunity to increase the species diversity within these simplified swards. This will happen naturally over time, but success depends on many factors such as management practice, fertiliser treatment and the supply of natural propagules of potential new colonisers.

Wathern (1976) studied 69 grasslands from 17 sites in the city of Sheffield, Yorkshire, ranging in age from 1 to approximately 200 years. At each site a number of random 1m² quadrats were placed to sample the range of grasslands present. Within each quadrat, rooting frequency was recorded using 10 × 10 subdivisions. As part of a larger study of the structure of these communities, Wathern used dominance-diversity curves to demonstrate differences among the various grassland types. Three types of grassland were

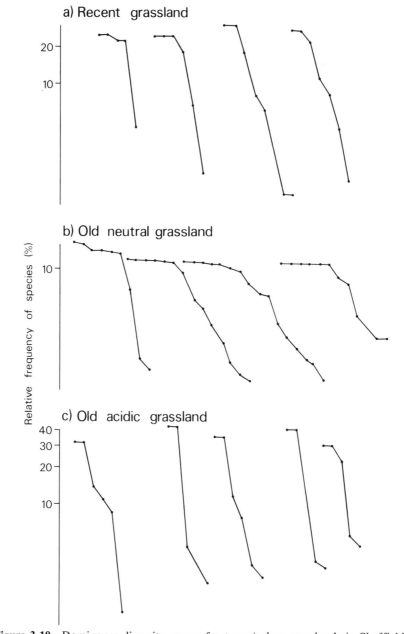

Figure 3.10 Dominance-diversity curves for town/urban grasslands in Sheffield, Yorkshire (Wathern, 1976)

identified in the survey, and dominance-diversity curves were drawn from samples of each (Figure 3.10). They show very clearly the simplified nature of the recently created swards (Figure 3.10a), containing between 5–8 species, compared with the 12–16 species in the old neutral grassland (Figure 3.10b). The curves for the recent swards are very much steeper indicating the one or two dominant sown species, together with only a few others. In contrast, those for the old neutral grassland show a much flatter curve, indicating a more even distribution of species as well as greater species richness. Interestingly, however, some of the old acidic grasslands which were also included in the study had curves which were much closer to those of the recent swards (Figure 3.10c).

These graphs demonstrate two important points. First, a number of factors control grassland composition and diversity, but nutrient status is one of the most critical of all. Second, although the most obvious characteristic of sown swards is their relatively low species diversity and unevenness, as shown by the steep dominance-diversity curves, some natural communities exhibit the same properties to an even more marked degree. Thus in terms of management and conservation goals, diversification of all swards is not necessarily always desirable. Careful consideration has to be given to the local environment, particularly nutrient status, and to the management strategies which will or will not be employed once the grassland is established.

4

Basic statistical analysis of vegetation and environmental data

Introduction

The aim of this chapter is to introduce a range of simple statistical techniques which are of value in examining data collected on vegetation and its environment. All techniques are concerned with situations where **only one or two variables** are being analysed. Understanding these methods and the possibilities for their application in plant ecology is a very important starting point for the more complex multivariate methods described in later chapters. Ultimately, the application of all statistical methods is part of **problem solving** and scientific method (Chapter 1). A valuable introduction to ideas of problem solving is presented in Chatfield (1988).

Exploratory and confirmatory data analysis

Most statisticians now believe that it is helpful to make a distinction between **exploratory** and **confirmatory** data analysis (Chatfield, 1988). Exploratory data analysis (EDA) is used to study a new set of data which has not been subject to previous analysis. A confirmatory analysis, however, is used to check for the presence or absence of phenomena observed in a previous analysis or to test hypotheses derived from existing results or established theories. In most reported scientific literature in biology and ecology, 'significant' results are derived from one-off exploratory data sets rather than from data sets specifically collected for confirmatory analysis. Furthermore, it is virtually universal for a researcher to carry out exploratory data analysis followed by hypothesis generation and confirmatory analysis on the **same** data set after having noted some very interesting feature of it in the exploratory analysis. As Nelder (1986) has pointed out however, exploratory data analysis should only really be used in the initial description of data and a new set of data should be collected for confirmatory analysis.

Exploratory data analysis (EDA)

Although much statistical analysis is concerned with hypothesis generation and testing, some reservations about the dominance of this approach have been expressed over the past 15 years (Tukey, 1969; Erikson and Nosanchuck, 1977; Sibley, 1987; Marsh, 1988). While rigorous statistical testing and **confirmation** is important, these authors make the point that a great deal of data analysis is **exploratory** in approach and is primarily concerned with the search for pattern and order in data. Sibley (1987) argues that the process of data exploration should be seen as **circular**, rather than as a series of steps. Also, it is a process of '**successive deepening**.' Examination of a set of data by one form of preliminary analysis will lead to further refinement of ideas and hypotheses and throw up new ideas and show patterns that had not previously been apparent. When these patterns are explored further, still more ideas are generated. Analogies have been made with peeling the layers off an onion; each successive layer increasing understanding about the patterns and properties of the data.

Given this basic idea of EDA, a whole set of **'robust'** techniques have been evolved which are designed to explore, rather than rigorously test data. An important aspect of these techniques is that they are much less rigorous in their assumptions both in terms of sampling and statistical properties. This is because of the removal of the need for confirmatory analysis based on hypothesis generation and testing and the probabilistic approach.

In the context of vegetation description and analysis, it is ironic that at a time when methods for the description and analysis of vegetation are being criticised for being insufficiently rigorous and researchers are being attacked for failing to generate and test hypotheses, there is also a plea from some statisticians for this less rigorous and more exploratory approach to simple data analysis. There is no contradiction, however. The reason is that both hypothesis generation and testing (confirmatory analysis) and exploratory data analysis have their place and, as Tukey (1977) and Chatfield (1986; 1988) stress, they should be placed side by side as suitable approaches for scientists to use.

Tukey and EDA are not without their critics. Chatfield (1985; 1986; 1988) recommends the use of EDA but criticises Tukey for introducing too much new statistical jargon and for failing to emphasise sufficiently the value of the combined approach of both exploratory and confirmatory analysis. Chatfield also introduces the term **initial data analysis** (IDA). He rejects the use of the word 'exploratory' because he sees 'exploratory' as only appropriate for describing the analysis of completely new data, whereas the methods are equally suited to existing data and are not used merely at the preliminary stage. EDA (or IDA) have also been criticised by other established statisticians as being 'common sense' and as already being practised by many researchers. Chatfield (1985; 1988) counters this, arguing that most students nevertheless have to be taught 'common sense'. Despite these criticisms, it would seem that exploratory data analysis has much to offer and represents an approach which would be of value to many vegetation scientists.

Descriptive and inferential statistical analysis

Descriptive statistical analysis (exploratory data analysis)

As explained in Chapter 1, a large amount of work in vegetation science is **inductive** and **descriptive** in approach and is concerned with the examination of variation within a set of data and the search for patterns and trends. Thus when collecting and analysing floristic data for the purpose of defining plant communities, there will often be no clear idea at the outset of what exactly the result will be. Instead, the aim is simply to show how vegetation varies from place to place or within a certain area or region and to present these data in summary form. All the multivariate methods of classification and ordination presented in Chapters 5, 6, 7 and 8 are usually used in this inductive manner. Once variation has been summarised and pattern displayed, then **hypothesis generation** can occur.

Hypothesis testing and inferential statistics (confirmatory data analysis)

The kinds of questions that emerge from vegetation analysis and which lead to hypothesis generation often take the form of 'why?' Why does the vegetation vary in the way that it does? Can a reasonable **explanation** be offered for any of the observed variations and differences? These then lead to a need for further understanding of the **biological and ecological processes** behind the observed variations. Once hypotheses have been generated, then the **deductive** approach to scientific enquiry is being applied and different approaches to data analysis have to be employed. This will involve the use of **inferential statistical analysis**, where data are collected and analysed with the aims of accepting or rejecting a hypothesis. Inferential statistics are concerned with **mathematical probabilities** and in the context of scientific investigation involve a search for principles that have a degree of generality. Using inferential statistics enables the vegetation scientist to apply results taken from a small sample of reality to a much larger environment or population. Making generalizations from a sample to a population is therefore known as statistical inference and is a vital part of scientific method.

Hypotheses

A hypothesis can be defined as a preliminary explanation of observed facts or phenomena, which is usually then tested for its validity. Most hypotheses are either concerned with observed **differences** in plant species, vegetation or environmental factors at different locations in space or time, or else they are concerned with **relationships** between phenomena — plant species, vegetation and environment or between species or environmental factors themselves. In practice, hypotheses are stated with varying degrees of precision. Many statistical texts distinguish three forms of hypothesis:

(a)   the **research hypothesis** or general ecological statement relating to plant species vegetation and/or environment.
(b)   the **null hypothesis** of no difference or no relationship (H_0).
(c)   the **alternative hypothesis** stating the nature of the difference or the relationship often in some detail (H_1).

As a simple example, if, having made a preliminary study of the vegetation of an area in terms of both vegetation cover and its environmental controls, an idea emerges that the distribution of one species (X) is closely related to soil drainage quality, a general **research hypothesis** may then be formulated: that is 'there is a relationship between the distribution of species X and the quantity of moisture in the underlying soil'. For all types of statistical analysis, the strategy is broadly similar. Next, the hypothesis can be stated in the negative, known as the **null hypothesis** (H_0). In the example, this would be that there is **no** relationship between the distribution of species X and soil moisture. Then the **alternative hypothesis** is stated, usually in more detail than the original general **research hypothesis** because it must be defined in relation to the particular statistical test being carried out.

The normal distribution and levels of significance

The sampling distribution of test statistics

A **sampling distribution** is a theoretical distribution defined as the distribution that would result by randomly taking all possible samples from a specified population. With inferential statistical testing, if it was possible to repeat any test using different repeat samples a large number of times, each time with a **test statistic** being calculated, the **distribution** of all those test statistics could be plotted. The result would often be a **normal** or **Gaussian** distribution as shown in Figure 4.1. Obviously

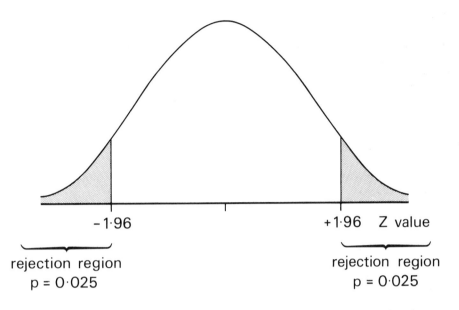

Figure 4.1 Rejection areas on the normal curve for the 0.05 significance level (redrawn
from Shaw and Wheeler (1985); with kind permission of David Fulton
Publishers)

it is not possible to repeat a test using a large number of different repeat samples,
so mathematicians have devised theorems which make use of known distributions
such as the normal distribution. Some distributions may be closer to other
theoretical distributions such as the Poisson distribution. These mathematically
calculated distributions provide either a close approximation to the presumed
distribution or in some cases, an exact probability distribution.

   The key point is that from a known probability distribution of a test statistic, it
is possible to calculate the probability with which any value of the test statistic will
occur. This characteristic underlies the whole basis of inferential statistics. Further
details of the normal and Poisson distributions are given in most elementary
statistical textbooks.

Directional and non-directional hypotheses

A further important aspect of hypothesis statements is whether the hypothesis is
stated in a **directional** or **non-directional** manner. In the case of soil moisture, a **directional** hypothesis states the anticipated direction of the relationship — perhaps that
the abundance of species X is **highest** where soil moisture values are lowest. A **non-
directional** hypothesis would not go so far, saying simply that some form of relationship exists between the two variables, the abundance of species X and soil moisture.
Whether the alternative hypothesis is stated in directional or non-directional terms is
very important in the interpretation of probability values at the end of the statistical
analysis. Where the direction of the test has been stated, a **one-tailed** significance test
is applied which assesses the likelihood of the result in terms of just one end of the
normal distribution as shown in Figure 4.1 (probability $p = 0.025$). Where the
hypothesis cannot be stated with such precision, then the relationship is non-
directional, a **two-tailed** test is used and the probability of the result occurring in
either of the tails of the distribution of Figure 4.1 is assessed (probability $p = 0.05$).

As explained above, each statistical method has its own test statistic and associated probability distribution. From the latter it is possible to assess the random probability of the calculated test statistic: that is what is the chance of this test figure occurring randomly or by chance? When the test statistic is shown to be unlikely to have occurred by chance, the null hypothesis can be rejected.

Usually, it can be assumed that the random probability of the test statistic is also the probability of H_0 being correct. Probability values are usually expressed in the range 0–1 or 0–100 per cent. Low random probabilities such as 0.01 or 1 per cent would imply that H_0 can be rejected, but a higher figure of 0.5 or 50 per cent would mean that it should be accepted. In rejecting H_0, it follows that H_1 must be accepted. Before the calculation of any test statistic, the rejection levels or confidence limits, beyond which the probability of the null hypothesis being correct is unacceptable, must be decided. This is a matter of tradition, with values of 0.01 (1 per cent) and 0.05 (5 per cent) typically being used. Figure 4.1 shows the rejection regions or confidence limits of the normal distribution for 0.05 (95 per cent). These are also known as **significance levels**.

Once a test statistic has been calculated, for **most** tests, if the value of the test statistic is **greater than** the critical value for a given confidence level, then there is only a small probability that the null hypothesis will be wrongly rejected, and the rejection of the null hypothesis is justified. However, if the test statistic is **less than** the critical value for a given rejection level, then the probability that the null hypothesis will be wrongly rejected is low, the null hypothesis cannot be rejected and H_0 must be accepted.

Degrees of freedom

Associated with all inferential statistical tests is the quantity known as **degrees of freedom**, which is defined as the number of observations in a particular statistical analysis which can take on any value but within the limitations imposed by any calculations on those values. As an example, suppose that the density of a plant species (X) has been measured in five different localities: a–e. The mean density is 36; then

$$\text{If } \overline{X} = 36; \text{ and } N = 5; \text{ then } \Sigma X = 36.0 \times 5 = 180$$

Given any four of the density values such as for areas a–d: 75, 25, 50 and 20 then:

$$(a + b + c + d + e) = 180$$
$$(75 + 25 + 50 + 20 + e) = 180$$
$$e = 180 - (75 + 25 + 50 + 20) = 10$$

Thus, when $(N-1)$ numbers are specified, the Nth is determined and the degrees of freedom in this case are the number of observations minus 1 $(N-1)$. The basic rule in applying degrees of freedom is that one degree of freedom is lost for every fixed value. The fixed or known value in this case is the mean, and knowledge of the mean always enables the value of the last observation to be determined. Hence one degree of freedom was lost.

Sampling and inferential statistics

Most inferential statistical tests make assumptions about the nature of the data being analysed and the manner in which they have been collected. Many of these assumptions relate to the idea of **random sampling** particularly if **parametric** methods are being used (see below). Very rarely are statistical analyses based on the total population of individuals. Instead, they are based on the premise that a subset of the population can be measured and that this subset may exhibit the same properties as the total population from which it was drawn. With random sampling, each sample measurement should be **independent** of any other, and on every sampling occasion every individual should have an equal chance of being selected. If these principles are not followed, the sample becomes **biased**. In order to prevent such bias, the nature of the overall population must be known but often problems arise in assessing how representative any sample is of its population. In general, the larger the sample, the more likely it is to be representative of the population from which it is drawn. Unfortunately, comparatively little vegetation data are collected at random (see Chapter 2), partly because of the practical difficulties involved in locating a truly random sample in the field and also because of the time involved.

Spatial autocorrelation

A further difficulty which affects the independence of samples is that of **spatial autocorrelation**. This is a difficult and complex subject which statisticians and geographers are only beginning to understand, although the concepts are fairly readily understood (Cliff and Ord, 1973; 1981; Cox, 1989). The problem is present in all spatial sampling (all vegetation sampling is, by definition, spatial) and is due to the inevitable relationships between points in space and in particular, their proximity to each other. For example, if four quadrats A–D are thrown along a transect at 50m intervals following an environmental gradient, it is impossible for those samples to be completely independent of each other. The observed species composition of quadrat B will be affected by the composition of quadrat A simply because of its proximity in space to quadrat A. Similar relationships and varying degrees of influence will exist between the other quadrats because of their proximity to each other. Geographers have begun to look at special forms of statistical analysis to tackle this problem, but satisfactory methods which can be widely applied have yet to be devised (Cliff and Ord, 1973; Johnston, 1978; Cliff and Ord, 1981). The same problem exists if samples are collected from one location but at successive times. This is known as **temporal autocorrelation**. Vegetation scientists usually choose to ignore these problems, since a great deal of their work is inductive and descriptive rather than deductive and does not involve the use of inferential statistics. Nevertheless, this does represent an important area for future research, and if plant ecologists are to be encouraged to apply more inferential statistical analysis and hypothesis testing, they will have to take greater notice of the problem (Sokal and Thompson, 1987; Fortin *et al.*, 1989; Legendre and Fortin, 1989).

Parametric and non-parametric statistics

Some of these problems of sampling in relation to statistical analysis may be solved by using **non-parametric** as opposed to **parametric** tests. Parametric tests make certain assumptions about the background populations from which samples are drawn. The most important of these is that the background population is approximately normally distributed (Figure 4.1) and the smaller the sample size being tested, the more important it is that the background population approximates to normality. For many variables, it may not be reasonable to assume that the background population is normally distributed, and in many circumstances it is thus not possible to test this assumption by collecting a very large sample. This problem can be overcome by using **distribution-free** (non-parametric) tests, which make no such assumptions about the distribution of the background population. Many of the tests described in the remainder of this chapter are distribution-free and non-parametric.

Siegel (1956) was the statistician who was originally responsible for the widespread dissemination of non-parametric methods. In addition to the removal of the assumption of normality in the population distribution, he also lists other advantages of non-parametric techniques. They can deal with very small samples; they can be used on a variety of measurement scales; suitable non-parametric tests exist for analysing samples drawn from several different populations and finally, the methods are usually easier to understand and to apply than parametric ones. A further important concept is the **power-efficiency** of a non-parametric test. In statistical terms, the power of a test is related to its ability to state correctly whether a hypothesis is true or false. Parametric tests are usually more powerful than non-parametric and the power of a test is influenced by the size of a sample. A non-parametric test of low power-efficiency requires a larger sample to achieve the same level of power as a parametric test which has a higher power-efficiency ratio. As an example, the parametric method for testing differences between the means of two independent sets of observations or samples on one variable is the t-test (100% efficient). The non-parametric equivalent is the Mann–Whitney U test which is around 95 per cent power-efficient.

Descriptive statistics and exploratory data analysis

First, **measures of central tendency** are described — the mean, mode and median. Second, **measures of dispersion** are dealt with — the variance and the standard deviation. In addition to the following discussion, good introductions to these topics are given in Bishop (1983), Clarke (1980), Ebdon (1985), Finney (1980), Hammond and McCullagh (1978), Lee and Lee (1982), Shaw and Wheeler (1985) and Silk (1979). Basic graphical methods of exploratory data analysis are also important, notably **histograms**, **stem and leaf plots** and **box plots** (Tukey, 1977 and Marsh, 1988).

Measures of central tendency

The arithmetic mean or average
The mean ($\overline{X}$) is found by summing all of the observed values on one variable in a set of data and dividing by the number of observations (n). Thus:

$$\overline{X} = \frac{\Sigma X}{n} \tag{4:1}$$

The mean is an important and widely used measure, but interpretation of means need to be made with care. They are particularly affected by unusual or extreme values.

The median
This is the value at the mid-point of a set of observations where half the scores lie above the median and half below. Data are first ranked, and if n is the number of observations, the median is $(n+1)/2$. This is easiest when there is an **odd** number of data points, for example if there are 51 data points, the median would be $(51+1)/2$ — = the 26th observation — and there would be 25 observations above

and 25 observations below it. If the number of data points is **even**, then the median is conventionally taken as the value halfway between the two middle data points. The median is an important measure in exploratory data analysis (EDA) (see boxplots, below) and is much less prone to the effects of unusual or extreme data values.

The mode

The **mode** is the value that occurs most frequently in a set of observations. Thus if there are 10 numbers:

72 70 72 79 77 72 68 69 72 73

the mode is 72 because it occurs four times and the others only occur once. In a **frequency distribution** or **histogram**, the mode is the value or class (modal class) with the highest number of observations. The mode is most useful for describing data measured on the **nominal** scale.

Frequency tables and histograms

Any set of observations can be displayed in grouped form as a grouped frequency table or histogram. These will enable the **distribution** of points to be seen. It is important that the data in either case are present as a **count** of the number of observations in each group (that is, frequencies). To demonstrate this point, the diagrams in Figures 2.1 and 2.2 are not histograms using frequencies but percentages.

Stem and leaf plots

These are similar to histograms and frequency tables in that they are a method for visually displaying the distribution of a set of observations. The advantage over frequency tables and histograms is that the original data values are retained in the display and stem and leaf plots are simple and quick to use. The method splits the data into two components, a stem and a leaf, and a display is constructed by placing one leaf for every data point at an appropriate point on the stem. For the two sets of observations on light under *Calluna* in Table 4.1, the stem and leaf plots can be constructed by taking stem units of 2.0 and leaf units of 1.0, as below:

% light under pioneer *Calluna* n = 25 Stem unit = 2.0 Leaf unit = 1.0	% light under mature *Calluna* n = 25 Stem unit = 2.0 Leaf unit = 1.0
1 4 9	2 3 77
3 5 01	5 3 899
7 5 2333	8 4 011
(6) 5 444445	(5) 4 22333
12 5 666677	12 4 44455
6 5 889	7 4 667
3 6 01	4 4 889
1 6 3	1 5 0

Table 4.1 Light at ground level as a percentage of that available at the surface under *Calluna* heathland at the pioneer and mature stages, Barden Moor, near Ilkley, Yorkshire, England

Pioneer		% light at the surface Mature	
54.7		37.5	
49.5		45.6	
56.2		39.4	
63.1		46.7	
54.8		42.8	
54.3		43.6	
55.7		48.2	
51.2		44.5	
56.6		49.0	
58.2		50.6	
54.5		41.8	
53.5		38.8	
54.2		39.7	
60.1		41.6	
59.7		42.1	
57.6		44.3	
56.2		43.2	
53.1		48.9	
53.1		37.1	
52.9	$\bar{X} = 55.71$	45.3	$\bar{Y} = 43.70$
56.4		46.4	
61.2	$n_x = 25$	47.1	$n_y = 25$
57.3		44.3	
58.4	$S_x^2 = 10.98$	43.7	$S_y^2 = 13.39$
50.3		40.4	

The first column is the **depth** of a line and tells how many leaves lie on that line or 'beyond'. Thus, in the first plot, the 7 on the third line from the top means that **there are 7 leaves on that line and above** it; the 6 on the third line from the bottom indicates that **there are 6 leaves on that line and below** it. The line with brackets contains the middle observation if the total number of observations, n, is odd, and the two middle values if n is even. The brackets enclose the number of leaves on that middle line. The stem unit is 2.0 and the leaf unit is 1.0. Thus in the first column of Table 4.1 there is one value of 49.5, which is represented by the 4 9 of the first row of the stem and leaf plot. The next values have to be in the range 50.0–51.9. There are two values in Table 4.1: 50.3 and 51.2, and these are shown as 5 01 and so on. Decimal points are not used in a stem and leaf display. Thus the numbers 370, 37, 3.7 and .37 would all be split into stem = 3 and leaf = 7. The **leaf unit** indicates where the decimal point belongs. For 370, the leaf unit would be 10; for 37, it would be 1; for 3.7, 0.1 and for .37, 0.01. The stem units can be altered to units of 2.0, 5.0 or 10.0.

Stem and leaf plots, histograms and frequency tables all show:

(a)   the typical values of the distribution;
(b)   the dispersion or spread of values;
(c)   the shape of the distribution, particularly if it approximates to normality — which is important if parametric tests are being used and
(d)   the occurrence of outliers or rogue values which may distort any statistics calculated on the data.

This last point is particularly important. Stem and leaf plots are explained at greater length in Tukey (1977) and Marsh (1988).

Measures of dispersion

The range and interquartile range (dQ)

The **range** is the difference between the highest and lowest observations in a set of data. It is, however, sensitive to extreme values. The **quartiles** are the **lower quartile** below which 25 per cent of the observations occur, the **median** below which 50 per cent of values lie and the **upper quartile**, below which 75 per cent of values are found. The **interquartile range** (dQ) is the difference between the values of the upper and lower quartiles.

Boxplots and outliers

Boxplots (sometimes called 'box and whisker plots') use the measures of the median and the quartiles to show the spread of data around the central value (the median) and are particularly useful to highlight **outliers** in a set of data.

In a boxplot diagram (Figure 4.2), the middle 50 per cent of the distribution, or the interquartile range (dQ), is shown by a box with the median as a +. To identify outliers, a value which is 1.5 and 3 times the interquartile range (dQ) is calculated. From this, two limits are defined:

(i)   inner fences — the upper and lower quartiles plus 1.5 dQ
(ii) outer fences — the upper and lower quartiles plus 3.0 dQ

'Whiskers' are then drawn, connected to either end of the box to the 'adjacent values', which are those data points that come nearest to the inner fences while still being inside or on them (Figure 4.2). Any points which occur outside this range are classified as **outliers** and are worth examining prior to any data analysis. Such values are very likely to affect calculations such as the mean and the standard deviation (see below) based on those data. They also represent data values which are potentially subject to error, perhaps in measurement or in sampling. The boxplot for the first set of data on per cent light under *Calluna* (Table 4.1) is shown below:

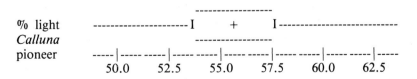

The effect of introducing two 'rogue' data points to that set of data is demonstrated below, where values of 40.1 and 43.7 have been substituted for the first two items of data:

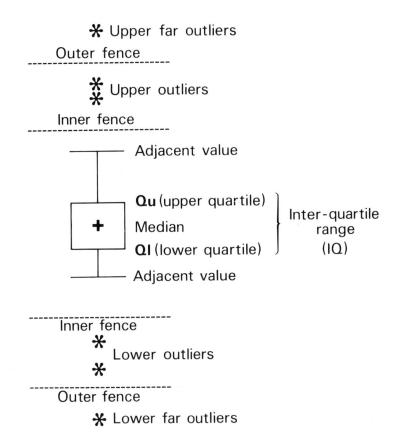

Figure 4.2 Principles of boxplots (redrawn from Marsh (1988); with kind permission of Polity Press)

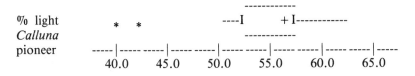

The values, symbolised by asterisks, show up very clearly as outliers, and deserve particular attention in interpretation of the data. Finding outliers does not automatically mean that they should be discarded.

Along with stem and leaf plots, boxplots are one of the most useful tools of exploratory data analysis.

The variance and standard deviation

The **variance and standard deviation** are very important measures upon which many other statistical concepts are based. The variance is determined by calculating the deviation of each value in a set of data from the mean, squaring these deviations, summing those squares and then dividing by the number of observations to give the mean square deviation. In most books, the variance is denoted by either the symbol

S^2 or σ^2. The variance equation is:

$$S^2_x \text{ or } \sigma^2_x = \frac{\sum_{i=1}^{n} (X_i - \bar{X})^2}{n} \qquad (4:2)$$

where S^2_x or σ^2_x = variance of the observations of X
X_i = the ith value of X
$\bar{X}$ = the mean or average of X
n = the number of observations

The **standard deviation** is simply the square root of the variance and is denoted by S or σ.

$$S_x \text{ or } \sigma_x = \sqrt{\left(\frac{\sum_{i=1}^{n} (X_i - \bar{X})^2}{n}\right)} \qquad (4:3)$$

When working by hand, it is easier to use the formula:

$$S_x \text{ or } \sigma_x = \sqrt{\left(\frac{\sum_{i=1}^{n} X_i^2}{n}\right) - \bar{X}^2} \qquad (4:4)$$

For small samples (<30), it is better to estimate the variance and the standard deviation by incorporating 'Bessel's correction', which divides by $n-1$, rather than n, to give what is known as the 'best estimate' of S^2 (σ^2) or S (σ) denoted by $\hat{S}^2$ ($\hat{\sigma}^2$) or $\hat{S}$ ($\hat{\sigma}$).

$$\hat{S}_x \text{ or } \hat{\sigma}_x = \sqrt{\frac{\sum_{i=1}^{n} (X_i - \bar{X})^2}{n-1}} \qquad (4:5)$$

or

$$\hat{S}_x \text{ or } \hat{\sigma}_x = S_x \frac{n}{n-1} \qquad (4:6)$$

The variance and standard deviation are strongly influenced by outliers.

Coefficient of variation (v)

The standard deviation is a measure of dispersion which is of little value if variations in one set of data are to be compared with another. As an example, a standard deviation of 2.77 around a mean of 8.43 shows a much greater spread than the same standard deviation around a mean of 540.3. The **coefficient of variation** enables these to be compared by converting the standard deviation to a percentage of the mean. Thus:

$$V = \frac{S_x}{\bar{X}} \times 100 \qquad (4:7)$$

Inferential statistical anlaysis (confirmatory data analysis)

Comparison of samples

In plant ecology, it is often necessary to test whether two samples of the same phenomenon are derived from the same parent population. For example, hypotheses may have been formulated that the abundance of one species of plant differ significantly between two rock types or slope aspects, or that the number of grazing animals varies between two different plant communities. Data will then be collected on the plant species in the two environments or on the different slopes or on the numbers of grazing animals on the two different vegetation types. Where the samples are **independent** of each other, then the **t-test** may be applied as a parametric test and the **Mann–Whitney U test** as a non-parametric test. Where the samples are not independent but **paired**, then the **paired t-test** is applied for data which are up to parametric standards and the **Wilcoxon signed rank test** for paired samples in the non-parametric case.

The t-test for difference between the means of independent samples

As a parametric test, certain conditions must be met by the data. The data must be interval or ratio in type. The background populations of the two samples must be distributed normally, as in Figure 4.1, although some deviation from this is permissible. Also the test has two forms; one where the variances of the two samples are assumed to be more or less equal; and another where they are not. The two forms of test are similar and the t-test statistic is derived from the difference between the sample means divided by the standard error of that difference:

$$t = \frac{\text{difference between the means}}{\text{standard error of the difference}} = \frac{\overline{X} - \overline{Y}}{\sigma_{\overline{X} - \overline{Y}}} \qquad (4:8)$$

where $\overline{X}$ and $\overline{Y}$ are the sample means and $\sigma_{\overline{X} - \overline{Y}}$ is the standard error of $\overline{X} - \overline{Y}$. The standard error of the difference is the standard deviation of a theoretical distribution of differences between the two means. The idea is very similar to that explained above in the section on the sampling distribution of test statistics.

If there are small differences between the means together with a wide spread of observations (Figure 4.3a), the t value will be small. The greater the differences between the means and the lower the variances or spread of observations, the higher will be the t value (Figure 4.3b).

As an example of the application of the method, Table 4.1 presents data on light levels under *Calluna vulgaris* (Scots heather or ling) in the pioneer and mature phases of the *Calluna* cycle (Watt, 1947; Gimingham, 1972). It is well known that *Calluna* heathlands in north-western Europe undergo cyclical succession with four phases:

Pioneer   — *Calluna* aged 3–10 years and 10 per cent cover
Building   — *Calluna* aged 7–13 years and 90 per cent cover
Mature   — *Calluna* aged 12–28 years and 75–100 per cent cover
Degenerate — *Calluna* aged 16–29 years and 40 per cent cover

                                                  (Gimingham, 1972)

As a result of the changes in the growth form of *Calluna* over its thirty-year life span, there are associated changes in other species and in the microclimate. These were studied in depth by Barclay–Estrup and Gimingham (1969) and Barclay–

a)

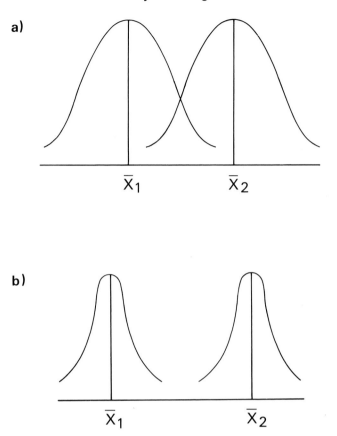

$\overline{X}_1$ $\overline{X}_2$

b)

$\overline{X}_1$ $\overline{X}_2$

Figure 4.3 (a) overlapping and (b) distinct sets of data in the t-test for differences between the means of two samples

Estrup (1971). One of the parameters measured was the light level under the *Calluna* bushes at different stages of the cycle. The data in Table 4.1 were collected at Barden Moor near Ilkley in North Yorkshire, England and show light measured underneath bushes as a percentage of available light at the surface above the bush for both the pioneer and mature phases.

The null hypothesis (H_0) is that there is no difference between the sample means for light in the pioneer and mature phases of the cycle. The alternative hypothesis (H_1) is that the means differ by an amount which is too great to be due to random sampling variation in both sets of samples. The test is also **one-tailed** because it is reasonable to suppose that light levels will be higher under pioneer *Calluna* than the larger and much more dominant bushes of the mature phase.

The first stage is to test for equality of variance between the two samples. This is done by calculating the variances of the two samples and then establishing the ratio of the greater to the lesser variance to give the variance ratio, which is known as the F statistic. Tables for the distribution of F are given in most statistical textbooks (Bishop, 1983; Shaw and Wheeler, 1988; Wardlaw, 1985). These tables are set out as a list of critical values for significance levels of 0.01 and 0.05. Once the significance level has been chosen (0.05 in this case), then the appropriate critical value is found for the degrees of freedom associated with the greater and lesser variances (v_1 and v_2 respectively, where $v = n-1$). If the F ratio exceeds this critical value, then the null hypothesis (H_0) of no difference between the variances is rejected and the t-test should not be applied.

In the case of the *Calluna* data of Table 4.1, the values for the two variances are 10.98 (X) and 13.39 (Y) which give the following:

$$F = \frac{13.39}{10.98} = 1.22 \quad \text{with 24 and 24 degrees of freedom}$$

The calculation of the two variances uses $n-1$ rather than n. This is known as Bessel's correction and is applied to the calculation of the variance and the standard deviation of small samples of under 30 observations. The variance or the standard deviation is then known as a 'best estimate'. From F tables, the critical value at the 0.05 level is 1.98. The observed value of F is 1.22, which is less than the critical value and thus the null hypothesis (H_0) that there is no significant difference between the variances is accepted.

Given that the variances can be assumed equal, the following equation is used to determine t:

$$t = \frac{\bar{X} - \bar{Y}}{\sqrt{\left(\dfrac{n_X S_X^2 + n_Y S_Y^2}{n_X + n_Y - 2}\right)} \sqrt{\left(\dfrac{n_X + n_Y}{n_X n_Y}\right)}} \tag{4:9}$$

where $\bar{X}$ = mean of variable X
         $\bar{Y}$ = mean of variable Y
         S_X^2 = variance of X
         S_Y^2 = variance of Y
         n_X = sample size of variable X
         n_Y = sample size of variable Y

The first part of the denominator is termed the 'pooled' or combined variance estimate and as such can only be used for samples of similar variances. An advantage of this test is that the terms of the equation have already been calculated in Table 4.1 for the F-ratio test. Thus:

$$t = \frac{55.71 - 43.70}{\sqrt{\left(\dfrac{25 \times 10.98 + 25 \times 13.39}{25 + 25 - 2}\right)} \sqrt{\left(\dfrac{25 + 25}{25 \times 25}\right)}}$$

$$t = \frac{12.01}{\sqrt{\left(\dfrac{274.5 + 334.8}{48.0}\right)} \sqrt{\left(\dfrac{50.0}{625.0}\right)}}$$

$$t = 12.01$$

The degrees of freedom for this test are $(n_X - 1) + (n_Y - 1)$; in this case $v = 48$. Tables of t are then consulted, which are available in most statistical textbooks. Assuming a significance level of 0.05 and remembering that the test is one-tailed, the critical t value is $+1.68$. The calculated t value of 12.01 far exceeds this figure and thus H_0 is rejected and the alternative hypothesis (H_1), that the samples come from different populations and the light levels under the pioneer phase are significantly higher than those under the mature phase is accepted. This result is then interpreted in terms of the effects of shading and light availability in the pioneer and mature phases of the *Calluna* cycle.

If the requirement of equal variances had not been met, then the following expression (with the same notation as formula 4:9) would have been used to determine t:

$$ t = \frac{\bar{X} - \bar{Y}}{\sqrt{\left(\dfrac{S_X^2}{n_{X-1}} + \dfrac{S_Y^2}{n_{Y-1}} \right)}} \qquad (4:10) $$

The two different formulae will only give similar results when the samples are large and the variances similar. It is thus very important to carry out the F-ratio test at the outset and to determine which of the two formulae is the correct one to use. Finally, it should be noted that although the two sets of observations in this case were identical in size, the test is equally applicable to samples of differing sizes.

The Mann–Whitney U test for independent samples

The Mann–Whitney U test is the non-parametric equivalent of the t-test for independent samples. The workings of the test are illustrated by a study of plant species richness in the hedgerows of lowland England. Two areas with very different farming regimes and environments were selected, and data on plant species numbers were collected by sampling a 32m length of 15 different hedgerows in Huntingdonshire and 13 similar lengths in Devon. The number of higher plant species in each of these lengths is shown in Table 4.2. The aim of the study was to see if there was a significant difference between the numbers of species in the hedgerows of the different counties. Thus the null hypothesis (H_0) was that there was no difference between the numbers of species in the samples of hedgerows from the two counties. The alternative hypothesis (H_1) was that there were more species in the hedgerows in Devon than in Huntingdonshire. The reason for this was that environmental and farming conditions in Devon (smaller fields, livestock farming and less use of herbicides) were better than in Huntingdon. Here the alternative hypothesis (H_1) is directional because it has been hypothesised that the hedgerows in Devon have higher plant-species numbers than those in Huntingdonshire, and the test is therefore one-tailed. Note also that the Mann–Whitney U test can be performed with two samples of differing sizes.

To calculate the U test statistic, the scores for both sets of samples are ranked together on a continuous scale from lowest to highest. Where ties occur, the mean rank of all the scores involved in the tie is given to those observations. The ranked scores for each set of observations are then summed to give values of Σr_1 and Σr_2. These are then entered to the following formula for the calculation of U and U_1:

Table 4.2   Data for the Mann-Whitney U test on higher plant species numbers in hedgerows in Devon and Huntingdonshire, England (Gilbertson *et al.*, 1985)

Number of higher plant species in a 32m-length of hedge		Rank n_1	Rank n_2
Devon (n_1)	Huntingdonshire (n_2)		
28	14	26	5
27	20	25	13.5
33	16	28	8.5
23	13	20	2.5
24	18	23	11
17	21	10	16
25	23	24	20
23	20	20	13.5
31	14	27	5
23	20	20	13.5
23	20	20	13.5
22	14	17	5
15	11	7	1
	16		8.5
	13		2.5
$n_1 = 13$	$n_2 = 15$	$\Sigma\, r_1 = 267$	$\Sigma\, r_2 = 139$

$$U = 13 \times 15 + \frac{13(13+1)}{2} - 267 = 19$$

$$U_1 = 13 \times 15 + \frac{15(15+1)}{2} - 139 = 176$$

$$U \text{ (lower value)} = 19$$

$$U = n_1 n_2 + \frac{n_1(n_1+1)}{2} - \Sigma\, r_1 \tag{4:11}$$

$$U_1 = n_1 n_2 + \frac{n_2(n_2+1)}{2} - \Sigma\, r_2 \tag{4:12}$$

where n_1 = the number of observations in the first sample
      n_2 = the number of observations in the second sample

The application of these formulae to the hedgerow data is shown in Table 4.2. The lower of the values for U and U_1 is taken to assess the significance of the difference between the two sets of samples.

In Table 4.2, the value of U is 19 and U_1 176. The lower score is thus 19 and the significance tables for Mann–Whitney U, available in most statistical textbooks are consulted at an appropriate significance level. It is important to realise that the

Table 4.3 Paired sample t-test on productivities of species in successional old-fields in central New York. Data for fields abandoned 17 years previously with fertilised (F) and unfertilised plots (U). Productivity in g/m^2 day obtained by dividing peak biomass of each species by time from last frost to that biomass (Abstracted from Table 2; Mellinger and McNaughton, 1975)

Species	Unfertilised (control) plot (U)	Fertilised (treated) plot (F)	Difference (d) (F − U)
Asclepias syriaca	0.034	0.247	0.213
Aster laevis	0.244	0.096	− 0.148
Aster lateriflorus	0.041	0.146	0.105
Aster novae-angliae	0.310	0.365	0.055
Aster simplex	0.062	0.088	0.026
Dactylis glomerata	0.001	0.055	0.054
Fragaria virginiana	0.441	0.385	− 0.056
Hieracium pratense	0.592	0.626	0.034
Phleum pratense	0.387	0.911	0.524
Picris hieracoides	1.369	1.510	0.141
Plantago lanceolata	0.260	0.208	− 0.052
Poa compressa	0.610	0.773	0.163
Poa pratensis	0.054	0.116	0.062
Solidago altissima	0.843	1.967	1.124
Solidago graminifolia	0.201	0.097	− 0.104
Solidago juncea	0.278	0.148	− 0.130
Solidago rugosa	0.156	0.197	0.041
Taraxacum officinale	0.100	0.151	0.051
	$n = 18$	$\bar{d} = 0.117$	$\sigma_d = 0.294$

greater the difference between the two sets of samples, the smaller the test statistic will be, that is the lower the value of U or U_1. Thus **unlike most other statistical tests, if the computed value is lower than the critical value in the statistical tables, then the null hypothesis (H_0) is rejected for the given significance level. If the computed value is larger than the critical value, then the null hypothesis is accepted.**

For the example, with a significance level of 0.05 and sample sizes of $n_1 = 13$ and $n_2 = 15$, the critical value for a one-tailed test is 61. The calculated value of 19 is obviously well below this figure; the null hypothesis (H_0) is rejected and the stated directional hypothesis (H_1) is accepted. The samples thus indicate that there are significantly more higher plant species in the hedgerows of Devon than in Huntingdonshire.

Finally, if the size of the samples exceeds 20, standard significance tables cannot be used; different tables should be used or a value called a 'z score' is calculated and a different set of significance tables is required (see Hammond and McCullagh, 1978, p. 208 and Silk, 1979, p. 189).

The t-test for paired samples

Samples are said to be paired when they have something in common with each other which they do not share with the remainder of the data, and each observation in one sample can be matched with one in the second sample. To illustrate this, Table 4.3 presents a set of data on the above-ground productivity (g m² day) of species in successional plots within old fields in central New York (Mellinger and McNaughton, 1975). The data represent productivities for two plots, both of which were abandoned 17 years prior to the experiment. One of the fields had subsequently been fertilised with 10-10-10 (N-P-K) dry pebble fertiliser. The other had been left unfertilised as a control. The data thus represent the paired productivities of the species in the two different plots.

In terms of statistical analysis, it is obviously interesting to establish whether there is a significant difference between the means of the two samples. The null hypothesis (H_0) is that no difference in productivities exists. The alternative directional hypothesis (H_1) is that the species in the fertilised fields (F) would be expected to have higher productivities than those in the unfertilised control plot (U).

To calculate the paired t-test, the test for equality of variances should again be carried out first and if the t-test is to be used the null hypothesis of no difference between the variances must be accepted. The variance for the species in the control plot, using Bessel's correction, is 0.1204, while that for the species in the fertilised plot is 0.2864. The variance ratio test is thus:

$$F = \frac{0.2864}{0.1204} = 2.38 \quad \text{with 17 and 17 degrees of freedom}$$

Examination of F tables available in most statistical texts shows that with 17 and 17 degrees of freedom, the F value has to exceed 2.28 to reject the null hypothesis of no difference between the variances at the 0.05 level (H_0). The calculated value of 2.38 just exceeds this, the null hypothesis (H_0) cannot be accepted and thus there is a doubt as to whether the paired t-test should be applied to these data. However, at the 0.01 level the calculated value has to exceed 3.25 to reject H_0. Accordingly the F test shows that we are just at the margins of the assumption of equality of variances in this application of the t-test. Because of this, the t-test is applied to these data, but also the same data are analysed later using the non-parametric Wilcoxon signed rank test.

Assuming that we can take the variances as approximately equal, the best estimate of the standard deviation of the difference between the pairs is calculated. This involves finding the difference between each of the pairs of values (F − U; column 3 in Table 4.3) and taking the mean of the difference ($\bar{d}$). The best estimate of the standard deviation of the differences is then found from:

$$\hat{\sigma}_d = \sqrt{\left(\frac{\Sigma(d - \bar{d})^2}{n - 1} \right)} \tag{4:13}$$

For the example this gives the foliowing:

$$\hat{\sigma}_d = \sqrt{\left(\frac{1.466}{17} \right)}$$

$$\hat{\sigma}_d = 0.294$$

Next calculate the standard deviation of the sampling distribution of mean differences (standard error):

$$\text{S.E.}_{\bar{d}} = \frac{\hat{\sigma}_d}{\sqrt{n}} = \frac{0.294}{\sqrt{18}} = \frac{0.294}{4.24} = 0.069$$

t is then calculated:

$$t = \frac{\text{the mean of the differences between the paired values}}{\text{the standard error of the difference}}$$

$$t = \frac{\bar{d}}{\text{S.E.}_{\bar{d}}}$$

$$t = \frac{0.1168}{0.069} = 1.69$$

With paired data, the number of degrees of freedom is $n-1$, where n is the number of pairs of observations. Using the t tables found in most statistical textbooks, for a one-tailed test, at the 0.05 level, with 17 degrees of freedom, the critical value is 1.74. The calculated value is 1.69. Thus the null hypothesis (H_0) of no difference between the productivities is accepted, and the alternative hypothesis (H_1) of a significantly greater productivity in the fertilised plots is rejected.

Final points on t-tests

One of the assumptions of the t-tests is that the background populations of the samples are normally distributed. In the case of the *Calluna* light data and the New York old field data, the distributions approximate to normality. However, where distributions are **skewed**, data can be normalised by **transformation**. The commonest transformations are logarithms or square roots of the original data where data are positively skewed. Negative skewness is much rarer. This topic is dealt with at greater length in Hammond and McCullagh (1978), Silk (1979), Finney (1980), Bishop (1983), Ebdon (1985), Barber (1988) and Fowler and Cohen (1990).

The Wilcoxon paired-sample test

The Wilcoxon paired-sample test is the non-parametric equivalent of the paired t-test. It is thus appropriate to situations where the parametric assumptions of the paired t-test cannot be met, particularly the assumption of normality of the background populations of the two samples and the requirement of equal variances. Here the same data on productivities of species in the 17-year-old abandoned field in central New York are used. When using the paired t-test, the F-ratio test of the null hypothesis of no difference between the variances (H_0) indicated that the variances were not equal at the 0.05 level, although they were at the 0.01 level. It is almost certainly safer to use the non-parametric alternative in this case.

The Wilcoxon test is based on differences between the pairs of data (X and Y) from the two samples when they are **ranked** or put on the ordinal scale. Significance levels are determined by the allocation of ranks into two groups: first, where the X value of a pair is greater than the Y and second, where the Y value is greater than the X. Obviously in ranking the data, some information is lost, but this is the price that is paid for the relaxation of parametric assumptions. The Wilcoxon test is carried out in two ways, depending on sample size. If a sample size is less than 25, then the following procedure is followed. Here the New York old-field data with a sample size of 18 are used (Table 4.4).

Table 4.4 Wilcoxon paired-sample test on productivities of species in successional old-fields in central New York. Data for fields abandoned 17 years previously with fertilised (F) and unfertilised plots (U). Productivity in g/m² day obtained by dividing peak biomass of each species by time from last frost to that biomass (abstracted from Table 2; Mellinger and McNaughton, 1975)

Species	Unfertilised (control) plot (F)	Fertilised (treated) plot (U)	Absolute difference $|F-U|$	Ranks of $|F-U|$	Assigned ranks	
					R_1 (F>U)	R_2 (F<U)
Asclepias syriaca	0.034	0.247	0.213	16	16	
Aster laevis	0.244	0.096	0.148	14		14
Aster lateriflorus	0.041	0.146	0.105	11	11	
Aster novae-angliae	0.310	0.365	0.055	7	7	
Aster simplex	0.062	0.088	0.026	1	1	
Dactylis glomerata	0.001	0.055	0.054	6	6	
Fragaria virginiana	0.441	0.385	0.056	8		8
Hieracium pratense	0.592	0.626	0.034	2	2	
Phleum pratense	0.387	0.911	0.524	17	17	
Picris hieracoides	1.369	1.510	0.141	13	13	
Plantago lanceolata	0.260	0.208	0.052	5		5
Poa compressa	0.610	0.773	0.163	15	15	
Poa pratensis	0.054	0.116	0.062	9	9	
Solidago altissima	0.843	1.967	1.124	18	18	
Solidago graminifolia	0.201	0.097	0.104	10		10
Solidago juncea	0.278	0.148	0.130	12		12
Solidago rugosa	0.156	0.197	0.041	3	3	
Taraxacum officinale	0.100	0.151	0.051	4	4	

$$\Sigma R_1 = 122 \quad \Sigma R_2 = 49$$

Number of pairs of observations (n) = 18
$|F-U|$ differences are taken regardless of sign (absolute difference)

First the null hypothesis must be stated (H_0) — that there is no difference in the productivities of the unfertilised and fertilised plots. Second, a directional alternative hypothesis (H_1) is put forward that the productivities in the fertilised plots (F) are greater than those in the unfertilised (U). A significance level of 0.01 is selected.

Next, the **absolute difference** between the matched pairs of observations is found ($|F-U|$, column 3 in Table 4.4). Any pairs which are equal in value and where $|F-U|$ = 0 are removed from the analysis. The values of $|F-U|$ are then ranked (column 4 in Table 4.4) **with the smallest value given the rank of 1**. Ties are given the average of the ranks that they would have occupied, as in the case of the Mann–Whitney U test. Ranks are then sorted into two columns (columns 5 and 6 in Table 4.4): R_1, where F is greater than U and R_2, where U is greater than F. R_1 and R_2 are then summed and whichever total is **smaller** is called T. In the example, ΣR_1 is 122 and ΣR_2 49. Thus T is the smaller value 49.

This T value can then be looked up in appropriate statistical tables, which give critical values of T for samples of differing sizes from n = 6 to n = 33. The test cannot be applied to samples of less than 6. An important point is that to achieve significance at a specified level, the T value must be **less than** the critical value, again as was the case with the Mann–Whitney U test. The column headed 0.01 for one-tailed tests is used. With 18 pairs of observations, T must be less than 32 to be significant and to reject the null hypothesis (H_0). The calculated value of T is 49, which is greater than the critical value. The alternative hypothesis (H_1), that productivities are greater in the fertilised than in unfertilised plots is rejected, and the null hypothesis (H_0) is accepted. It is interesting to note that even if the significance level had been lowered to 0.05, the test would not have been significant, since for a one-tailed test, the value of T must be less than 47.

It is worth noting that if H_0 is correct, ΣR_1 and ΣR_2 should be approximately equal because the differences between the paired values will be distributed randomly among the sets of ranks, where X is greater than Y and where Y is greater than X. Hence, the greater the difference between ΣR_1 and ΣR_2, the less likely it is that H_0 is correct. Also, since $(R_1 + R_2)$ remains constant for a sample of given size, the larger the difference between ΣR_1 and ΣR_2, the smaller will be the lower value (T) and the greater will be the other. Thus the smaller the value of T, the less probable it is that H_0 is correct.

With larger samples ($n > 25$), the procedure is very similar to that described above, except when examining values of T in the table of critical values. In most statistical tables, these are only given for sample sizes up to 33. For larger samples, T is converted into a z score. This is T minus the mean of the sampling distribution, divided by the standard deviation of the sampling distribution. Thus z is calculated as:

$$z = \frac{T - \frac{1}{4}n(n+1)}{\sqrt{\{n(n+1)(2n+1)\}/24}} \qquad (4{:}14)$$

A value of p is then found from z tables in most statistical texts, which give one- and two-tailed probabilities associated with values of z in the normal distribution. The logic of this is that the smaller the value of T, the further it is from the mean of the sampling distribution and therefore the greater its z score; hence the smaller the possibility (p) of its occurrence under H_0. When p is **less than** the rejection level, then H_0 can be rejected and H_1 accepted. If it is **more than** the rejection level, H_0 may not be rejected.

Finally, where a large number of ties occur, a correction factor is introduced as below:

$$z = \frac{T - \frac{1}{4}n(n+1)}{\sqrt{\dfrac{n(n+1)(2n+1)}{24} - \dfrac{\Sigma u^3 - \Sigma u}{48}}} \qquad (4{:}15)$$

where   u = the number of ties on each rank where ties occur
$\qquad$ Σu = the sum of u over all sets of tied ranks

Correlation and regression analysis

An appreciation of methods for correlation and regression is very important for straightforward analysis of floristic and environmental data and also as a basis for understanding of the more complex methods of ordination and classification described in Chapters 5, 6 and 8. **Correlation analysis** is a set of methods which

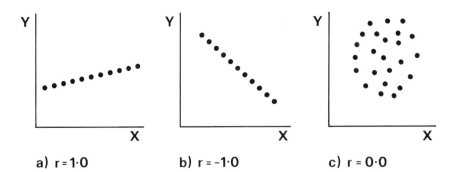

Figure 4.4   Scattergrams of (a) perfect positive correlation; (b) perfect negative correlation and (c) no correlation between variables X and Y

is used to determine the **strength of relationships** between variables. The result of a correlation analysis is a statistic lying between -1.0 through 0.0 to $+1.0$ which describes the degree of relationship between the two variables. **Regression analysis** takes this a stage further by measuring and describing the **form** of the relationship between two variables and allowing **prediction** of values of one variable in terms of variation in the other. Correlation and regression are closely related techniques.

Both parametric and non-parametric methods exist for correlation and regression. For correlation, the major parametric method is **Pearson's product–moment correlation coefficient** and for regression, the equivalent is the **least-squares technique**. However, non-parametric alternatives, which are often better suited to the quality of data generated by vegetation scientists, also exist. For correlation, there are several possibilities but the most widely used is **Spearman's rank correlation coefficient**. There is also the very useful method of **point-biserial correlation** which allows one binary or dichotomous $(+/-$ or 0/1) variable to be correlated against a continuous variable. For regression, a simple non-parametric alternative for describing the form of the regression relationship is the **method of semi-averages**, although it does not enable any significance testing to be carried out. An even more useful non-parametric regression method, which is derived from techniques of exploratory data analysis (Tukey, 1977), is **the fitting of resistant lines**.

Once again, the decision as to whether to use parametric or non-parametric tests depends on the quality of the data. In correlation, the Pearson product–moment correlation coefficient (r) is the most powerful, while Spearman's rank correlation coefficient (r_s) is only 91 per cent as power-efficient as Pearson's r. This means that if Pearson's r is significant in a sample of 100 cases taken from two normally distributed variables, it will require a sample of 110 cases of the same data to achieve the same level of significance from the rank order coefficient r_s.

Scattergrams

At the start of any correlation and regression analysis, it is always worthwhile to plot a graph of the relationship between the two variables (Dunn, 1989). Values of correlations vary from -1.0 through 0.0 to $+1.0$. The values $+1.0$ and -1.0 represent a perfect relationship between the two variables. Thus Figure 4.4a shows a perfect positive relationship (r $= +1.0$) between two variables X and Y which means that an **increase** in the amount of X is directly matched by an **increase** in

Table 4.5 Calculation of the product-moment correlation coefficient for data on plant species richness and site age from 26 vacant urban lots in Chicago (Crowe, 1979)

1 Lot	2 Species numbers (Y_i)	3 Lot age (months) (X_i)	4 Y_i^2	5 X_i^2	6 X_iY_i
1	11	3	121	9	33
2	9	3	81	9	27
3	16	7	256	49	112
4	27	15	729	225	405
5	21	18	441	324	378
6	32	20	1024	400	640
7	22	20	484	400	440
8	16	20	256	400	320
9	15	20	225	400	300
10	26	22	676	484	572
11	25	25	625	625	625
12	15	30	225	900	450
13	30	30	900	900	900
14	45	40	2025	1600	1800
15	22	50	484	2500	1100
16	56	65	3136	4225	3640
17	47	70	2209	4900	3290
18	20	70	400	4900	1400
19	45	80	2025	6400	3600
20	44	90	1936	8100	3960
21	54	100	2916	10000	5400
22	47	100	2209	10000	4700
23	37	100	1369	10000	3700
24	69	113	4761	12769	7797
25	46	120	2116	14400	5520
26	47	150	2209	22500	7050

$$\sum_{i=1}^{n} Y_i \ 844 \qquad \sum_{i=1}^{n} X_i \ 1381 \qquad \sum_{i=1}^{n} Y_i^2 \ 33838 \qquad \sum_{i=1}^{n} X_i^2 \ 117419 \qquad \sum_{i=1}^{n} X_iY_i \ 58159$$

the amount of Y. Figure 4.4b shows a perfect negative relationship with an **increase** in the amount of X being matched exactly by a **decrease** in the amount of Y. It is also important to note that in either case, the relationship is **linear**. Figure 4.4c shows the graph where there is **no correlation**. The closer the value is to $+1.0$ or -1.0, the **stronger** is the relationship.

However, the fact that two variables have a strong relationship does not necessarily mean that one variable is **causing** the variation in the other. **A causal relationship** cannot be deduced from a correlation coefficient alone. Cause and effect can only be determined through other evidence and the judgement of the researcher.

In order to demonstrate the use and calculation of correlation and regression analysis, an example of exploration of the relationship between plant species richness and site age is taken for 26 vacant urban lots in Chicago (Crowe, 1979). Crowe recorded the number of plant species growing on abandoned urban plots, ranging in age following abandonment from 3 to 150 months. The data are shown in columns 2 and 3 of Table 4.5. In Figure 4.5, these data are plotted as a scattergram. Inspection of this graph shows evidence of a positive linear relationship between the two variables.

Correlation analysis

The Pearson product–moment correlation coefficient

As with all parametric tests, the product–moment correlation coefficient is based on certain assumptions about the data to which it can be applied. First, data must be continuous and measured on either the interval or ratio scales. Second, the background populations of each set of data should fit the normal distribution (Figure 4.1). The computational formula for the product–moment correlation coefficient is as follows:

$$r = \frac{\sum_{i=1}^{n} X_i Y_i - (\sum_{i=1}^{n} X_i)(\sum_{i=1}^{n} Y_i) / n}{\sqrt{[\sum_{i=1}^{n} X_i^2 - (\sum_{i=1}^{n} X_i)^2 / n]} \sqrt{[\sum_{i=1}^{n} Y_i^2 - (\sum_{i=1}^{n} Y_i)^2 / n]}} \qquad (4{:}16)$$

where

r = product–moment correlation coefficient
n = the number of pairs of observations

$\sum_{i=1}^{n} X_i$ = sum of observations on X

$\sum_{i=1}^{n} Y_i$ = sum of observations on Y

In the Chicago example, the general research hypothesis is that there is a relationship between species numbers and age of urban lot. The null hypothesis (H_0) is that there is no such relationship. The alternative directional hypothesis (H_1), derived from examination of Figure 4.5, is that there is a positive correlation between the two variables, so that plant-species richness increases as lot age increases.

Using the above formula allows the calculation to be broken down into a number of simple steps, as shown in Table 4.5. Using the sub-totals from the table, the following equation results:

$$r = \frac{58159 - (1381)(844) / 26}{\sqrt{[117419 - (1381)^2/26]} \sqrt{[33838 - (844)^2/26]}}$$

$$r = \frac{58159 - 44829.38}{209.92 \quad 80.25} = \frac{13329.62}{16846.08} = 0.791$$

Thus the existence of a positive relationship between species richness and age is demonstrated with a correlation coefficient of 0.791. However, this must then be tested for significance.

Testing the significance of r

When sampling from bivariate populations, there is always the possibility that an entirely spurious correlation coefficient may be derived, particularly where small samples have been collected. The significance test is designed to calculate the probability that for the given sample size, the correlation coefficient could have been derived by chance. The test is based on the use of t tables printed in most statistical textbooks. Since a directional alternative hypothesis (H_1) has been stated, the significance test is one-tailed.

The value of t is found by the formula:

$$t = r \sqrt{\frac{n-2}{1.0 - r^2}} \qquad\qquad (4:17)$$

thus

$$t = 0.791 \sqrt{\frac{26-2}{1.0 - 0.626}} = 0.791 \sqrt{\frac{24}{0.374}} = 6.336$$

Degrees of freedom for correlation coefficients are the number of pairs of observations less two ($n-2$). Reference to tables of Student's t shows that with 24 degrees of freedom t must exceed 2.80 to be significant in a two-tailed test at the 0.01 level and would have to exceed 2.49 in a one-tailed test. The calculated value was 6.336. The hypothesised positive relationship between the two variables is thus highly significant.

However, although this ecologically interesting relationship between plant species richness and age of site in Chicago has been found, it is important to be very careful in interpretation of the result and in assigning causality. Undoubtedly, the length of time during which colonisation has been possible is an important causal variable, but there must be others. A measure of how much of the variation in the species richness data is **explained** by the variation in lot age is obtained by squaring the correlation coefficient to give r^2 — in this case $r^2 = 0.791^2 = 0.63$. This means that 63 per cent of the variation in species richness in the 26 urban lots is accounted for by variation in lot age. This value is also known as the **coefficient of explanation or the coefficient of determination**. However, 37 per cent of the variation still remains unexplained and may be attributable to other factors, for example lot size, isolation, substrate/geology or degree of human interference. The implications of this are discussed further in the case study at the end of the chapter.

Spearman's rank correlation coefficient

This is the most widely used non-parametric coefficient although others, notably Kendall's Tau, exist (see Siegel, 1956; Hammond and McCullagh, 1978 and Lee and Lee 1982). The data from Chicago are again used to illustrate the calculation of Spearman's rank correlation coefficient. The following formula is used:

Table 4.6 Calculation of Spearman's rank correlation coefficient between plant species richness and site age from 26 vacant urban lots in Chicago (Crowe, 1979)

1 Lot	2 Species numbers	3 Rank	4 Lot age (months)	5 Rank	6 Difference (d)	7 Difference^2 (d^2)
1	11	2.0	3	1.5	0.5	0.25
2	9	1.0	3	1.5	0.5	0.25
3	16	5.5	7	3.0	2.5	6.25
4	27	13.0	15	4.0	9.0	81.00
5	21	8.0	18	5.0	3.0	9.00
6	32	15.0	20	7.5	7.5	56.25
7	22	9.5	20	7.5	2.0	4.00
8	16	5.5	20	7.5	2.0	4.00
9	15	3.5	20	7.5	4.0	16.00
10	26	12.0	22	10.0	2.0	4.00
11	25	11.0	25	11.0	0.0	0.00
12	15	3.5	30	12.5	9.0	81.00
13	30	14.0	30	12.5	1.5	2.25
14	45	18.5	40	14.0	4.5	20.25
15	22	9.5	50	15.0	5.5	30.25
16	56	25.0	65	16.0	9.0	81.00
17	47	22.0	70	17.5	4.5	20.25
18	20	7.0	70	17.5	10.5	110.25
19	45	18.5	80	19.0	0.5	0.25
20	44	17.0	90	20.0	3.0	9.00
21	54	24.0	100	22.0	2.0	4.00
22	47	22.0	100	22.0	0.0	0.00
23	37	16.0	100	22.0	6.0	36.00
24	69	26.0	113	24.0	2.0	4.00
25	46	20.0	120	25.0	5.0	25.00
26	47	22.0	150	26.0	4.0	16.00

$$\sum_{i=1}^{n} d^2 \quad 620.50$$

$$r_s = 1.0 - \frac{6 \; \Sigma d^2}{n^3 - n} \qquad (4:18)$$

where d = difference between paired ranks (see text)
      n = number of pairs of observations

For the Chicago example, the hypotheses are laid out in exactly the same way as for the product–moment coefficient, and the calculation is shown in Table 4.6. The two variables are ranked separately and consistently (that is from high to low or low to high), and each observation is shown by the ranked values in Table 4.6. Where two or more values have the same value, they are said to be 'tied'. In this case, the rank scores that would have been given to the values are taken and averaged and that value is then given to all those observations with the same value in the original data. As an example, in the species richness data, there are three lots with the score of 47 species: numbers 17, 22 and 26 (Table 4.6). These would be ranked 21, 22 and 23. When these ranks are summed [21 + 22 + 23 = 66] and averaged [66/3 = 22], the average rank of 22 is given to all three scores of 47 species.

Once the ranks have been assigned, the difference between each pair of ranks is taken (d) and this value is then squared (d^2). The sum of the d^2 values is then calculated to give Σd^2. Using the figure for Σd^2 from the table, the following calculations are made:

$$r_s = 1.0 - \frac{6 \times 620.5}{26^3 - 26}$$

$$r_s = 1.0 - \frac{3723}{17550} = 0.787$$

The value of r_s (0.787) is less than that of the product–moment correlation coefficient (0.791) but only marginally so. This demonstrates the **power-efficiency** of the test.

Significance testing

As before, the r_s value must be tested for significance using the t statistic and the tables available in most statistical textbooks. Exactly the same formula as for the product–moment coefficient is applied:

$$t = r_s \sqrt{\frac{n-2}{1.0 - r_s^2}} \tag{4:17}$$

thus

$$t = 0.787 \sqrt{\frac{26-2}{1.0 - 0.619}}$$

$$t = 0.787 \sqrt{\frac{24}{0.381}} = 6.249$$

The value of t has to exceed 2.49 in a one-tailed test at the 0.01 level. This is achieved with the t value of 6.249 and once again, the positive correlation is highly significant.

The point-biserial correlation coefficient (r_p)

Data are often collected on the distribution of a species in a series of quadrats, together with information on several environmental variables, with the aim of establishing correlations between the species distribution and an environmental variable. A range of statistical tests is available but not all are appropriate. As an example, the use of the Pearson product–moment correlation coefficient to examine correlations between species abundance — perhaps measured on a five-point Domin scale or as percentage cover, against variables such as pH, per cent of soil moisture content or the levels of soil nutrients — is usually invalid since the statistic requires continuous data for both variables and that the background populations of the two variables be normally distributed. Unfortunately, most species abundance data are not strictly continuous, nor are they normally distributed. Furthermore, such data for many species will also contain a substantial number of zeros or absences; thus it may be possible to use Spearman's rank correlation coefficient but only if there

Table 4.7 Partitioned data for the calculation of the point-biserial correlation coefficient between the present or absence of bracken (*Pteridium aquilinum*) and % soil moisture

Presence (+) or absence (−) of *Pteridium*	% soil moisture	Partitioned data	
		p (*Pteridium* present)	q (*Pteridium* absent)
−	85.2		85.2
+	55.9	55.9	
−	95.2		95.2
−	58.1		58.1
+	54.9	54.9	
−	89.4		89.4
+	68.2	68.2	
+	46.8	46.8	
−	54.8		54.8
−	92.2		92.2
−	90.7		90.7
+	30.2	30.2	
+	35.2	35.2	
−	90.3		90.3
−	95.6		95.6
−	91.5		91.5
−	55.4		55.4
+	35.5	35.5	
+	42.1	42.1	
−	49.3		49.3
$S_x = 23.05$		Mean p = 46.10	Mean q = 78.98

are not a large number of zeros or absences in the data.

This problem of absences may be partially overcome by the use of a method of correlation which has rarely been used in vegetation studies — the point-biserial correlation coefficient (r_p) which is applied in situations where one variable is in presence/absence or binary form, while the other variable is continuous. It thus becomes possible to correlate the presence or absence of a species which is present or absent in a series of quadrats against a continuous environmental variable.

As an example of the use of the coefficient, data from 20 quadrats on the presence/absence of bracken (*Pteridium aquilinum*) and the per cent of soil moisture in a series of quadrats are presented in Table 4.7. The hypothesis is set up that a relationship exists between the soil moisture content and the presence and absence of bracken. The null hypothesis (H_0) is that no such relationship exists, while the alternative hypothesis (H_1) can be phrased in directional terms: the soil moisture content of quadrats where bracken is absent will be higher than in those quadrats where it is present, thus indicating a negative correlation.

The procedure is as follows. First, both variables are tabulated (Table 4.7) and the continuous variable is then partitioned into two sets on the basis of the presence and absence of bracken (p = bracken present; q = bracken absent). The mean of each group of the partitioned continuous variable is calculated, plus the standard deviation for the whole of the continuous variable. Also the proportion of all values in each group is found. The point-biserial correlation coefficient is then calculated from the formula:

$$r_p = \frac{|M_p - M_q|}{S_x} \times \sqrt{(p \times q)} \qquad (4{:}19)$$

where: r_p = point-biserial correlation coefficient
         M_p = mean of the first group of values
         M_q = mean of the second group of values
         $|M_p - M_q|$ = the absolute difference between the means regardless of sign
         p   = proportion of observations in the first group
         q   = proportion of observations in the second group
         S_x = the standard deviation of all observations on the continuous variable

In the example (Table 4.7), group p includes 8 out of the total of 20 observations and thus p is 0.4, while q contains 12 observations and is 0.6. Using the above formula and the data in the table, r_p is computed as follows:

$$r_p = \frac{|46.1 - 78.98|}{23.05} \times \sqrt{(0.4 \times 0.6)}$$

$$r_p = \frac{32.88}{23.05} \times 0.489 = 0.698$$

Testing for significance

The same formula is again used as for the other correlation coefficients (4:17) and t is calculated as follows:

$$t = r_p \sqrt{\frac{n-2}{1.0 - r_p^2}} \qquad (4{:}17)$$

thus

$$t = 0.698 \sqrt{\frac{20-2}{1.0 - 0.487}} = 0.698 \sqrt{\frac{18}{0.513}} = 4.134$$

Reference to a table of t values shows that with $n - 2 = 18$ degrees of freedom, a t value of greater than 2.55 is significant at the 0.01 level for a one-tailed test. Thus the hypothesis that *Pteridium* is absent where percentage soil moisture values are high and present where they are low is accepted with a high degree of confidence.

Interpretation of correlations

Great care should be taken in the interpretation of correlation coefficients. A significant result in a correlation analysis does **not** necessarily mean that there is a **causal** relationship between the variables. While this may appear to be the case for the examples given in the case of urban lot age and species richness in Chicago and the soil moisture/*Pteridium* relationship from Britain, **other** factors may also be important. Many correlations are examples of **size relationships**. A correlation between plant size or height and productivity which gave a significant positive correlation would be a good demonstration of this. Still others are best described as **mutual interaction** between the variables. Rather than one **causing** variation in

Table 4.8 Closed and open systems in correlation analysis (based on Silk, 1979, p. 209)

(a) open number system — habitat types in units of 100km^2

Region	Habitat type		
	Woodland (1)	Scrubland (2)	Grassland (3)
A	5	2	1
B	7	4	2
C	8	6	4
D	10	8	5

$$r_{1,2} = 0.99; \; r_{1,3} = 0.96; \; r_{2,3} = 0.89$$

(b) closed number system — habitat types expressed as % (region = 100%)

Region	Habitat type		
	Woodland (1)	Scrubland (2)	Grassland (3)
A	63	25	12
B	54	31	15
C	44	33	23
D	43	35	22

$$r_{1,2} = -0.97; \; r_{1,3} = -0.98; \; r_{2,3} = 0.89$$

the other, it is more realistic to talk of these variables 'varying together' rather than a one-way causal relationship.

Finally, there is the problem of **closed number systems**. This relates to the use of **percentage data or proportions** in correlation. As an example, take a situation where the extent of certain habitat types is being measured in four different regions of a country. The regions are A, B, C and D. The habitat types are woodland, grassland and scrub. Correlation is used to describe the degree of relationship or similarity in distribution of the three habitat types across the four regions. In Table 4.8a, the areas of the habitats are shown as hypothetical values in units of 100km^2. Here the number system is open and if the areas of each habitat type across the four regions are correlated with each other, the correlations in Table 4.8a are found. However, if the areas of habitat are now expressed as percentages, taking the total area of each region as 100 per cent, the values in Table 4.8b result. These values occur within a **closed number system**, and the corresponding correlation coefficients can be seen to be very different. It has been demonstrated that where there are three variables involving values that sum to a fixed total (in this case percentages summing to 100 per cent), two of the correlations will always be negative and one positive, regardless of what the correlations are between the set of open numbers from which the closed numbers were obtained. Thus great caution should be taken when applying correlation to numbers based on percentages in different categories which sum to 100 per cent or with proportions which may sum to 1.0. Further discussion of closed number sets is presented in Johnston (1978) and Silk (1979).

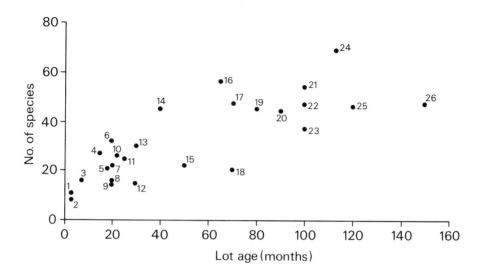

Figure 4.5 Scattergram of the two variables, species richness (Y) and abandoned urban lot age (X) for 26 sites in Chicago (Crowe, 1979; redrawn with kind permission of *Journal of Biogeography*)

Regression analysis

The purpose of simple (two-variable or bivariate) regression analysis is to determine the nature of the relationship between two variables by fitting a mathematical function to the set of data. All the methods described below are for **linear** regression, which aims to fit a straight line to a graph or scattergram showing the relationship between two variables, as in Figure 4.5. As with correlation, there are both parametric and non-parametric methods for regression. **Least-squares regression** is the major parametric technique for linear regression. Several alternative non-parametric and essentially descriptive methods also exist, for example **semi-averages** and various techniques for deriving **resistant lines**. When developing a research hypothesis involving the use of regression, it is important to decide on **dependent** and **independent** variables. In the example of the Chicago data, used to explain correlation (Table 4.5), the dependent variable is species richness and the independent variable is urban lot age. It is most unlikely that lot age could be dependent on species richness. Normally, the dependent variable is plotted on the Y axis of a scattergram and the independent variable on the X axis (Figure 4.6). Unfortunately, in some situations, it is less easy to determine which variable is the dependent. Nevertheless, it is still possible to use regression to describe the relationship.

Least-squares regression

Least-squares regression is a parametric method and thus makes a number of assumptions about the data, notably that they are measured on the interval or ratio scale and that the background population of each variable is normally distributed. The method is best described in two sections: first, the fitting of the regression line and testing for significance; second, the use of the method for prediction and the analysis of residuals.

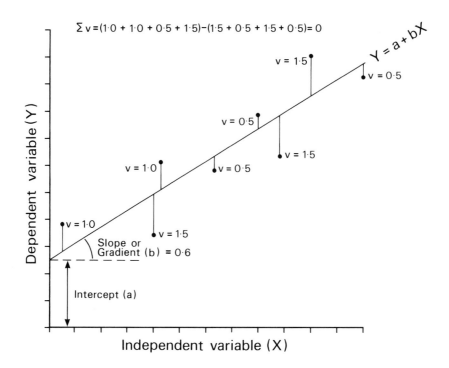

Figure 4.6 Properties of the linear regression line (Redrawn from Shaw and Wheeler, 1985; with kind permission of David Fulton Publishers)

The linear regression model

Any straight-line graph drawn on X and Y axes can be represented by an equation of the form:

$$Y = a + bX \tag{4:20}$$

where a is the **intercept term** and represents the point on the graph where the straight line intersects the Y axis (Figure 4.6) and b is the **slope term** or **gradient** of the line. The gradient determines the rate at which the line rises or falls as X increases. In Figure 4.6, the line has a slope of 0.6. This means that for every unit (1.0) increase of X, there is an equivalent increase of 0.6 in the value of Y.

The regression line is fitted through a scattergram of points in such a way that the positive and negative deviations of individual points, measured in terms of the Y axis, from the line must sum to zero ($\Sigma v = 0$), and the sum of the squared deviations of the individual points from the line must be smaller than for any other line (Σv^2 = minimum). In Figure 4.6, both of these conditions are met for the eight points in the scattergram. It is important to realise that **the line is fitted in terms of predicted values of Y for given values of X**.

The differences between actual points, in terms of Y and the points on the calculated line in terms of Y ($\hat{Y}$), are known as the **residuals** and the least-squares method involves finding the line such that the sum of the squares of the residuals (that is, of differences between the actual and predicted line values) is minimised (Figure 4.6).

Computation of the line requires values of a (intercept) and b (slope term). b is determined from the formula:

$$b = \frac{\sum_{i=1}^{n} X_i Y_i - [(\sum_{i=1}^{n} X_i)(\sum_{i=1}^{n} Y_i)] / n}{\sum_{i=1}^{n} X_i^2 - (\sum_{i=1}^{n} X_i)^2 / n} \qquad (4:21)$$

where   n = number of pairs of observations

$$\sum_{i=1}^{n} X_i = \text{sum of observations on } X$$

$$\sum_{i=1}^{n} Y_i = \text{sum of observations on } Y$$

and a from:

$$a = \overline{Y} - b\overline{X} \qquad (4:22)$$

where $\overline{X}$ = mean of X
      $\overline{Y}$ = mean of Y

Using these formulae and the Chicago data from Table 4.9, the least-squares line can be calculated as follows:

$$b = \frac{58159 - [(1381)(844)] / 26}{117419 - (1381)^2 / 26}$$

$$b = \frac{58159 - 44829}{117419 - 73352} = \frac{13330}{44067}$$

$$b = 0.302$$

$$a = 32.46 - (0.302 \times 53.12) = 16.42$$

Thus the regression equation is:

$$Y = 16.42 + 0.302X$$

The resulting line has been plotted on the scattergram of Figure 4.5 in Figure 4.7. There are two limitations to the use of the regression equation:

(a) the line is a line of **best fit** only within the range of the values of X in the analysis. The line should therefore never extend beyond the lowest and highest X values on the graph, since the relationship defined by the regression may not hold good outside the range of X values.

(b) the regression line giving estimated values of Y (species richness) on X (urban-lot age) is not reversible. The line **cannot** be used to predict or estimate values of X for a given value of Y. This is because unless the line just happens to be at 45°, the sum of squares of the Y (vertical) variation is different from the sum of squares of the X (horizontal) variation (Figure 4.8). Thus, if it was

Table 4.9 Calculation of least-squares regression for data on plant species-richness and site age from 26 vacant urban lots in Chicago (Crowe, 1979)

1 Lot	2 Species numbers (Y_i)	3 Lot age (months) (X_i)	4 Y_i^2	5 X_i^2	6 X_iY_i	7 $\hat{Y}_i$	8 $Y_i - \hat{Y}_i$	9 Standard-ised residuals $\dfrac{Y_i - \hat{Y}_i}{\hat{\sigma}_e}$
1	11	3	121	9	33	19.72	−8.72	−0.66
2	9	3	81	9	27	19.12	−10.12	−0.87
3	16	7	256	49	112	21.23	−5.23	−0.26
4	27	15	729	225	405	24.56	2.44	0.63
5	21	18	441	324	378	22.75	−1.75	−0.08
6	32	20	1024	400	640	26.07	5.93	0.99
7	22	20	484	400	440	23.05	−1.05	−0.05
8	16	20	256	400	320	21.23	−5.23	−0.66
9	15	20	225	400	300	20.93	−5.93	−0.76
10	26	22	676	484	572	24.26	1.74	0.30
11	25	25	625	625	625	23.96	1.04	0.11
12	15	30	225	900	450	20.93	−5.93	−1.07
13	30	30	900	900	900	25.47	4.53	0.46
14	45	40	2025	1600	1800	30.01	14.99	1.68
15	22	50	484	2500	1100	23.05	−1.05	−0.96
16	56	65	3136	4225	3640	33.33	22.67	2.03
17	47	70	2209	4900	3290	30.61	16.39	0.96
18	20	70	400	4900	1400	22.44	−2.44	−1.79
19	45	80	2025	6400	3600	30.01	14.99	0.45
20	44	90	1936	8100	3960	29.70	14.30	0.04
21	54	100	2916	10000	5400	32.73	21.27	0.77
22	47	100	2209	10000	4700	30.61	16.39	0.04
23	37	100	1369	10000	3700	27.59	9.41	−1.01
24	69	113	4761	12769	7797	37.27	31.73	1.96
25	46	120	2116	14400	5520	30.31	15.69	−0.72
26	47	150	2209	22500	7050	30.61	16.39	−1.70

$$\sum_{i=1}^{n} Y_i \; 844 \qquad \sum_{i=1}^{n} X_i \; 1381 \qquad \sum_{i=1}^{n} Y_i^2 \; 33838 \qquad \sum_{i=1}^{n} X_i^2 \; 117419 \qquad \sum_{i=1}^{n} X_iY_i \; 58159$$

necessary to estimate values of X from Y, then the regression would have to be recalculated and a different line would result.

Explained and unexplained variation in the Y values
The total variation in

$$Y = \sum_{i=1}^{n} (Y_i - \bar{Y}_i)^2$$

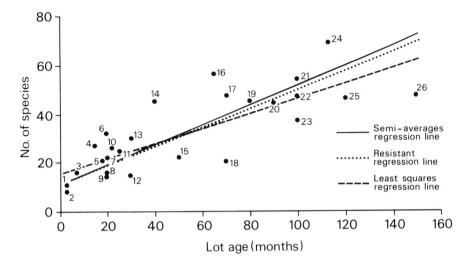

Figure 4.7 Least-squares, semi-averages and resistant regression lines plotted on the scattergram of the Chicago urban-lot data (Crowe, 1979; redrawn and adapted with kind permission of *Journal of Biogeography*)

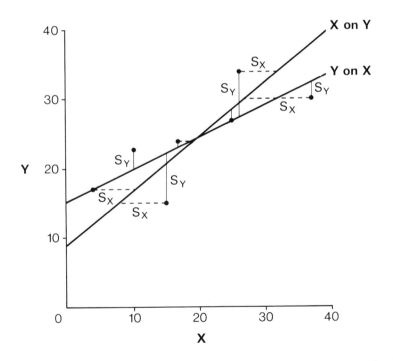

Figure 4.8 Vertical and horizontal deviations from a regression line and regressions of X on Y and Y on X (Hammond and McCullagh (1978); redrawn with kind permission Oxford University Press)

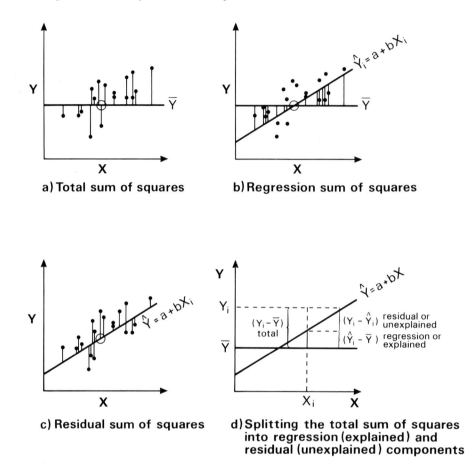

a) Total sum of squares

b) Regression sum of squares

c) Residual sum of squares

d) Splitting the total sum of squares into regression (explained) and residual (unexplained) components

Figure 4.9 Components of sums of squares and explained (regression) and unexplained (residual) variation (Silk, 1979; redrawn with kind permission of Chapman and Hall)

(Figure 4.9a), or the sum of the squared deviations from the mean of Y. This total variation comprises two parts:

(a) the **explained variation** (Figure 4.9b) which is

$$\sum_{i=1}^{n} (\hat{Y}_i - \overline{Y})^2$$

representing the deviation of the regression line from $\overline{Y}$. This indicates the extent to which the prediction or estimation of the line has been improved by using $\hat{Y}_i$ instead of Y_i, which would have been the best estimate of Y values if there were no knowledge of X and the regression equation.

(b) the **unexplained variation** which is

$$\sum_{i=1}^{n} (\hat{Y}_i - Y_i)^2$$

(Figure 4.9c). This represents the deviation of the original Y values (Y_i) from the estimated or calculated line ($\hat{Y}_i$). Table 4.9, column 7, shows the estimated values of Y ($\hat{Y}_i$) for the original values of X. These are the values of $\hat{Y}$ calculated along the regression line for each value of X. The difference between Y_i and $\hat{Y}_i$ (column 8) is the **residual or unexplained variation** in each case. The sum of squares of these values is the total unexplained variation or the residual variation (Figure 4.9c).

For a single observation (X_i/Y_i), the contributions to total variation, explained (regression) and unexplained (residual) variation are shown in Figure 4.9d.

The coefficient of determination or explanation

The explained and unexplained variation is used to calculate the coefficient of determination (explanation) which is given by the formula:

$$\frac{\text{Explained variation}}{\text{Total variation}} = \frac{\sum_{i=1}^{n} (\hat{Y}_i - \bar{Y})^2}{\sum_{i=1}^{n} (Y_i - \bar{Y})^2} \tag{4:23}$$

In the case of the Chicago data:

$$\text{Coefficient of determination} = \frac{4032.0}{6440.5} = 0.63 \text{ (63 per cent)}$$

Significance testing of least-squares regression lines

The $\hat{Y}$ values obtained from the regression line are only estimates based on samples drawn from a much larger and usually unknown population. As such, they are subject to sampling variations, and it is important to identify the reliability of sample estimates for the regression coefficients a and b and the estimated values of $\hat{Y}$. This can be done by using analysis of variance (ANOVA) on the explained and unexplained components of the variation described above.

An F (variance ratio) test is used to examine the ratio between the explained (regression) variance and the unexplained (residual) variance. The variance has to be related to degrees of freedom, as in Table 4.10.

A null hypothesis can be erected which is that there is no explanation in the variability of Y (the dependent variable) in terms of X (the predictor). It follows that greater F ratios are provided by higher proportions of explained variance. The associated degrees of freedom and predetermined significance level are used to determine critical F values from published statistical tables. Table 4.11 shows the figures for the Chicago case. Here the F statistic is calculated as:

$$\begin{array}{l} \text{Explained variance} \\ \text{(Regression)} \end{array} = \frac{\sum_{i=1}^{n} (\hat{Y}_i - \bar{Y})^2}{k} = \frac{4032.0}{1} = 4032.0 \tag{4:24}$$

where k = the number of predictors (always 1 in simple regression)

Table 4.10 Regression equation analysis of variance

Source of variation	Sums of squares	General description	Degrees of freedom
Explained (Regression)	$\displaystyle\sum_{i=1}^{n} (\hat{Y}_i - \bar{Y})^2$	Sum of squared deviations of predicted (estimated) values from sample mean	k
Unexplained (Residual)	$\displaystyle\sum_{i=1}^{n} (\hat{Y}_i - Y_i)^2$	Sum of squared differences between observed and predicted (estimated) values	$n - k - 1$
Total	$\displaystyle\sum_{i=1}^{n} (Y_i - \bar{Y})^2$	Sum of squared deviations of observations from sample mean	$n - 1$

where: n = number of observations
$\bar{Y}$ = mean of observed Y values
Y_i = individual Y values
$\hat{Y}_i$ = estimated (predicted Y values)
k = number of predictors (always 1 in simple regression)

Table 4.11 Regression analysis of variance table from the Chicago study

Source of variation	Sums of squares	Degrees of freedom	Mean square	F
Explained (Regression)	4032.0	1	4032.0	40.18
Unexplained (Residual)	2408.4	24	100.4	
Total	6440.4	25		

$$\text{Unexplained variance (Residual)} = \frac{\displaystyle\sum_{i=1}^{n} (\hat{Y}_i - Y_i)^2}{n - k - 1} = \frac{2408.4}{24} = 100.4 \tag{4:25}$$

$$F = \frac{4032.0}{100.4} = 40.18$$

Examination of statistical tables for critical values of F at the 0.01 level shows that with 1 and 24 degrees of freedom (v_1 is the greater variance estimate; v_2 is the lesser variance estimate), the F value must exceed 7.82. Thus the regression is highly significant.

Confidence limits of least-squares regression lines

Regression lines are used for prediction of $\hat{Y}$ values for given values of X. However, the $\hat{Y}$ values are only estimates based on a sample of a much larger population. It is therefore necessary to calculate the standard deviation of the estimated values ($\hat{Y}$) from the observed values (Y). This is the same as the standard deviation or error of the residuals. The formula is:

$$S_{YX} = \sqrt{\frac{\sum_{i=1}^{n}(Y_i - \hat{Y}_i)^2}{n}} \qquad (4:26)$$

where S_{YX} = standard deviation or error of the residuals

An alternative formula if the product–moment correlation coefficient has been calculated is:

$$S_{YX} = \sigma Y \sqrt{1 - r^2} \qquad (4:27)$$

where σY = the standard deviation of y
     r = the product–moment correlation coefficient between X and Y

In the Chicago example:

$$S_{YX} = \sqrt{\frac{2408.4}{26}} = 9.62$$

Where there are a small number of observations (<30), as in the Chicago example ($n=26$), an alternative formula is applied to allow for underestimating of variances. This 'best estimate' ($\hat{\sigma}_e$) is found from:

$$\hat{\sigma}_e = S_{YX} \sqrt{\frac{n}{n-k-1}} \qquad (4:28)$$

where S_{YX} = the standard deviation of the residuals
     n = the number of observations
     k = the number of predictors (always 1 in simple regression)

In the Chicago case:

$$\hat{\sigma}_e = 9.62 \times \sqrt{26/24} = 10.01$$

This value is a measure of the spread of the observed points about the regression line. It is then possible, if desired, to plot confidence limits to the regression line. However, these will be curved, owing to sources of error in the estimation of both the regression coefficients a and b. The methods for this and allowing for sampling errors in a and b are described well in Shaw and Wheeler (1985), Silk (1979) and Lee and Lee (1982).

Problems in the application of least-squares regression

A number of assumptions underlie the application of the **linear regression model**. These are only briefly mentioned here but are explained in much greater detail in Silk (1975), Johnston (1978), Lee and Lee (1982), Shaw and Wheeler (1985), Barber (1988) and Dunn (1989).

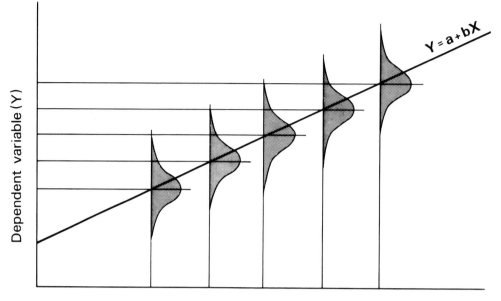

Figure 4.10 Representation of the principle of homoscedasticity where observations have constant variance along the regression line (Redrawn from Shaw and Wheeler, 1985; with kind permission of David Fulton Publishers)

(a) the data for each variable must be normally distributed.

(b) the method assumes that the data are best fitted to a linear model — in the case of the Chicago data, transformation of the data or curve-fitting could improve the degree of fit of the relationship (see case study).

(c) autocorrelation — this can be either spatial or temporal in nature. A further assumption of least-squares regression analysis is that each observation is independent of all others. The idea of spatial autocorrelation was touched on in Chapter 2 on sampling. The spatial positioning of sample points with respect to each other almost inevitably results in spatially autocorrelated data. Similar problems occur when data are collected from successive points in time.

(d) lack of measurement error — it is assumed in least-squares regression that both X and Y are measured without error. If this is not the case, then the coefficients of the regression equation may be biased. It is, however, extremely difficult to determine the magnitude of measurement error in most instances.

(e) homoscedasticity — which means 'equally scattered' and refers to the standard error of the residuals in a regression analysis. Ideally, this standard error describing the scatter of observations around the regression should remain the same along the whole length of the line, as in Figure 4.10. If there is considerable variation in the values of

$$\sum_{i=1}^{n} (Y_i - \hat{Y}_i)^2 / n$$

for each value of X, then the coefficients of the regression equation may be severely biased.

(f)   the means of the conditional distributions should be zero — for every value of X, the mean of $(Y_i - \bar{Y}_i)$ should be zero. If not, the coefficients of the regression may be biased estimates.

Prediction

Once the least-squares regression line has been calculated and confidence limits have been set, the equation and the line can be used for **prediction**. It is important to remember that predictions can only be made for values of Y in terms of X from the regression line of Y on X. Also no predictions should be made beyond the limits of X.

The analysis of residuals

Analysis of residuals, the differences between the observed values of Y and those predicted by the regression equation $(\hat{Y})$, is a very important further step in regression analysis which nevertheless is often neglected. Study of the residuals can assist in understanding the unexplained variance between X and Y and enable information to be gained on possible new and additional variables causing variation in Y. This in turn may help with both refinement and modification of existing hypotheses and the generation of new ones.

Usually, the study of residuals involves standardisation of the residual values because otherwise they would be expressed in units of the original Y variable, which would depend on the particular measurement units for the Y variable. Given that one of the assumptions of the regression model is that the residuals are normally distributed about the regression line (homoscedasticity) with uniform variance along the line (Figure 4.10), the process of standardising the residuals is achieved by making a best estimate of the standard error of the residuals ($\hat{\sigma}_e$), which was described in equation (4:26). An individual residual value can then be standardised by the formula:

$$\text{Standardised residual} = \frac{Y_i - \hat{Y}_i}{\hat{\sigma}_e} \qquad (4:29)$$

The standardised residuals for the Chicago data are shown in Table 4.9, column 9. Examination of these values shows that urban lot number 16, in particular, is worth close examination with a standardised residual of 2.03. Other sites with large standardised residuals are numbers 14, 18, 24 and 26, all the values between 1.5 to 2.0 times the standard error. The position of these points on the scattergram in Figure 4.6 also shows their deviant nature. All four sites would almost certainly merit more detailed ecological survey to ascertain other possible factors affecting species richness. Further analysis of residuals is possible using residual plots and analysis for autocorrelation (Silk, 1979; Shaw and Wheeler, 1985).

As a parametric method, least-squares regression involves making a number of assumptions and has many limitations that are often conveniently ignored by many researchers. Having used the Chicago data set as an example, some questions could be asked as to its suitability for least-squares regression analysis. As a response to these kinds of difficulties, various other non-parametric methods have been devised which, although not widely used, are frequently better suited to the quality of data generated by vegetation scientists.

Non-parametric regression

The following methods of non-parametric regression do not make the rigorous assumptions of the least-squares (parametric) approach described above.

Semi-averages regression

The simplest means of finding a best-fit line through a scattergram of points is by using **semi-averages** regression. Taking the Chicago data as an example, the procedure is as follows:

(a)   calculate the overall means (averages) for both variables.

> Overall mean of X (urban lot age) = 53.12
> Overall mean of Y (species numbers) = 32.46

(b)   calculate the first semi-average.

> X coordinate = mean of all values below the overall mean of X = 21.53
> (15 values)
> Y coordinate = mean of all values below the overall mean of Y = 20.47
> (15 values)

(c)   calculate the second semi-average.

> X coordinate = mean of all values above the overall mean of X = 96.18
> (11 values)
> Y coordinate = mean of all values above the overall mean of Y = 48.81
> (11 values)

These three sets of coordinates (x:y — 21.53:20.47; 53.12:32.46; 96.18:48.81) for the lower semi-average, the overall mean and the upper semi-average are then plotted on the scattergram together with the original data points (Figure 4.7). The three points should very nearly form a straight line and the 'best-fit' line can then be drawn as close as possible to the three points, although the line need not actually join them.

Resistant lines and robust regression

The numerous assumptions of the least-squares regression model and the evolution of techniques for exploratory data analysis have led to the development of methods for **resistant** or **robust** lines. Such methods are based on ranking of the regression data pairs on the basis of the X variable, partitioning the data into two or three groups, calculating the **medians** of those groups and using those values to derive a regression line. Various authors have suggested this approach (Bartlett, 1949; Brown and Mood, 1951; Quenouille, 1959). The technique described below and variations are presented in Daniel (1978), Sibley (1987) and Marsh (1988), with further comment on resistant lines in Besag (1981).

An important aspect of this approach is, however, that it is exploratory in nature and is concerned as much with studying variation in residuals as the fitting of the regression line. In least-squares regression, correlated residuals violate the assumption of the model that the error term is a random component. Also, any exceptions or '**rogue**' data points can influence the positioning of the line such that it becomes a somewhat meaningless summary of the overall relationship between the variables. In the Chicago example, analysis by least-squares regression is considerably influenced by extreme data values such as urban lots 16 and 26. The problem of these 'rogue' points can be dealt with in several ways. In least-squares regression, it is possible to calculate the regression line with all the data and then to recalculate the equation with the 'rogue' points omitted. Any differences between the two lines are then a measure of the influence of the 'rogue' points on the overall relationship between X and Y. Another approach is to weight observations with large residuals to reduce their influence on the line. However, neither of these alternatives is entirely satisfactory, particularly given the assumptions of the least-squares regression model. Also in both cases, the definition of a 'rogue' point is somewhat subjective. The use of resistant lines provides a much better alternative and will generally give a fit far less influenced by extreme values.

The technique is based on dividing the X data into three roughly equal groups and calculating the medians of X and Y for each group. Either the two outer points or all three points are then used to derive a 'best-fit' line. It is important to realise that the aim is still to fit a **linear** relationship and the equation for a straight line: $Y = a + bX$ still applies. The procedure is as follows:

(a) rank the X values in the order in which they occur from lowest to highest along the X axis.

(b) split the rank order of X into three groups, left (L), middle (M) and right (R) as nearly equal as possible. If even groups are not possible, then use the following to achieve a balanced distribution:

| | Size of group (k = group size) | | |
Group	$n = 3k$	$n = 3k + 1$	$n = 3k + 2$
Left	k	k	$k + 1$
Middle	k	$k + 1$	k
Right	k	k	$k + 1$

Where ranks are tied, however, all X tied values are allocated to the same group. This can be a serious problem if there are a large number of ties and the method becomes inaccurate. Exactly how many ties are needed to invalidate an analysis is uncertain, however.

(c) calculate the summary points which are the coordinates of the medians of X and Y in each group. Thus for left, middle and right groups these are $(X_L:Y_L)$; $(X_M:Y_M)$ and $(X_R:Y_R)$.

(d) the slope of the line is then given by:

$$b = \frac{Y_R - Y_L}{X_R - X_L} \qquad (4:30)$$

and the intercept by:

$$a = 1/3 \{(Y_L - bX_L) + (Y_M - bX_M) + (Y_R - bX_R)\} \qquad (4:31)$$

(e) on the assumption that a linear fit is appropriate, the line can be plotted. For the Chicago data this gives the following calculation based on Table 4.12. 26 observations give 3 groups of X values with 9 observations in the first group, 8 in the second and 9 in the third.

Table 4.12   Derivation of three groupings from the Chicago data for calculation of medians in resistant regression (n = 26)

Group L 9 observations			Group M 8 observations			Group R 9 observations		
X	Y	Y(ordered)	X	Y	Y(ordered)	X	Y	Y(ordered)
3	11	9	22	26	15	70	20	20
3	9	11	25	25	22	80	45	37
7	16	15	30	15	25	90	44	44
15	27	16	30	30	26	100	54	45
18	21	16	40	45	30	100	47	46
20	32	21	50	22	45	100	37	47
20	22	22	65	56	47	113	69	47
20	16	27	70	47	56	120	46	54
20	15	32				150	47	69

Medians

$X_L = 18.0$ $Y_L = 16.0$ $X_M = 35.0$ $Y_M = 28.0$ $X_R = 100.0$ $Y_R = 46.0$

Slope (b) is calculated as follows:

$$b = \frac{Y_R - Y_L}{X_R - X_L} = \frac{46.0 - 16.0}{100.0 - 18.0} = \frac{30.0}{82.0} = 0.366$$

The intercept is calculated from:

$$a = 1/3 \{(Y_L - bX_L) + (Y_M - bX_M) + (Y_R - bX_R)\}$$

$$a = 1/3 \{9.41 + 15.19 + 9.40\} = 11.33$$

The equation for the resistant regression line is:

$$\hat{Y} = 11.33 + 0.366X.$$

This line has been plotted on Figure 4.7. The last very important step in the analysis is to calculate and examine the residuals which can be calculated from the predicted values:

$$r_i = Y_i - (a + bX_i) \tag{4:32}$$

Analysis of residuals

As with the least-squares method, analysis of the residuals from resistant regression lines can be extremely useful for further hypothesis generation and modification. The values of the residuals from the Chicago data are shown in Table 4.13 and these can be inspected for pattern. The same points — 14, 16, 18, 24 and 26 — are shown to have the highest deviations from the computed regression line.

Table 4.13 Residuals from resistant line regression of the Chicago data

Lot number	Species number Y_i	Predicted value $\hat{Y}_i$	Residual $Y_i - \hat{Y}_i$
1	11.0	12.4	−1.4
2	9.0	12.4	−3.4
3	16.0	13.9	+2.1
4	27.0	16.8	+10.2
5	21.0	17.9	+3.1
6	32.0	18.7	+13.3
7	22.0	18.7	+3.3
8	16.0	18.7	−2.7
9	15.0	18.7	−3.7
10	26.0	19.4	+6.6
11	25.0	20.5	+4.5
12	15.0	22.3	−7.3
13	30.0	22.3	+7.7
14	45.0	26.0	+19.0
15	22.0	29.6	−7.6
16	56.0	35.1	+20.9
17	47.0	37.0	+10.0
18	20.0	37.0	−17.0
19	45.0	40.6	+4.4
20	44.0	44.3	−0.3
21	54.0	47.9	+6.1
22	47.0	47.9	−0.9
23	37.0	47.9	−10.9
24	69.0	53.1	+15.9
25	46.0	55.3	−9.3
26	47.0	66.2	−19.2

Case study

Multiple correlation and regression analysis and urban plant ecology in Chicago (Crowe, 1979)

The study of the vegetation of towns and cities is now a very important area of plant ecology. Apart from the pioneering work of Shenstone (1912) in London, until recently most ecologists ignored the great variety of plants which have adapted to the highly modified urban ecosystems of the world. However, since the mid-1970s, the vegetation of urban environments has begun to be studied in more detail (Davis, 1976; Nature Conservancy Council, 1979; Haigh, 1980; Whitney and Adams, 1980; Kunick, 1982; Whitney, 1985; Emery, 1986; Gilbert, 1990; Sukopp et al., 1990). Urban ecosystems were estimated to cover over 28m ha of the land surface of the United States in 1978 (Grey and Deneke, 1978).

Many aspects of urban plant ecology merit study and philosophically, there are important reasons for studying wildlife in the city (Harrison et al., 1987). The recognition and definition of plant communities is an interesting topic, as are surveys into the origins of different urban species. Some species are found in the surrounding countryside, others occur only in city environments, while still others are garden escapes. Further work can include diversity, succession and productivity, similar to Wathern's case study (1976) in the previous chapter. More specialist topics are also possible such as the study of the flora of walls (Darlington, 1981) or of pavements.

The study by Crowe (1979) of urban lots in Chicago, which was used to introduce techniques of correlation and regression, is a good example of urban plant ecology. Crowe was interested in ideas of island biogeography (MacArthur and Wilson, 1967) and in particular the idea that species richness is related to island area. Crowe makes the analogy between true oceanic islands and the abandoned urban lots of big cities, arguing that they represent islands for potential colonisation within the 'ocean' or 'sea' of concrete or tarmac in the urban environment. Concepts of island biogeography and the problems of making these extrapolations of the theory to terrestrial environments are discussed in Kent (1987). Crowe was particularly interested in the higher plant species richness within abandoned urban lots in Chicago and the factors which determine species numbers. He examined 26 urban lots and in addition to collecting data on species richness, several variables relating to the area and distance effects predicted by island biogeographic theory were measured. The simple correlation and regression analyses already described had demonstrated that lot age (time since last major disturbance) was very important and accounted for 63 per cent of the variation in species richness. However, other island biogeographic variables could be significant as well and help to explain some of the remaining 37 per cent of the variance. Thus, measurements were taken of lot area, distance to the nearest older lots, the number of other lots within 1 km, distance to the oldest lot, distance to the nearest other lot and distance to the largest lot. Clearly all of these distance measures would affect potential colonisation, while area is related to extinction and the maximum number of species which a site can hold at one time.

One of the first interesting features of Crowe's paper is that having found a correlation of 0.79 ($r^2 = 0.63$) for species richness and age, he was able to increase this to 0.85 ($r^2 = 0.72$) by **transforming** the data and taking the logarithms of both the data on species richness and age. Crowe also calculated correlation coefficients between the other variables to give a **correlation matrix**. The correlation of species richness and lot area was also highly significant with a coefficient of 0.48 ($r^2 = 0.23$) for untransformed data, significant at the 0.05 level but increasing to 0.67 ($r^2 = 0.45$), highly significant, at the 0.01 level when logarithmic transformations were applied although the strength of the relationship was much less strong than for lot age. A well-established part of island biogeographic theory predicts a log–log relationship between species and area, and Crowe was able to demonstrate this for the Chicago data. The fact that it applied for age as well, and age was even more highly correlated with species richness, was also a very interesting result.

The obvious question to ask next is, 'what are the combined effects of age and area?' Clearly, the joint explanation is not the sum of the two coefficients of determination (r^2 age = 0.63; r^2 area = 0.23; sum = 0.86) because there is correlation between age and area (r = 0.288) and thus they overlap in some of their explanation of the variance.

In order to examine the joint effects of more than one variable, **multiple correlation and regression** need to be applied. There is a very large literature on these methods, and they have to be used very carefully. In multiple correlation, one variable, in this case species richness, is correlated with two or more others and the degree of correlation with the other variables **in combination** is determined. In the case of the Chicago data, if the scores for species richness are correlated against **both** age and area with no transformation, the correlation coefficient rises to 0.83 and the coefficient of determination (r^2) to 0.69. With transformation, this rises to 0.92 ($r^2 = 0.84$).

Multiple regression analysis similarly regresses the dependent variable (Y) on more than one independent (X) variable (X_1; X_2; X_3; X_n). With the Chicago data, using both lot age (X_1) and lot area (X_2) as independent variables and species richness (Y) as the dependent, the regression equation becomes:

$$Y = 14.5 + 0.273X_1 + 0.00281X_2$$

The figure of 14.5 is the **intercept** of **constant** and 0.273 is the **partial regression coefficient** for lot age (X_1) and 0.00281 is the **partial regression coefficient** for lot area (X_2). The regression coefficients then have to be tested for significance. This is done using t-tests and is described in Johnston (1978) and Shaw and Wheeler (1985). Results are presented in the following table:

Predictor	Coefficient	Standard deviation	t-ratio	p
Constant	14.46	3.091	4.68	0.00
Age (X_1)	0.273	0.046	5.91	0.00
Area (X_2)	0.0028	0.0013	2.24	0.03

The p values indicate that all three coefficients are significant at the 0.05 level. Also a table is drawn up in a similar manner to Tables 4.10 and 4.11 for testing the significance of r^2 (0.69) (Table 4.14). The F ratio is looked up with 2 and 24 degrees of freedom in appropriate statistical tables and at the 0.01 level, the value has to exceed 5.66 to be significant. Clearly, the calculated value of 25.95 is highly significant and the hypothesised relationship between species richness and the combined effects of age and area is accepted. This is a very interesting result in terms of island biogeographic theory and the possible analogies which have been made between urban lots and oceanic islands. It also shows how both time (age) and space (lot area) are important in successional processes and in determining species richness in these superficially uninteresting urban areas.

Again, the multiple regression can also be calculated with the three variables **transformed** using logarithms. In this case, the regression equation becomes:

$$Y = 0.759 + 0.350X_1 + 0.203X_2$$

and the table for testing regression coefficients is:

Predictor	Coefficient	Standard deviation	t-ratio	p
Constant	0.7595	0.2978	2.55	0.02
Age (X_1)	0.3504	0.0464	7.56	0.00
Area (X_2)	0.2029	0.4834	4.20	0.00

Table 4.14 Multiple regression analysis of variance table from the Chicago study

Untransformed data

Source of variation	Sums of squares	Degrees of freedom	Mean square	F
Explained (Regression)	4462.0	2	2231.3	25.95
Unexplained (Residual)	1177.9	23	86.0	
Total	6440.4	25		

Transformed data

Source of variation	Sums of squares	Degrees of freedom	Mean square	F
Explained (Regression)	6.2343	2	3.1171	60.31
Unexplained (Residual)	1.1888	23	0.0517	
Total	7.4231	25		

The table for the F-ratio test is given in Table 4.14 and the multiple regression is even more highly significant (F = 60.3 with 2 and 23 degrees of freedom). **Transformation** also assists with normalising the variables thereby bringing the data closer to the assumptions of the multiple regression model. Multiple correlation and regression are complex methods which need to be handled with care. More detailed explanations are given in Johnston (1978), Shaw and Wheeler (1985) and Jongman *et al.* (1987), although the latter is heavily mathematical. There are many assumptions in the use of the multiple regression model. All of those discussed previously for simple (two variables) regression apply plus the assumption of absence of **multicollinearity** between the independent variables. This means that there should not be correlation between the independent variables or at least those correlations should be very low.

Multiple correlation and regression analyses always need to be carried out by computer; it is best to use one of the well-known packages such as MINITAB (Ryan *et al.*, 1985) or the Statistical Package for the Social Sciences (SPSS) (Nie, 1983) see Chaper 9.

Regression and multiple regression are important methods for plant ecologists and are related to some of the more recent methods of ordination discussed in Chapter 6. Austin (1971) provides a very useful review and example of its application in plant ecology. He uses multiple regression to study the causes of variation in abundance of *Eucalyptus rossii* in relation to a number of environmental factors in the Southern Tablelands of New South Wales, Australia.

Conclusion

This is a very important chapter of this book. The description and analysis of vegetation and environmental data have tended to concentrate on the methods of ordination and classification described in the following chapters and to ignore the more simple but equally important methods of statistical analysis in relation to hypotheses and data exploration. The need to ask more specific questions about vegetation and its environment and to generate and test hypotheses about individual species and their relationships to each other and to their environment has already been stressed (Keddy, 1987; Kent and Ballard, 1988). The limited range of methods for exploratory data analysis which have been introduced here will be of value to many vegetation scientists and their students in addition to the more traditional confirmatory approach of hypothesis generation and testing. The techniques presented here are by no means the only ones appropriate to the simple analysis of vegetation data, and reference to the many textbooks mentioned will show how all these ideas can be taken further. Nevertheless, from the authors' experience, these are the methods which are likely to prove most useful.

Various statisticians have recently reviewed the problems of applying methods of statistical analysis and the advanced student should read the articles published by Nelder (1986), Cormack (1988) and McPherson (1989) and the book by Chatfield (1988) which deal with the role of statistical analysis in general and in the biological and environmental sciences in particular.

5

Ordination methods I, 1950–70

The word 'ordination' means 'to set in order' and was first used by Goodall (1954), who showed that it is derived from the German word *Ordnung* applied to this approach by Ramensky (1930). Here, ordering means the arrangement of vegetation samples in relation to each other in terms of their similarity of species composition and/or their associated environmental controls. Ordination methods are also part of **gradient analysis**. In gradient analysis, variation in species composition is related to variation in associated environmental factors which can usually be represented by **environmental gradients** (Chapter 1). The emphasis in ordination methods is thus on **individual** samples or species and their degrees of similarity to each other and on determining how the order of the individuals is correlated with underlying environmental controls.

Gradient analysis and ordination techniques are a group of methods for **data reduction and exploration** leading to **hypothesis generation** (Figure 5.1). The methods are essentially descriptive and enable researchers to formulate ideas about plant community structure as well as possible **causal** relationships between variation in vegetation and its environment.

Within plant ecology, gradient analysis and ordination methods can help with one or more of the following areas of research:

(a) summarising plant community data and providing an indication of the true nature of variation within the vegetation of the area under study.

(b) enabling the distribution of individual species within different communities to be examined and compared.

(c) providing summaries of variation within sets of vegetation samples which can then be correlated with environmental controls to define environmental gradients.

Direct and indirect gradient analysis or ordination

Direct gradient analysis is used to display the variation of vegetation in relation to environmental factors **by using environmental data to order the vegetation samples**. As the name suggests, the environmental data are used **directly** to organise the information on vegetation (central pathway — Figure 5.1). The methods necessarily assume that the underlying environmental gradients are known and can be seen as quite distinct from the methods of indirect gradient analysis or ordination described below.

The term **indirect ordination methods or gradient analysis** is applied to techniques which operate on a set of vegetation by first examining the variation within it (left pathway — Figure 5.1). This is done independently of the environmental data. A second stage of analysis is then performed once the major sources of variation in the vegetation data have been described and summarised. Only then are the environmental data compared and correlated with the summarised vegetation data in order to detect possible environmental gradients. The environmental interpretation is thus **indirect**. These methods can be used in situations where the underlying environmental gradients are unknown or are unclear, although they are equally applicable where the environmental gradients are known (Whittaker, 1967).

THE REAL WORLD

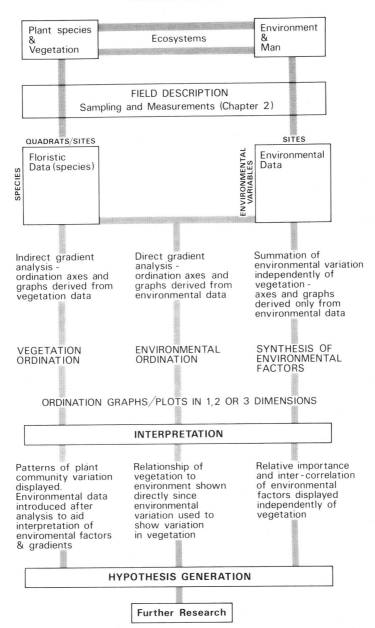

Figure 5.1 Approaches to ordination in plant ecology. (a) Indirect ordination, where vegetation data are analysed independently of environmental data and environmental data are introduced only after the ordination diagram has been produced (vegetation ordination). (b) Direct ordination, where the vegetation samples are ordinated by using environmental data to construct the ordination diagram (environmental ordination). (c) Summarisation of environmental variation, where the variability of the environmental data is analysed independently of the floristic vegetation data (synthesis of environmental factors)

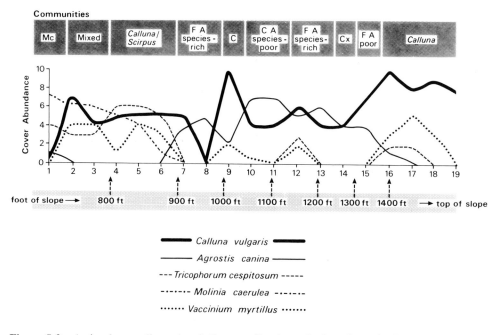

Figure 5.2 A simple one-dimensional direct ordination of selected species in 1m quadrats along a continuous transect up a hillslope near Creetown, Kirkcudbrightshire, Scotland; F = *Festuca ovina*; A = *Agrostis capillaris*; C = *Calluna vulgaris* (redrawn with kind permission of Longman from Tivy, 1982; after Mitchell, 1977)

Historically, direct ordination methods are older and are particularly prominent in America, where they have largely been developed and where there is a much stronger tradition of seeing vegetation as a **continuum** rather than as a collection of distinct communities. Elsewhere in the world, particularly in Europe, indirect methods have assumed much greater importance.

Some confusion has arisen over the direct/indirect gradient analysis terminology. Gauch (1982b), for example, reserves the word 'ordination' solely for indirect gradient analysis. However, most ecologists would probably agree with the definitions given here. Austin (1968) has concurred with this by calling methods for direct gradient analysis **environmental ordination** and those for indirect gradient analysis **vegetation ordination**.

A third approach to ordination involves the separate analysis of the environmental data collected in a set of quadrats. This can be described as summarisation of environmental variation (right pathway — Figure 5.1). Once the patterns of environmental variation have been analysed, species data may be introduced to examine relationships between species distributions and environmental factors or gradients.

Direct gradient analysis or ordination

The simplest form of direct gradient analysis is a graph of species response to one environmental factor. A good example is provided in Tivy (1982), where species

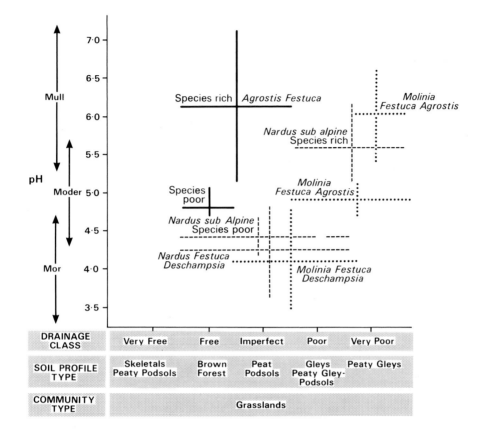

Figure 5.3 Two-dimensional direct ordination of grass-moorland community types in
Scotland. Dominant species: 1. *Agrostis/Festuca*; 2. *Nardus stricta*; 3. *Molinia
caerulea* (adapted and redrawn from Burnett, 1964 and Tivy, 1982; reproduced
with kind permission of Longman)

abundances along a transect up a hillslope have been plotted with distance up the
hillslope corresponding to increase in altitude (Figure 5.2). This graph demonstrates
all the basic principles of gradient analysis and ordination — the nature of
community variation up the slope is summarised, individual species distributions are
displayed and relationships between both the community composition and the
environmental factor of altitude are shown. As such it is an example of a
coenocline. However, it does represent a very basic form of gradient analysis. The
gradient itself — altitude — is an example of a **complex gradient** within which many
other more specific environmental factors such as temperature, exposure, slope,
drainage quality and even biotic factors are operating.

Mosaic diagrams of two factors

Burnett (1964) provides an example of a direct ordination based on two factors. He
was studying variation in the upland pastures of Scotland and believed that variation in species composition of those pastures was a function of drainage quality and
pH (acidity) of the soil. He collected a large number of vegetation samples together

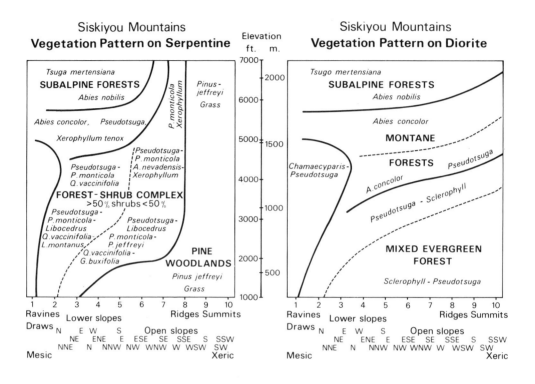

Figure 5.4   Mosaic diagrams of the plant communities of the Siskiyou mountains, southwestern Oregon, USA. Different diagrams were produced for the two major rock types (serpentine and diorite) with separation of the plant communities on the basis of altitude and topographic position (Whittaker, 1960; redrawn with kind permission of *Ecological Monographs*)

with data on soil moisture content and pH at each site. Burnett was then able to construct the ordination diagram of Figure 5.3 by placing each quadrat on the graph defined by soil moisture content and pH. On the graph, the points representing quadrats formed a number of groups, and by examining the species composition of the quadrats in each group he was able to define the particular types of grassland which characterise the uplands of Scotland and examine their tolerance ranges and degrees of overlap. This approach is similar to that of Whittaker (1960) in describing the vegetation of the Siskiyou Mountains in southwestern Oregon (Figure 5.4). Whittaker carried out a number of other similar studies, for example in the Great Smokey Mountains (1948; 1956). The success of the approach depends entirely on the choice of environmental axes. In Whittaker's studies (Whittaker, 1967) with altitude and geology/terrain type, the gradients were pronounced and obvious. However, in many other situations, the environmental gradients will be less clear

Table 5.1 Derivation of the continuum indices for upland hardwood forests in Wisconsin (Curtis and McIntosh, 1951; reproduced with kind permission of *Ecology*)

(a) Average importance values (IV) and constancy (%) of the four trees selected as the leading dominants in the four groups of stands recognised as the first stage of the analysis (for species with the highest importance potential only — 80 stands)

| Species | Leading dominant in stand | | | |
	Q. velutina	*Q. alba*	*Q. rubra*	*Acer saccharum*
Q. velutina				
Average IV	165.1	39.6	13.6	0.0
Constancy (%)	100.0	72.3	38.3	0.0
Q. alba				
Average IV	69.9	126.8	52.7	13.7
Constancy (%)	100.0	100.0	97.1	66.7
Q. rubra				
Average IV	3.6	39.2	152.3	37.2
Constancy (%)	25.0	94.5	100.0	76.3
Acer saccharum				
Average IV	0.0	0.8	11.7	127.0
Constancy (%)	0.0	5.6	29.4	100.0

(b) Average importance value (IV) and constancy (%) of associated trees in the four groups of stands with the above four species as leading dominants

| Species | | Leading dominant in stand | | | |
		Q. velutina	*Q. alba*	*Q. rubra*	*A. saccharum*
Q. macrocarpa	IV	15.6	3.5	4.2	0.1
	Constancy (%)	50.0	38.9	20.6	4.8
Prunus serotina	IV	21.4	21.8	5.9	1.4
	Constancy (%)	87.5	89.0	64.8	19.0
Carya ovata	IV	0.3	8.8	5.2	5.9
	Constancy (%)	12.5	61.2	38.3	33.3
Juglans nigra	IV	1.5	1.2	2.2	1.9
	Constancy (%)	12.5	11.1	20.6	23.8
Acer rubrum	IV	3.9	2.3	2.4	1.0
	Constancy (%)	12.5	33.3	23.5	4.8
Juglans cinerea	IV	0.0	2.7	1.7	4.8
	Constancy (%)	0.0	11.1	20.6	47.6
Fraxinus americana	IV	0.0	1.9	5.1	7.6
	Constancy (%)	0.0	11.1	20.6	42.8
Ulmus rubra	IV	4.6	7.7	8.3	32.5
	Constancy (%)	25.0	27.8	53.3	85.7
Tilia americana	IV	0.3	5.9	19.0	33.0
	Constancy (%)	12.5	16.7	73.5	100.0
Carya cordiformis	IV	2.5	5.8	4.1	8.2
	Constancy (%)	12.5	33.3	41.2	66.7
Ostrya virginiana	IV	0.0	2.4	5.5	16.2
	Constancy (%)	0.0	22.2	41.2	95.3

Table 5.2 Climax adaptation numbers for the tree species found in stands of upland hardwood forests in Wisconsin (Curtis and McIntosh, 1951; by kind permission of *Ecology*)

	Climax adaptation number
Quercus macrocarpa	1.0
Populus tremuloides	1.0
*Acer negundo	1.0
Populus grandidentata	1.5
Quercus velutina	2.0
Carya ovata	3.5
Prunus serotina	3.5
Quercus alba	4.0
Juglans nigra	5.0
Quercus rubra	6.0
Juglans cinerea	7.0
*Ulmus thomasi	7.0
*Acer rubrum	7.0
Fraxinus americana	7.5
*Gymnocladus dioica	7.5
Tilia americana	8.0
Ulmus rubra	8.0
*Carpinus caroliniana	8.0
*Celtis occidentalis	8.0
Carya cordiformis	8.5
Ostrya virginiana	9.0
Acer saccharum	10.0

* The climax adaptation number of these species is tentative owing to their low frequency of occurrence in the survey.

and the application of this method will be more difficult.

The continuum index approach and weighted averages ordination (Curtis and McIntosh, 1950; 1951)

Some of the earliest developments in ordination were made by Curtis and McIntosh. They studied the distribution of forest trees in southern Wisconsin. For each stand or forest sample, an **importance value** was calculated with three components, as follows:

1. $\text{Relative density} = \dfrac{\text{Number of individuals of species}}{\text{Total number of individuals}} \times 100$

2. $\text{Relative dominance} = \dfrac{\text{Dominance* of a species}}{\text{Dominance of all species}} \times 100$

3. $\text{Relative frequency} = \dfrac{\text{Frequency of a species}}{\text{Frequency of all species}} \times 100$

> Importance value for each species = Relative density + relative dominance + relative frequency

*Dominance is defined as the mean basal area per tree times the number of trees of the species

Using the importance values for each species, stands were arranged in a sequence, so that, as far as possible, the amount of each species showed a single peak or increased or decreased monotonically along the series. Then each tree species was allocated a weight according to the position of the peak and these weightings were then used to provide a more accurate position of the stands or samples on the axis corresponding to the initial ordering. The weightings were calculated by initially grouping the stands according to the four 'leading dominant' species (the most abundant species) and arranging those groups in order (Table 5.1a). The average importance values of the other species (Table 5.1b) were then taken to calculate weightings or **climax adaptation numbers** for all species as in Table 5.2, where 1 = pioneer species and 10 = climax species.

Then a **weighted average** was calculated for each stand using the abundance of each tree species in the stand and the species weight. Thus if A_{ij} is the abundance of the tree species i in stand j, and W_i is the species weight or climax adaptation number (Table 5.2), an ordination score for each sample (S_j) on a single axis can be computed using the formula:

$$S_j = \frac{\sum_{i=1}^{n} A_{ij}W_i}{\sum_{i=1}^{n} A_{ij}} \qquad (5:1)$$

where $\sum_{i=1}^{n}$ = summation across all species in the stand or sample

$\quad$ n $\quad$ = number of species

The result was a single axis ordination which positioned each stand or sample along the gradient. This could then be used to show individual species distributions or values of environmental factors.

There were two problems with the Curtis and McIntosh technique. First, the method only derives one axis of variation and thus only works well where there is one primary environmental gradient, as was the case in the woodlands of southern Wisconsin where a marked successional series was evident. Second, the precise weight given to each tree species (the climax adaptation number) was based on a subjective assessment of its position in a successional series, even though that was based on information provided by the importance values. The climax adaptation numbers would thus vary from one worker to another and depend on the familiarity and experience which the worker has of the vegetation and its environmental controls. Further refinement of the method to overcome some of these problems was achieved by Brown and Curtis (1952) and Goff and Cottam (1967).

Indirect gradient analysis

Indirect ordinations are based on analysis of floristic data independently of any

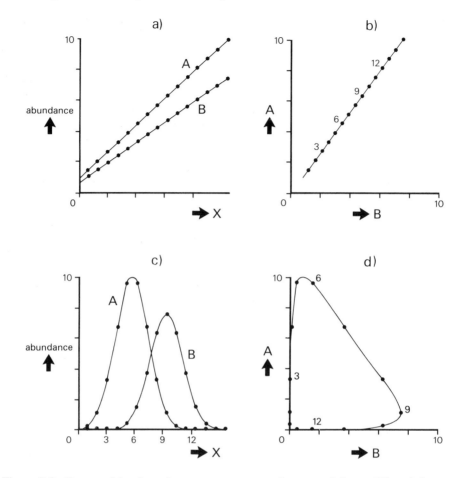

Figure 5.5 Two models of species response to an environmental factor (X) and the resulting graphs for joint distribution of the species. (a) a rectilinear response model of two species (A and B) in relation to X. (b) the resulting line when species are plotted together. (c) a unimodal bell-shaped curve response model of two species (A and B) in relation to X. (d) the resulting contorted curve when both species are plotted together (ter Braak, 1987a; redrawn with kind permission of PUDOC — Centre for Agricultural Publishing and Documentation, Wageningen, The Netherlands)

preconceived notions of controlling environmental factors or successional sequences. The concern is with the internal variability of the data and the assumption is made that examination of variability in floristics will inevitably reflect variation in environment; controlling factors and the relationships between vegetation and its environment can be explored **after** the floristic variation has been analysed and displayed. Thus environmental data are not used at any stage of analysis and are only introduced at the interpretation stage.

Methods of indirect ordination can be seen most simply as a means of summarising the information in a data matrix, such as Tables 3.1–3.4, in scatter diagrams (Figures 5.6 and 5.8). As such, they are known as methods of **matrix approximation**. However, Prentice (1977) and ter Braak (1987a) take another, more sophisticated view, that there is an underlying or **latent structure** in the data matrix such that the pattern of species entries in the matrix is determined by a few environmental variables (the latent variables). The purpose of ordination is then to recover that latent structure and in the process show how species are responding to prevailing environmental factors. In this case, ordination becomes more like regression, except that in ordination, the explanatory variables are not known environmental variables but theoretical variables (latent variables) that are constructed to explain the species data best. Thus, as with regression, each species represents a response variable, but in ordination the response variables are analysed simultaneously. ter Braak (1987a) illustrates this point by taking two examples of species-response models: first, a model which assumes rectilinear (straight line) relationships between two species A and B in response to an environmental factor — X. If the abundances of A and B are recorded in a series of quadrats, these can be plotted against X (Figure 5.5a). If the relationship between the two species is rectilinear (that is a straight line) then the graph showing the relationship between the two species will be as in Figure 5.5b. If, however, a different species-response model is taken, corresponding to bell-shaped unimodal curves (Figure 5.5c), then the joint plot of the species A and B produces the contorted curve of Figure 5.5d. In indirect gradient analysis, the aim is to make an inference about the relationships with the latent environmental variables (Figure 5.5a,c), using only the species data (Figure 5.5b,d). The form of the relationship clearly has a very substantial effect on the resulting graphs.

Methods of indirect gradient analysis are much more widely used than those of direct gradient analysis. The first reason for this is that species data are usually very much easier to collect than environmental data. Within any specific research problem, there are many environmental variables which could be measured and it is often not easy to predict in advance which ones are going to be of importance. Also many environmental variables do not lend themselves easily to measurement, for example biotic pressures such as grazing and trampling. Many environmental measures are indirect in terms of their effects on plants. A good example of this is soil pH, which is easy to measure with relative accuracy and is widely used as a general index of soil nutrient status. However, in terms of detailed variation in species composition it may be a rather crude variable with more detailed measurements of soil chemistry being of greater importance. Finally, environmental measurement is often expensive in terms of time, resources and money. For all these reasons, data on species composition is much easier and more rapid to collect and to analyse.

Indirect methods are also favoured because plant ecologists are more often concerned with community structure in relation to environment rather than with the response of individual species which are often very unpredictable. Indirect ordination methods represent a more efficient means of displaying variation in plant community structure than attempting to correlate single species distributions with environmental factors.

Ordination diagrams — the end product of indirect ordinations

Indirect quadrat ordinations

Indirect ordinations examine the similarity or dissimilarity of floristic composition of vegetation samples. This similarity or dissimilarity is expressed in graph form with plots of points in one, two or three dimensions where each point represents a vegetation sample or quadrat (Figure 5.6). **The distances between the points on**

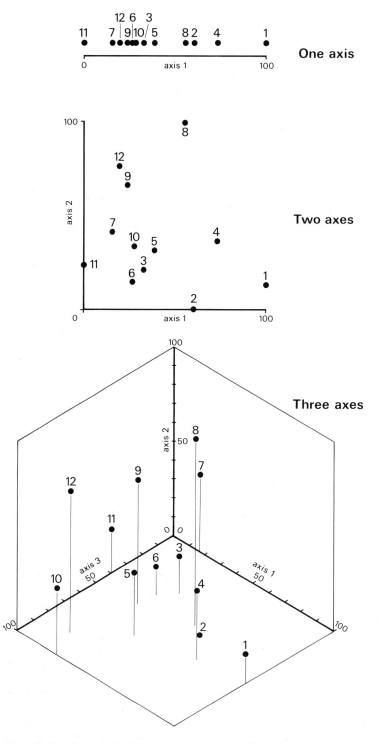

Figure 5.6 Ordination graphs in one, two and three dimensions from a polar quadrat ordination of the New Jersey salt marsh data (Table 3.1)

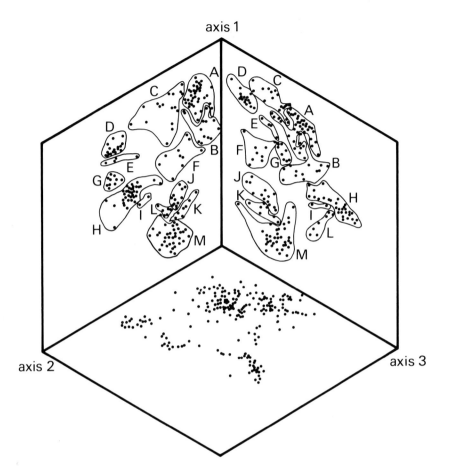

Figure 5.7   A three-dimensional ordination plot of savanna vegetation data from northern Nigeria; graphs are drawn in pairs — left axis 1 v. axis 2; right axis 1 v. axis 3; bottom axis 2 v. axis 3; letters denote species groupings. (Kershaw, 1968; reproduced by kind permission of *Journal of Ecology*)

the graph are taken as a measure of their degree of similarity or difference. Points which are close together will represent quadrats that are similar in species composition; the further apart any two points are, the more dissimilar or different the quadrats will be. Most published ordination diagrams are two-dimensional, simply because such graphs are easiest to read on a flat page and a two-dimensional graph will show more information than a one-dimensional one. Three-dimensional plots are produced and can be drawn in pairs in perspective on a flat page (Figure 5.7) or as a stereo pair which will give a 3D image when viewed through a stereoscope (Figure 5.8).

As a general rule, the axes of the graphs produced by an ordination method come out in **descending order of importance,** with the first axis summarising more variation than the second, the second more than the third and so on. However, this is not a universal rule, as will be explained later. It is clearly also possible to plot ordination axes in pairs as two-dimensional graphs, axis 1 versus axis 2; axis 1 versus axis 3 and axis 2 versus axis 3. In most analyses, only the first three or four

G2 VEGETATION PATTERNS AND PROCESSES - YES TOR VEGETATION DATA 1989

QUADRAT ORDINATION

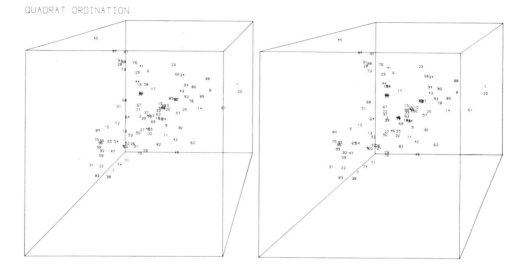

Figure 5.8 A stereo-pair of data from a vegetation survey of an area near Yes Tor, north-west Dartmoor, England

axes are of significance, although in a quadrat ordination it is normally possible to carry on extracting axes mathematically up the the number of species in the analysis. However, the summarising power of higher axes is minimal, and they make a trivial contribution to the analysis of overall variation in the vegetation data.

Figure 5.9 is a two-dimensional ordination plot for the vegetation data from Gutter Tor, Dartmoor (Table 3.4). This is known as a **quadrat, site or sample ordination**. Four groups of quadrats can be recognised. Group 1: quadrats 18, 3, 20, 13, 14 and perhaps 24; Group 2: 15, 8, 23 and 19; Group 3: 21, 16, 10, 6, 4, 11, 17 and 22 and Group 4: 1, 2, 12, 9 and 25. In each of these groups, quadrats are closer to each other than to quadrats in any other group. Thus within each group, quadrats must be relatively similar floristically. However, quadrats 5 and 7 do not belong to any of the four groups and are transitional between groups 1, 2 and 3. Figure 5.9a shows the nature of vegetational variation in the data. There are four reasonably distinct community types, but some quadrats are located between these types. Also the diagram demonstrates that Group 3 is the most diffuse group. It could be argued that quadrats 16, 21 and perhaps 10 could be separated from 6, 4, 11, 17 and 22. However, this is an entirely subjective decision. This point introduces the most important principle of interpretation which is that although the method is objective in the sense of repeatability, the interpretation still involves a considerable degree of subjectivity, based on the expertise and ecological knowledge of the researcher.

Species ordinations

Ordination can also be carried out for species, producing a one- two- or three-dimensional graph, where each point represents a species and the distances between the points are an expression of how similar the species are in their distribution across the quadrats (Figure 5.10). Such an ordination is called a **species ordination**.

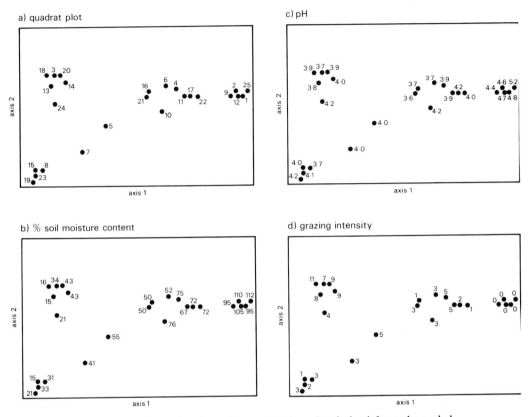

Figure 5.9 (a) A two-dimensional quadrat ordination plot derived from detrended correspondence analysis for the Gutter Tor data (Table 3.4). (b) % soil moisture (gravimetric method) plotted on the quadrat ordination diagram. (c) Soil pH plotted on the quadrat ordination diagram. (d) Grazing intensity plotted on the quadrat ordination diagram

The two-dimensional species ordination plot for the Gutter Tor data is presented in Figure 5.11.

Methods of indirect ordination

A number of different techniques for indirect ordination have been devised and applied over the past 30 years. Although the methods differ in their approach to analysis, it is important to realise that they all result in ordination diagrams of the type described above, with graphs in one, two or three dimensions.

Table 5.3 summarises the major methods of indirect ordination. An important feature of this table is the process of evolution in methodology. The earliest method was that of Bray and Curtis (polar) ordination, devised originally in 1957. This was then followed by principal component analysis (Orlóci, 1966; Austin and Orlóci, 1966; Gittins, 1969), reciprocal averaging and correspondence analysis (Benzécri, 1969, 1973; Hill, 1973b, 1974) and then detrended correspondence analysis (Hill, 1979a; Hill and Gauch, 1980). However, other methods also exist, although they have not been extensively used; for example non-metric multi-dimensional scaling (Fasham, 1977; Prentice, 1977) and canonical correspondence analysis (ter Braak,

Quadrat Ordination

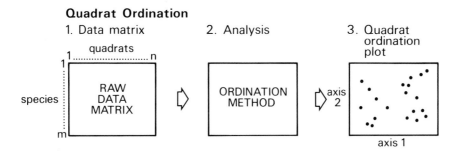

On the ordination plot, each point represents a **quadrat** and the greater the distance between any two points, the greater the difference in floristic composition of the quadrats which they represent.

Species Ordination

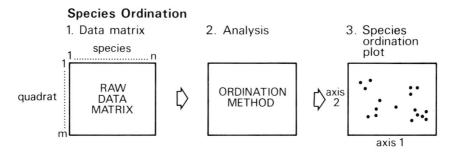

On the ordination plot, each point represents a **species** and the greater the distance between any two points, the greater the difference in the distribution of the species across all the quadrats

Figure 5.10 The concept of species ordination

1986a; 1987a,b; 1988a,b). The last is, however, a cross between direct and indirect methods.

For convenience, Bray and Curtis polar ordination (PO) and principal component analysis (PCA) are described in this chapter, representing developments up to 1970. Methods devised after that date are presented in Chapter 6.

Interpretation of ordination diagrams

Regardless of which technique is used, the process of interpretation of ordination diagrams is similar. Also, the procedure is the same whether the diagram is drawn in one, two or three dimensions.

The species ordination diagram

Most interpretation is carried out in relation to the **quadrat ordination diagrams** but the **species ordination diagram** is always worth examination as well. As an example, the species ordination from detrended correspondence analysis (DCA) of the

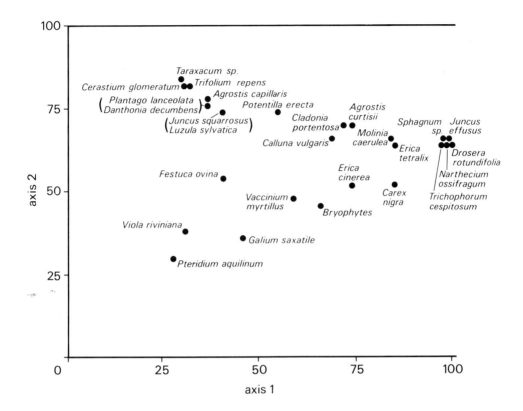

Figure 5.11 A two-dimensional species ordination plot derived from detrended correspondence analysis of the Gutter Tor data (Table 3.4)

Gutter Tor data (Table 3.4) is presented in Figure 5.11. Each point on the graph corresponds to a species, and the distances between points on the graph are an approximation to their degree of similarity in terms of distribution within the quadrats. Thus two species occurring with exactly the same abundances in the same quadrats would occupy the same point. As species distributions diverge, so the distances between points on the species ordination diagram increase. For purposes of interpretation, it is useful to look for groupings of species. Thus in Figure 5.11, four general groupings of species emerge: first, a group to the top left centre of the diagram containing species typical of improved upland pastures, for example *Agrostis capillaris*, *Trifolium repens* and *Cerastium glomeratum*; second, a bracken-invaded pasture group in the lower half, containing *Pteridium aquilinum*, *Galium saxatile*, *Viola riviniana* and *Festuca ovina*; third, a heathland group on the right-hand side of the diagram which can be split into dry heath — *Calluna vulgaris*, *Agrostis curtisii* and *Erica cinerea* and wet heath — *Molinia caerulea*, *Erica tetralix*, *Juncus effusus*, grading into a fourth group, valley bog, which includes *Narthecium ossifragum*, *Drosera rotundifolia* and *Sphagnum species*.

The quadrat ordination diagram

One of the principal aims of ordination is **to define the underlying environmental**

Table 5.3   A summary of methods of indirect ordination

Method	Author	Date	Comment	Application (Kent and Ballard, 1988)
Bray and Curtis (polar ordination) (PO)	Bray and Curtis	1957	Originally calculated and drawn using compass construction	Widely used between 1960 and 1970. Now superseded by more sophisticated techniques
Principal component analysis (PCA)	Gittins, Orlóci	1966	Relatively complex, requiring computing facilities for calculation	Widely used from 1966–present. However, now not recommended due to distortion ('horseshoe') effects
Reciprocal averaging/correspondence analysis (RA/CA)	Benzécri Hill	1969 1973	Simple calculation for one axis. Requires computer programs for full analysis	Used extensively from 1973–1985 – now largely replaced by detrended correspondence analysis
Detrended correspondence analysis (DCA)	Hill & Gauch	1979	'Improved' version of reciprocal averaging/correspondence analysis. Requires computer program DECORANA for analysis	Widely used 1980–present.
(Detrended) canonical correspondence analysis ([D]CCA)	ter Braak	1986 1988	Not strictly indirect ordination since it is a revised version of (detrended) correspondence analysis with ordination axes constrained by multiple regression with environmental factors. Requires computer program CANOCO/CANOPLOT/CANODRAW for analysis	Becoming widely used
Non-metric multi-dimensional scaling (NMDS)	Fasham, Prentice	1977	An alternative method rarely used but with considerable promise	Never widely used

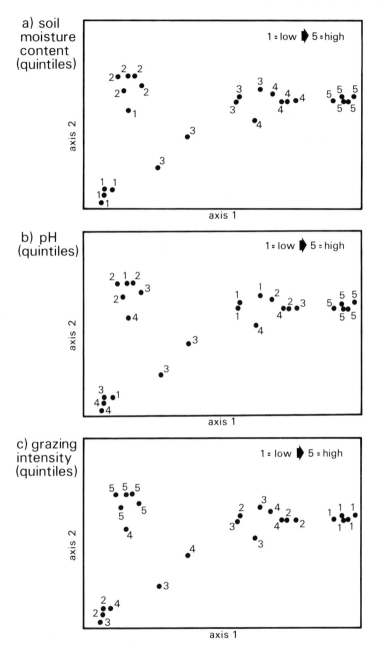

Figure 5.12 The quadrat ordination plots from Gutter Tor for (a) % soil moisture; (b) soil pH and (c) grazing intensity, plotted as quartiles to enable trends and patterns to be seen more clearly

gradients within a set of vegetation/environmental data. In indirect ordination, the analysis is carried out using **only** the vegetation/floristic data. Environmental relationships are determined once the ordination is complete and the graphs have been constructed. Most interpretation is carried out on two-dimensional graphs by superimposing environmental data collected at the same time as the floristic data in each quadrat on to the quadrat ordination diagram.

As an example, the two-dimensional quadrat ordination resulting from detrended correspondence analysis of the Gutter Tor data is presented in Figure 5.9a. From the data in Table 3.4, environmental plots have been produced for per cent of soil moisture (Figure 5.9b); soil pH (Figure 5.9c) and grazing intensity (Figure 5.9d). Either raw data can be plotted, as in Figure 5.9b–d or the data can be recoded as quartiles or quintiles as in Figure 5.12 which makes interpretation of trends very much easier.

By visual inspection, it is clear that soil moisture percentage values are generally lower on the left of the quadrat ordination diagram and higher on the right. A soil moisture gradient along the first axis is indicated. This trend is confirmed by the quintile plot (Figure 5.12a). Inspection of the plot of soil pH values, however, shows a more complicated picture (Figure 5.9c), although the quintile plot (Figure 5.12b) indicates that the highest values are all found in the group at the right of the diagram. This group would thus merit further investigation. Grazing pressure (Figures 5.9d and 5.12c) also shows a pattern, with high values in the top-left group and lower values in the right-hand group. This is another pattern which almost certainly deserves further study.

Inspection of these graphs enables trends in the data to be recognised and hypotheses to be generated concerning vegetation and plant community variation in relation to environment. If there are clear gradients across the graph, then environmental gradients may be assigned to axes. However, such patterns are rarely clear-cut. Alternatively, patterns of high or low distribution may be observed in different parts of the diagram and interpreted accordingly. If trends are detectable, particularly if parallel to the axes, then rank correlation coefficients may help to display and explore any possible relationships.

In Table 5.4, Spearman's rank correlation coefficients (Chapter 4) have been calculated between the positions of quadrats on the first two axes of the detrended correspondence analysis (DCA) of the Gutter Tor data and the environmental data collected for each of the 25 quadrats on soil moisture percentage, soil pH and grazing intensity. The first point to note is that the ordination axes are uncorrelated with each other, indicating that they may reflect different sources of variation within the vegetation data. Second, the first axis is very highly correlated with percentage soil moisture, confirming the visual relationship described above. However, both of the other environmental factors are also significantly correlated, although pH only at the 0.05 level. The second axis is only significantly correlated with grazing intensity, which is already more highly correlated with the first axis. This demonstrates that application of correlation methods may be useful, but they will only find significant linear relationships parallel to each axis. A diagonal trend in an environmental factor across an ordination plot will thus result in significant correlations on **both** axes, as is the case with grazing intensity.

An important concluding point is that such interpretations are only meant to be a beginning. From the descriptive trends identified here, hypotheses concerning variation in the vegetation and between the vegetation and environmental factors should emerge, resulting in ideas for more detailed and specific research work involving the collection of new data. As an example, while the importance of soil moisture as a primary gradient on Gutter Tor has been clearly demonstrated,

Table 5.4 A matrix of Spearman's rank correlation coefficients between the quadrat scores on the first two axes of the detrended correspondence analysis of the Gutter Tor data (Table 3.4) and the environmental data for % soil moisture, pH and grazing estimates for those quadrats

	Ordination axes 1	Ordination axes 2	% soil moisture	pH	Grazing intensity
Ordination axes 1	—	0.22	0.95	0.46	− 0.58
2	0.22	—	0.24	− 0.31	0.35
% soil moisture	0.95	0.24	—	0.48	− 0.52
pH	0.46	− 0.31	0.48	—	− 0.61
Grazing intensity	− 0.58	0.35	− 0.52	− 0.61	—

A coefficient greater than $+/-$ 0.33 is significant at the 0.05 level (one-tailed test)
A coefficient greater than $+/-$ 0.47 is significant at the 0.01 level (one-tailed test)

questions emerge as to the nature and intensity of grazing pressure on the different communities and how this may be reflected in the soils and their pH. A separate study of grazing habits could be valuable, as well as a study of nutrient turnover by animals and expressed in their dunging patterns.

Bray and Curtis (polar) ordination

Bray and Curtis or polar ordination (PO) was the first widely used method of indirect ordination devised in the 1950s before the advent of modern computers (Bray and Curtis, 1957). Although it has now been superseded, current research indicates that it may still be a relatively efficient method of ordination (Beals, 1973, 1984). More importantly, however, of all methods, it is the one which is easiest to understand and to teach, since it is based on simple calculation and drawing-compass construction.

The method

The method will be described using the New Jersey salt marsh data (Table 3.1).

(1) The raw data matrix, containing either presence/absence or abundance data is tabulated, usually with quadrats in columns and species in rows (Table 3.1).
(2) For a quadrat (site) ordination, a **dissimilarity half-matrix** is computed between all quadrats in all combinations. Several coefficients are available but the most commonly used are the **Czekanowski** coefficient (p. 93) and the coefficient of **squared Euclidean distance** (p. 94) for quantitative data and the **Jaccard or Sørensen** coefficient for qualitative data (p. 91). Bray and Curtis used the **Czekanowski** coefficient in their original method. This is a **similarity coefficient (S_C)** which calculates values between 0 and 1. If two quadrats are identical, then the coefficient = 1.0. If two quadrats have nothing in common the score is zero. These similarity coefficients can be converted into **dissimilarity coefficients (D_C)** by subtracting the values from 1.0, so that identical quadrats then have a score of 0.0, and quadrats which have no species in common then have a score of 1.0 (maximum dissimilarity). The dissimilarity measures are then used as **distances** in the construction of the ordination diagram.

Thus $D_C = 1.0 - S_C$ (5:2)

where S_C = Czekanowski similarity coefficient
D_C = Dissimilarity coefficient

For the New Jersey salt marsh data, the matrix of dissimilarity coefficients for quadrats is given in Table 3.6.

(3) The ordination diagram is then constructed geometrically on the basis of the dissimilarity matrix, using a drawing compass and ruler. The two quadrats with the highest dissimilarity value are taken. If two quadrats have no species in common, then the highest possible value will be 1.0. In the case of the New Jersey data, no two quadrats are totally different, but quadrats 1 and 11 and 1 and 12 both have scores of 0.99. Either of these pairs of quadrats can be taken as **reference quadrats** which are placed at either end of the first axis. Taking quadrats 1 and 11, the first axis itself is scaled to the dissimilarity value between 1 and 11 (0.99). Thus on a piece of graph paper, if a scale of 1cm = 0.1 dissimilarity is taken, a line 9.9 cm long representing the first axis would be drawn with quadrat 1 at one end and quadrat 11 at the other (Figure 5.13). It makes no difference which quadrat is placed at which end of the axis. The same plot will be produced, except that depending on which quadrat is at which end, one plot will be a mirror image of the other. The distances between the points representing quadrats will be the same.

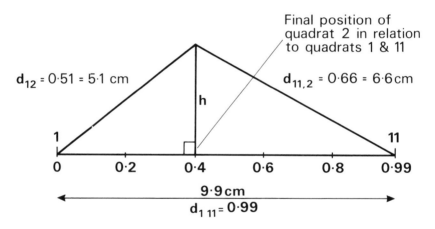

scale cm = 0.1 dissimilarity

Figure 5.13 Location of the end points (quadrats 1 and 11) and quadrat 2 on the first axis of the Bray and Curtis (polar) quadrat ordination of the New Jersey salt marsh data

(4) The remaining 10 quadrats are then positioned on the first axis using compass construction. Quadrat 2 is located on the first axis with respect to quadrats 1 and 11 by first taking the dissimilarity values between quadrats 1 and 2 (0.51, Table 3.6) and between quadrats 2 and 11 (0.66, Table 3.6). With the drawing compass point located at the end of the axis where quadrat 1 is located, an arc is drawn with its radius proportional to the dissimilarity between quadrats 1 and 2 (0.51 = 5.1 cm, Figure 5.13). Similarly, a second arc is drawn with the compass point positioned at the other end of the axis where quadrat 11 is located

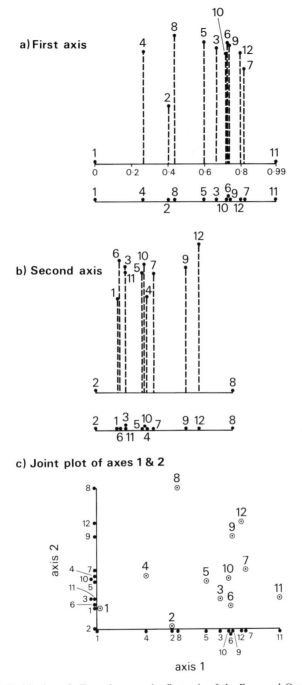

Figure 5.14   (a) Positioning of all quadrats on the first axis of the Bray and Curtis (polar) ordination of the New Jersey salt marsh data using compass construction. (b) Positioning of all quadrats on the second axis of the Bray and Curtis (polar) ordination of the New Jersey salt marsh data using compass construction. (c) The final ordination graph. Note that this is the same as Figure 5.6b, but the endpoints have been inverted on the first axis. This makes no difference to the results and is only a 'mirror-image'. The relative distances between the points are preserved.

with a radius proportional to the dissimilarity between quadrats 2 and 11 (0.66 = 6.6 cm). At the intersection of the two arcs, a perpendicular is dropped to the axis or baseline, and the point at which it intersects the line represents the final position of quadrat 2 on the first axis (Figure 5.13).

(5) The other quadrats are then positioned with respect to quadrats 1 and 11 in exactly the same manner, resulting in the single axis ordination shown in Figure 5.14a.

(6) The second axis is now constructed. Two reference quadrats for the second axis are chosen. The general rule for those quadrats which qualify is that they must be close to the centre of the first axis but still have a high dissimilarity value between them. In the New Jersey case, quadrats 2 and 8 are near to the centre of the first axis and yet have a dissimilarity value of 0.75. Thus, they are suitable as reference quadrats for the second axis.

(7) To construct the second axis, a line is drawn proportional to the dissimilarity between quadrats 2 and 8 (0.75 = 7.5 cm) and the other 10 quadrats are positioned by compass construction in the same manner as for the first axis (Figures 5.14b).

(8) The last stage involves the plotting of the points representing quadrats with respect to their coordinates on both axes. These are usually placed at right angles to each other (Figure 5.14c). Orlóci (1974; 1978) pointed out that the true angle between the axes is usually oblique and he presented an analytical method which calculates the true angle between the axes and which corrects the rotated coordinates on the second axis back to rectangular coordinates for plotting (Kent, 1977). The differences attributable to the correction are usually marginal, however.

(9) Construction of a third axis to give a three-dimensional ordination is possible. The procedure for the selection of reference quadrats is the same as on the second axis, except that selection is based on combined assessment of quadrats in the middle of both the first and second axes which still have relatively high dissimilarity values between them. Quadrats 7 and 10 would be suitable being relatively close together, near the centre of the second axis, and still having a dissimilarity value of 0.79. The resulting three-dimensional ordination diagram is shown in Figure 5.6c.

A problem in using Bray and Curtis ordination is that occasionally the positioning of a quadrat by the intersection of the two arcs may be impossible, because the arcs do not intersect. The reason for this is the failure of the Czekanowski dissimilarity coefficient to fulfil all the properties of a metric coefficient of similarity. The true distances between the quadrats are distorted when they are projected into a space of fewer dimensions. To position points accurately in relation to each other requires two dimensions of space, and when these distances are projected into one dimension on a particular axis, the distances are foreshortened. However, relatively, the distances are still preserved and this is why the ordination still works. If axes do not intersect, it is legitimate to select a different pair of reference quadrats and to reconstruct the ordination using those. The choosing of different pairs of reference quadrats only has the effect of 'viewing' the scatter of points and their relative dissimilarities from a different 'angle'. This section also introduces the idea of 'distortion' of the true relationships between points on an ordination diagram owing to the computational process of the ordination itself.

A further solution is to use a different similarity coefficient such as the coefficient of squared Euclidean distance (p. 94) or one of several recommended by Orlóci (1974; 1978).

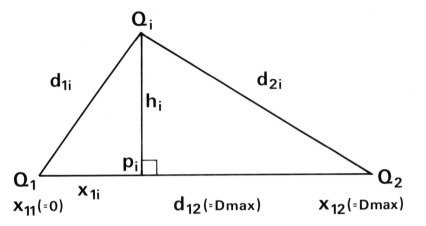

Figure 5.15 General principles for calculating the position of a given quadrat with respect to two reference quadrats (Q_1 and Q_2) which define the axis (Formulae 5:3–5:6) (Causton, 1988; reproduced with kind permission of Chapman and Hall)

For computerised analysis, the compass construction can be replaced by the following formulae (Causton, 1988):

In Figure 5.15, for triangle $Q_1Q_iP_i$, Pythagoras's theorem gives:

$$h_i^2 = d_{1i}^2 - x_{1i}^2 \tag{5:3}$$

For triangle $Q_2Q_iP_i$, Pythagoras gives:

$$h_i^2 = d_{2i}^2 - (d_{12} - x_{1i})^2 \tag{5:4}$$

Multiplying out the bracket in (5:3) and eliminating h_i^2 between the two equations gives:

$$d_{1i}^2 - x_{1i}^2 = d_{2i}^2 - d_{12}^2 + 2d_{12}x_{1i} - x_{1i}^2 \tag{5:5}$$

Re-arrangement, following cancellation of x_{1i}^2 terms gives:

$$x_{1i} = \frac{d_{1i}^2 - d_{2i}^2 + d_{12}^2}{2d_{12}} \tag{5:6}$$

The advantage of this is that the calculation of quadrat position on the axis (x_{1i}) is defined entirely from the dissimilarity values and thus can be computed from the original dissimilarity matrix.

In conclusion, Bray and Curtis (polar) ordination is still seen as an effective ordination method which compares favourably with some subsequent methods such as principal component analysis and reciprocal averaging or correspondence analysis (Gauch and Scruggs, 1979). However, polar ordination is not without its problems, the principal of which is the range of choice of pairs for endpoints of each axis. In small data sets this does not matter, but as amounts of data become larger the chances of obtaining consistent ordinations from different endpoint pairs is greatly reduced. The main advantage of polar ordination is its value in teaching, since if carried out manually using compass construction, students learn a great deal about the ordination process.

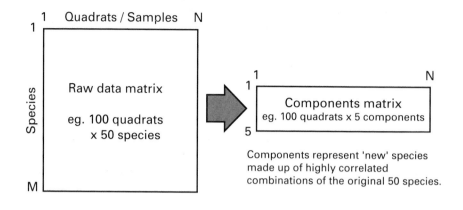

Figure 5.16 Reduction of many species into a few components

Principal component analysis

The method of principal component analysis (PCA) was first described by Karl Pearson (1901). However, a fully practical computational method was only devised much later by Hotelling (1933). Even then, the process of calculation was very daunting other than for very small data sets because it had to be carried out by hand. Only when computers became widely available did the technique come into general use in plant ecology. Although Goodall (1954) first applied the method under the erroneous title of factor analysis, PCA only became popular after the publication of a paper by Orlóci in 1966.

Although PCA is no longer recommended as an ordination method for floristic data (see p. 203 and Figure 5.1a), the method is important because it has been widely used in the past and there are many examples of its application in the literature. Even more important, however, is the the use of PCA to synthesise environmental data and to produce an ordination of quadrats based on **environmental variables alone** (right pathway — Figure 5.1). Here it is still one of the most effective methods of analysis.

PCA was the first ordination technique in which the ordination axes were computed from the data matrix alone and the researcher did not have to supply weights, endpoints or other subjective information during the computation process. For most non-mathematical students, PCA is best explained geometrically rather than by matrix algebra. The basic idea of PCA is that if data are collected to form a matrix of n quadrats and m species or m environmental variables, as in Figure 5.16, there will be a large amount of duplication or correlation in the variability of the species or environmental variables across the quadrats. Thus an original data matrix of, for example, 100 quadrats and 50 species/variables can be reduced to 100 quadrats and 5 or even fewer components. These components can be regarded as new 'super-species' or 'super-variables', made up of highly correlated combinations of the original 50 species or environmental variables (Figure 5.16).

One of the most important features of the components is that whereas all the former species/variables were almost certainly highly intercorrelated, the new components are completely uncorrelated and are said to be **orthogonal** and uncorrelated. The problems of intercorrelation and duplication within the original species/variables in the data set are removed by the analysis.

Another important concept is that just as in the original species/variable matrix, where each quadrat had a score for each species or environmental variable, so in the new components matrix, each individual has a score for each component which taken together are known as the **component scores**.

The core of any PCA are **eigenvectors** and **eigenvalues**.

Eigenvectors

Eigenvectors are sets of scores, each of which represents the **weighting** of each of the original species or variables on each component. The **eigenvector scores** are scaled like correlation coefficients and range from $+1.0$ through 0.0 to -1.0. For **each** component, **every** species or variable has a corresponding set of eigenvector scores and the **nearer the score is to $+1.0$ or -1.0**, that is the furthest **away from zero, the more important that species or variable is in terms of weighting that component.**

As an example, if a set of data collected in n quadrats contained 6 species or environmental variables A–F, following PCA, the eigenvector scores for the first two components could be:

Species/Variable	Eigenvectors Component I	Component II	
A	0.91	0.21	
B	0.79	0.19	Note that
C	0.75	0.29	conventionally
D	0.01	− 0.90	component numbers
E	− 0.12	− 0.86	are denoted by
F	− 0.01	− 0.93	Roman numerals

For component I, the highest weightings (eigenvector scores) are for species/variables A, B and C and for Component II, the highest weightings are for species D, E and F. It is on the basis of these eigenvector scores that the interpretation of a component is made. In this case, component I would be a 'new' species/variable with properties of the original species/variables A, B and C, and component II would be interpreted as a combination of the original species/variables D, E and F. If the similarity or correlation matrix between the species/variables was calculated, it would be clear that species/variables A, B and C would be very similar in their distribution among the quadrats, as would D, E and F. Note that in some analyses, eigenvectors and their scores are known as **component or factor loadings**.

Eigenvalues

Eigenvalues (as opposed to eigenvectors) are values that represent the relative contribution of each component to the explanation of the total variation in the data. There is one eigenvalue for each component, and the size of the eigenvalue for a component is a direct indication of the importance of that component in explaining the total variation within the data set.

A geometrical explanation of principal component analysis

The following geometrical explanation is adapted from Gould and White (1974; 1986):

The raw data matrix and standardisation

The starting point of PCA is a normal data matrix with quadrats and species as below. The data could be either the cover of each of 7 species measured on a 10-point Domin scale for 10 quadrats in a meadow or for 7 environmental variables which have been scaled between 0 and 10. This matrix has been deliberately contrived to assist with explanation of the method:

			Quadrats									
			1	2	3	4	5	6	7	8	9	10
S	V	A	1	5	7	9	10	8	6	4	3	2
p	a	B	8	10	7	9	6	5	4	3	1	2
e	r	C	3	6	7	9	10	8	5	4	2	1
c o	i	D	4	8	7	10	9	6	5	3	2	1
i r	a	E	10	9	7	8	6	5	4	3	1	2
e	b	F	2	6	7	9	10	8	5	4	3	1
s	l	G	1	6	8	9	10	5	7	4	3	2
	e											
	s											

Number of quadrats = 10
Number of species/variables = 7

In this example, the aim is to produce an ordination diagram for quadrats or samples. This is also known as a **normal** or R-analysis. The opposite of this is known as a Q-analysis or **inverse** analysis and provides the species ordination (if the data are species) or an ordination of the environmental factors (if the data are environmental variables). Other researchers also use the term **transposed** analysis.

In the application of PCA to vegetation/environmental data, the data may be initially **standardised**. This may be done for either or both species/variables and quadrats. The commonest means of standardisation is **zero mean, unit variance**. As the name suggests, each species/variable or quadrat is rescaled with a mean of zero and a variance of 1.0. If it is carried out for species/variables, then each species/variable score is expressed in units of standard deviations away from a mean of 0. The general formula is:

$$SS_i = \frac{(S_i - \overline{S})}{\sigma S} = \frac{\text{Each species/variable score} - \text{mean species/variable score}}{\text{Standard deviation of the species/variable}} \quad (5:7)$$

where SS_i = the standardised score of species/variable S in quadrat i
$\quad\quad S_i$ = the original scores of species/variable S in quadrat i
$\quad\quad \overline{S}$ = the mean score of species/variable S across all quadrats
$\quad\quad \sigma S$ = the standard deviation of species/variable S across all quadrats

Taking species/variable A in the above example, its mean across the 10 quadrats is 5.5 and the standard deviation is 3.03. The original score of 1 in quadrat 1 becomes -1.49, while the score of 9 in quadrat 4 becomes $+1.16$. All other scores on all other species/variables can be standardised in exactly the same way, using the mean and standard deviation for each species/variable in turn. If the data are not standardised, then the analysis will be biased towards those species/variables which have the highest variances. If the correlation coefficient is used as a measure of similarity, then standardisation is automatic. Quadrats or stands can be standardised in exactly the same way, except that the original scores are related to the mean and standard deviation of **each quadrat**. It is clearly possible to standardise both species/variables and then quadrats, and this is known as **double standardisation**.

The similarity or correlation matrix

The next stage of the R-analysis is to calculate a similarity or correlation matrix between all species/variables. The most commonly used coefficient is the product-moment correlation coefficient. For the above example, the product-moment correlation matrix is:

Species
(variable)

A	1.00						
B	0.35	1.00					
C	0.95	0.58	1.00				
D	0.83	0.79	0.93	1.00			
E	0.20	0.96	0.47	0.67	1.00		
F	0.98	0.49	0.99	0.90	0.36	1.00	
G	0.93	0.42	0.88	0.85	0.26	0.90	1.00
	A	B	C	D	E	F	G

Species/variable

The cosine of a right-angled triangle equals the correlation coefficient

Fundamental to the geometrical explanation of PCA is the idea that the correlation coefficient can be expressed as the cosine of a right-angled triangle.

In a right-angled triangle, the cosine of an angle is defined as the ratio of the length of the side adjacent to the angle, to the length of the hypotenuse. In a triangle with sides of lengths of the proportions 3,4,5, the cosine of α is 4/5 = 0.8, which in cosine tables gives an angle of 37°.

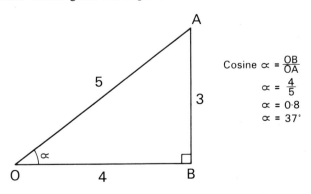

The next stage is to consider point O as a hinge, with side OA being lowered on to OB. The result is that the angle α becomes smaller and smaller and the two sides OA and OB become more and more similar in length, provided that the right angle OBA is maintained, until eventually, when $\alpha = 0°$, OA and OB are identical.

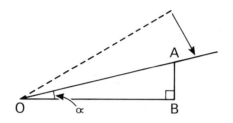

Again, side OA can be moved away from OB. As OA gets nearer to the vertical, the distance OB becomes very small, until when OA is at right angles to OB, the distance OB = zero; the ratio OA/OB = 0.0 and the angle $\alpha = 90°$ (Cosine 90° = 0.0).

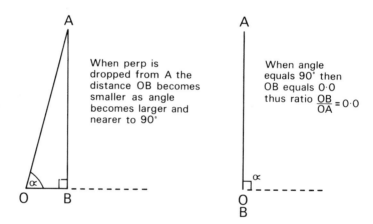

If OA is kept swinging around pivot O, the angle α then becomes greater than 90° and the cosine becomes negative, eventually approaching -1.0 when the angle $\alpha = 180°$.

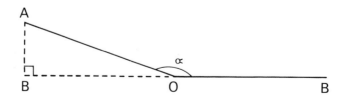

This idea can now be related to correlations between species or environmental variables. According to the above, the cosine expressed as a ratio between two sides of a right-angled triangle varies from $+1.0$ through zero to -1.0, as does the correlation coefficient. It thus becomes possible to represent correlations between species/variables geometrically.

Each species or variable in the analysis can be considered a **vector**. A vector is a line which possess properties of both length and direction.

——————————————⟶ A vector

The length of a vector representing a species/variable is directly related to its variance, but since the first step of PCA is usually to standardise all the species/variables to a variance of 1.0, vectors representing all species/variables in the analysis become the same length, that is 1.0.

Thus if there are two species/variables A and B, these can be expressed as two vectors of the same length. Also, using the cosine principle, the angle between the two vectors at their origin can be taken as an expression of their correlation.

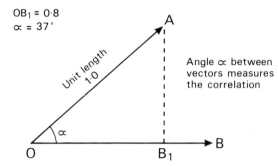

$OB_1 = 0.8$
$\alpha = 37°$

Angle α between vectors measures the correlation

If the perpendicular is dropped from the tip of vector A to intersect vector B, then depending on the angle α, the length of O to the point of intersection alters (B_1). Thus if OA = 1.0 (standardised) and angle $\alpha = 37°$, OB_1 must be 0.8 (cosine 37° = 0.8).

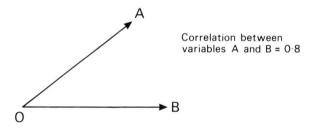

Correlation between variables A and B = 0.8

If the correlation between species/variables A and B, represented by vectors A and B, is 0.8, then this is shown geometrically by positioning the vectors at an angle of 37° to each other.

This logic can then be followed through for all other correlations between any other pair of species/variables represented as vectors. If the correlation between two species/variables is 0.0, the vectors must be at an angle of 90° to each other.

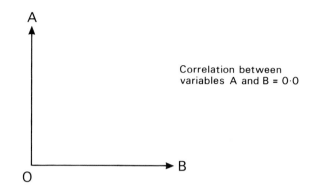

If the correlation between two species/variables is 1.0, then the two vectors will be on top of each other and the angle between them will be zero (0°).

If the correlation between two species/variables is −0.8, then the angle between the two vectors will be 143°.

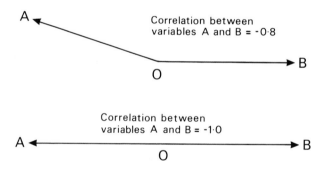

If two species/variables are perfectly negatively correlated (−1.0), then the two vectors will point in opposite directions and the angle between them will be 180°. This idea is now used to represent the intercorrelations between the seven species/variables as shown on p. 189. Each species/variable is represented as a unit vector, and the angles between them correspond to the degree of correlation. Strictly, these should be represented in several dimensions — one for each species/variable — but obviously this is impossible to visualise. Thus for purposes of explanation, the correlations have been carefully chosen, in order to be depictable in the two-dimensional plane of the page.

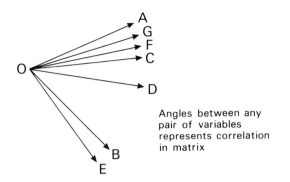

Angles between any
pair of variables
represents correlation
in matrix

The principal axis

Having obtained the geometric picture of the correlations between all seven
species/variables, the next stage is to collapse all the species/variables into one or
two major trends or components. On the geometric diagram, those species/variables
which are very much intercorrelated — A, G, C and F — tend to thrust together
in one particular direction. The aim of component analysis is to find this overall
directional thrust of the vectors by passing a line or axis through their common
origin, enabling each vector representing a species/variable to make a right-angled
projection on to this axis as in the following diagram. This line or axis is known
as the **principal axis**. Since the vectors are of unit length, their projections on to
the axis are no more than the cosines of the angles between the vectors and the
principal axis.

The important questions are: how is this principal axis located? How does one
know when it shows the direction of strongest thrust of all the vectors?

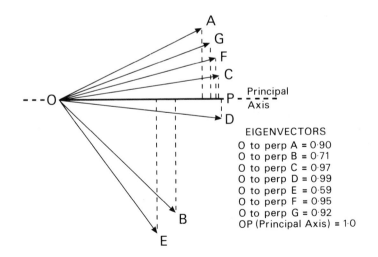

Principal
Axis

EIGENVECTORS
O to perp A = 0·90
O to perp B = 0·71
O to perp C = 0·97
O to perp D = 0·99
O to perp E = 0·59
O to perp F = 0·95
O to perp G = 0·92
OP (Principal Axis) = 1·0

In mathematical terms, the criteria can be very complex, but in geometrical terms,
the criteria are relatively simple. The principal axis is swung slowly around the
origin (O) and at each instant, the lengths on the principal axis OP from O to the
perpendiculars from the tips of all the vectors on to the principal axis are measured,
as in the above diagram. These projections are actually the **eigenvectors or loadings**

of all the species/variables on to the principal axis and these can be seen to be equivalent to the degree of correlation of each species/variable with the principal axis.

The position of the principal axis at the point which represents the maximum direction of thrust of all the vectors is found by taking the eigenvectors or loadings for any position of the principal axis, squaring them and then summing the squared values. The optimum position for the principal axis is found when **the sum of the squared values is at a maximum**. This value, which is the sum of the squared correlations between all species/variables and the axis, is the **eigenvalue** of that principal axis or first component. In practice, mathematically, there are rapid ways of finding the eigenvalues of any correlation matrix. The optimum position of the principal axis for the example is shown in the above diagram.

In the example, the eigenvector scores at the shown position of the principal axis are:

Species/Variable	Eigenvector
A	0.90
B	0.71
C	0.97
D	0.99
E	0.59
F	0.95
G	0.92

The eigenvalue is thus: $(0.90)^2 + (0.71)^2 + (0.97)^2 + (0.99)^2 + (0.59)^2$
$+ (0.95)^2 + (0.92)^2 = 5.33$

The eigenvalue represents the highest possible degree of correlation of all the species/variables with the principal axis and thus is a measure of the amount of variation in the data set accounted for by the first axis. The principal axis represents the first component which has varying contributions from each of the original species/variables, and those species/variables whose vectors are closest to the principal axis are the most important in attempting to explain or interpret the first axis.

Derivation of a second axis

Earlier, it was stated that in the simplest models of PCA, the components extracted were always **orthogonal** to each other which means that they are uncorrelated. In vector representation, if two species/variables are completely uncorrelated, that is correlation = 0.0, then the vectors representing those species will be at right angles to each other (90°). By the same logic, if one of the basic tenets of PCA is orthogonality, then the second axis or component must be at right angles to the first. This is shown in the following diagram.

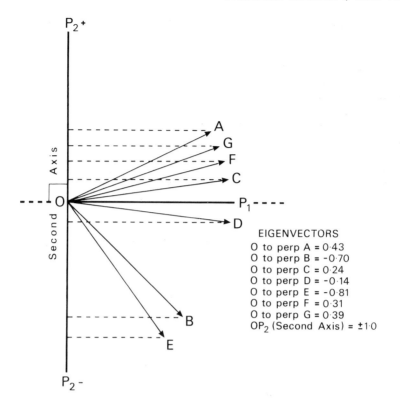

EIGENVECTORS
O to perp A = 0·43
O to perp B = -0·70
O to perp C = 0·24
O to perp D = -0·14
O to perp E = -0·81
O to perp F = 0·31
O to perp G = 0·39
OP₂ (Second Axis) = ±1·0

Once the second axis is in position, then the eigenvectors and the eigenvalue for that axis can be derived in exactly the same manner as for the first axis. The projections of the tips of the vectors representing each species/variable on to the second axis at right angles gives the eigenvectors for the second axis, and the sum of the squared eigenvector scores gives the eigenvalue. Thus:

Species/Variable	Eigenvector
A	0.43
B	− 0.70
C	0.24
D	− 0.14
E	− 0.81
F	0.31
G	0.39

The eigenvalue is thus: $(0.43)^2 + (-0.70)^2 + (0.24)^2 + (-0.14)^2 + (-0.81)^2 + (0.31)^2 + (0.39)^2 = 1.66$

Note that the second eigenvalue is much lower than the first. One of the features of PCA is that **axes are extracted in descending order of importance in terms of their contribution to the total variation in the data set**. Also note that those species/variables which had high eigenvector scores on the first axis tend to have lower eigenvector scores on the second.

Subsequent axes can be extracted in exactly the same manner, by erecting further axes at right angles to each other. Again a set of eigenvector scores can be determined, and the sum of squared eigenvector scores will give the eigenvalue although this value will be lower than on the first or second axis. Mathematically, the process of extracting further axes can go on until the number of axes or components equals the number of species or variables, but most higher axes are trivial and contribute little to explanation. For small data sets, two or three axes will summarise most of the variation in the data and with larger sets four or five will suffice. As a guideline, sometimes axes are only extracted up to the point where the eigenvalue associated with a particular axis or component has a value of $\geqslant 1.0$. An eigenvalue of less than 1.0 means that the axis is contributing less to the overall explanation of the variability in the data than any one of the original species or variables.

Calculation of component scores

So far, the analysis has produced two components, each of which has a set of eigenvectors or loadings and a corresponding eigenvalue which can be seen as summarising the amount of overall variation in the data accounted for by each component. The final stage of analysis is the calculation of the component scores for each quadrat in the original matrix. It is these scores which make up the new matrix of quadrats × components (Figure 5.16).

To obtain the component scores for each of the original 10 cases, each of the original values for each species or variable within a quadrat is multiplied by the eigenvector score for that species/variable on a given component. These values are then summed for each quadrat to give the component score for that individual. Thus:

			1	2	3	4	5	6	7	8	9	10	Component I	Component II
							Quadrats						Eigenvectors	
S	V	A	1	5	7	9	10	8	6	4	3	2	0.90	0.43
p	a	B	8	10	7	9	6	5	4	3	1	2	0.71	−0.70
e (o)	r	C	3	6	7	9	10	8	5	4	2	1	0.97	0.24
c (r)	i	D	4	8	7	10	9	6	5	3	2	1	0.99	−0.14
i	a	E	10	9	7	8	6	5	4	3	1	2	0.59	−0.81
e	b	F	2	6	7	9	10	8	5	4	3	1	0.95	0.31
s	l e s	G	1	6	8	9	10	5	7	4	3	2	0.92	0.39

To obtain the component score for quadrat 1 on component I:

Score = $(1 \times 0.90) + (8 \times 0.71) + (3 \times 0.97) + (4 \times 0.99)$
$+ (10 \times 0.59) + (2 \times 0.95) + (1 \times 0.92) = \underline{22.17}$

To obtain the component score for quadrat 10 on component II:

Score = $(2 \times 0.43) + (2 \times -0.70) + (1 \times 0.24) + (1 \times -0.14)$
$+ (2 \times -0.81) + (1 \times 0.31) + (2 \times 0.39) = \underline{-0.97}$

This process is thus carried out for every quadrat on every component extracted. In this example, where only two components have been extracted, this gives a new matrix of 10 quadrats × 2 components as below:

Components matrix

Components						Quadrats					
		1	2	3	4	5	6	7	8	9	10
I		22.17	41.80	42.90	54.40	53.80	39.50	31.40	21.70	13.40	9.10
II		−12.10	−7.60	−2.00	−1.80	−3.40	1.40	1.30	0.50	2.10	−0.97

Thus the original matrix of 10 quadrats and seven species/variables is reduced to 10 quadrats with two orthogonal (uncorrelated) components, with a high proportion of the total variation explained. The exact amount accounted for by each component can be calculated as follows below. The procedure is best explained by taking an extreme example. The highest possible explanation of any one component in PCA would be if 100 per cent of the variation was accounted for by one axis. For this to occur, all the species/variables in the analysis would have to be perfectly correlated with each other — that is all the correlations in the matrix would be 1.0, and all the species/variables would be identical in their distribution in the quadrats. In vector form, this would be represented as below, with all the vectors lying exactly on top of each other along the principal axis.

The eigenvector scores would thus also all be 1.0, and the eigenvalue, which is the sum of the squared eigenvector scores, would be equal to the number of species/variables, which in the example would be 7.0:

that is $(1.0)^2 + (1.0)^2 + (1.0)^2 + (1.0)^2 + (1.0)^2 + (1.0)^2 + (1.0)^2 = 7.0$

7 is the number of species/variables.

However, in a typical data set, where species or variables are intercorrelated to varying degrees, the angles between the vectors representing the species/variables start to appear, as in the previous geometrical diagrams, and thus the eigenvector scores fall below 1.0.

Since the eigenvector scores fall below 1.0, so the eigenvalue, which is the sum of the squared eigenvector scores, must also fall. It follows that the eigenvalue for any component is an exact measure of the proportion of the total variation in the data explained by that component and can be expressed as a fraction of the eigenvalue over the number of species/variables in the analysis and is usually then expressed as a percentage.

In the example, the percentage of the total variation explained by the first component is:

$$\frac{5.33 \ \text{(the eigenvalue)}}{7.00 \ \text{(the number of species/variables)}} \times 100 = 76.20\%$$

The percentage explanation of the second axis is:

$$\frac{1.66 \quad \text{(the eigenvalue)}}{7.00 \quad \text{(the number of species/}} \times 100 = 23.65\%$$
variables)

The cumulative explanation of the two components is:

76.20 + 23.65 = 99.85%

Thus all except for 0.15 per cent of the variation in the original matrix of 10 quadrats and seven species/variables is explained by the two components, and there would be little value in extracting any further components or axes, although in theory this could be calculated up to the seventh.

Production of the final quadrat ordination plot

The quadrat ordination plot is derived by taking the component scores for the quadrats on the two axes and plotting them as a graph, as below. This plot can

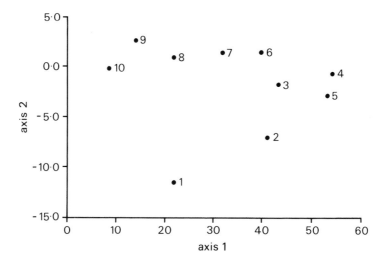

then be interpreted in exactly the same way as for any other ordination method based on the principle that the distance between any two points representing quadrats on the graph is an approximation to their similarity of species composition. Thus quadrats 4 and 5 are similar to each other, as are 9 and 10, but 4 and 5 are both very different from 9 and 10.

Clearly, this example has involved considerable simplification. Nevertheless, the basic principles of all applications of PCA are derived from the ideas presented here. For the mathematician, the method is more clearly explained by matrix algebra. Good examples using the matrix algebra are in Orlóci (1978), Digby and Kempton (1987), Jongman *et al.* (1987) and Causton (1988).

Data centring

A further important distinction in PCA is whether a **centred** or **non-centred** analysis is performed (Orlóci, 1967a; Gauch, 1982b; Causton, 1988). The idea of data centring is best explained by taking a simple case of two species or variables. When

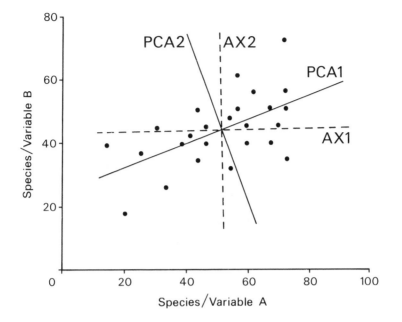

Figure 5.17 Data centring and derivation of principal components (adapted from Gauch, 1982b; redrawn with kind permission of Cambridge University Press)

plotted jointly, they give the swarm of points on the graph in Figure 5.17. There are three different sets of coordinates for these points. First, there are the original axes of the graph marked species/variable A and species/variable B. Second, there is a pair of axes which are placed on the **centroid** for the two species/variables where the centroid is simply the average of the scores for each of the two species/variables. These are labelled AX1 and AX2. This shifting of the coordinate system for the swarm of points to the centroid is known as centring. Third, there is another pair of axes PCA1 and PCA2, which represent the rotation of the first or principal axis around the centroid according to the principles of PCA so that the projection distances from the points to the axis are minimised, and the axis lies along the direction which summarises maximum variance as explained above. The second PCA axis in Figure 5.17 is defined as being at right angles or orthogonal to the first. Clearly, data centring, prior to the locating of axes, will alter the final position of those axes when compared to a situation where the data have not been centred. Ordinations of non-centred data tend to produce a trivial general first component or axis. However, differences between centred and non-centred analyses are usually not great unless there are major discontinuities in the data.

As an example of this, three PCA quadrat ordinations of the Garraf data set (Table 3.3), are presented in Figure 5.18. In Figure 5.18a, the data have not been centred while in Figure 5.18b, they have. In this case, the effects of centring are minimal. However, if the data are both centred and standardised (Figure 5.18c), quite striking differences are observable between the plots. Thus standardisation of data is much more likely to have a significant impact on the ordination. In most ecological applications, data are analysed using centred and standardised PCA, and if a variance-covariance matrix is used, data are centred in the process. If the correlation coefficient is used, data are both centred and standardised. The arguments for and against centring and standardisation are given by Dagnelie (1960;

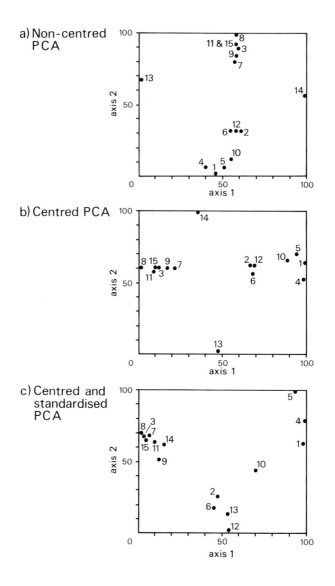

Figure 5.18   The effects of data centring and standardisation on the Garraf data set

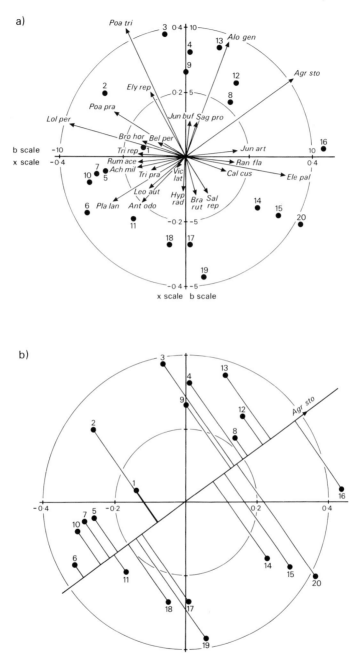

Figure 5.19 Biplot from principal component analysis of the dune meadow data (Jongman *et al.*, 1987) (a) species are represented by arrows; quadrats by dots; the b scale applies to species and the x scale to quadrats/sites. Species close to the origin are not plotted. (b) interpretation of the biplot in Figure 5.19(a) for *Agrostis stolonifera* (ter Braak, 1987a; redrawn with kind permission of PUDOC — Centre for Agricultural Publishing and Documentation, Wageningen, The Netherlands)

1978); Noy-Meir (1973); Noy-Meir *et al.* (1975); Orlóci (1967a; 1978), Greig-Smith (1980; 1983), Hinch and Somers (1987) and Causton (1988).

R and Q ordinations

Some confusion has arisen in the literature over this terminology, which is normally only applied to the use of PCA for ordination. An R ordination is the same as a quadrat ordination, where the component scores derived from a correlation or similarity matrix calculated between species is used to give ordination plots for quadrats. However, a further aspect of PCA is that the eigenvector scores (which represent the weightings of each species on each component) can be plotted to give a species ordination as well. With the R approach, the underlying geometrical model is thus one of quadrats in terms of species space. It is clearly possible to carry out PCA on the matrix inversely, so that the component scores would represent species, and the eigenvector scores on each component would be used to give an ordination of quadrats (Figure 5.10). This is known as the Q approach. In this case, the underlying geometrical model is of species defined in quadrat space. If the same centring and standardisation is applied, then an R analysis will give the same result as a Q analysis.

The biplot method

The fact that both component scores and eigenvectors scores from either an R or Q analysis can give ordination information on both quadrats and species has led to the development of the **biplot method** of displaying results. The method was first proposed by Gabriel (1971; 1981) and Bradu and Gabriel (1978). Their use in PCA for floristic ordinations is explained by ter Braak (1983) and ter Braak (1987a). Both quadrat scores and species are plotted on the same graph but using different scales because the scaling of quadrat (component) scores (in an R analysis) will be different from that for species which are represented by the eigenvector scores. In Figure 5.19a, following a centred PCA on the Dune Meadow data (Jongman *et al.*, 1987) the quadrats have been plotted in relation to the first two components in a standard quadrat plot. However, superimposed on this, with the axes aligned and suitably scaled, is the plot of species eigenvector scores (loadings) on the first two components. Arrows are then drawn from the joint centred ordination axes (0,0) to the points representing each species. The direction of the arrow indicates the direction in which the abundance of a species increases most rapidly. The length of the arrows shows the rate of change in abundance in that direction. Thus a long arrow indicates gradual rate of change in abundance, while a short arrow represents very rapid change. Figure 5.19b shows how this can be used to interpret the abundance of one species in relation to the quadrat plot. Those quadrats containing *Agrostis stolonifera* have been extracted, and lines have been constructed from each quadrat point to intersect the oblique arrow or axis representing *Agrostis stolonifera* on the original biplot. The sequence of quadrats from the top of the arrow through the origin to the tail is an approximation to the rank order of abundance of the species in the quadrats. Thus in Figure 5.19b, the abundance of *Agrostis stolonifera* will be greatest in quadrat 16, second highest at quadrat 13 and lowest at site 6, which is at the other end of the axis. All quadrats to the right of the origin will have abundance above the overall mean of the species, while those to the left will be below.

To conclude, PCA represents a very complex method of ordination with a range of variations within it. Although in many situations the problems of choice over standardisation and centring are quite minor, in others they may be very important. As a result, for many inexperienced users, interpretation may be difficult. However, PCA is now widely acknowledged as having serious limitations as a method for the ordination of floristic data. The assumption of linearity of correlations between species is rarely met, and the constraint of having ordination axes orthogonal to each other and thus uncorrelated means that the underlying mathematical model is very different from the ecological one, where major environmental gradients themselves will often be intercorrelated; yet axes are commonly taken to represent environmental gradients. If components are taken to represent new 'super-species' made up of highly correlated groups of the original species, interpretation can become difficult since the components represent mathematical constructions and abstractions, rather than real species.

The effects of this and the distortion often caused by applying PCA to floristic data are discussed in the next chapter. However, PCA is still extremely important as a method of summarising variation in environmental data. Provided the variables are standardised, usually to zero mean/unit variance, the results of PCA may be extremely valuable when interpreted alongside a floristic ordination of data from the same quadrats, carried out by some other ordination method. An excellent example of this is shown in the case study of research by Goldsmith (1973a,b) described below.

Case studies

The use of Bray and Curtis (polar) ordination to assist with site description prior to a study on the production ecology of an Alaskan Arctic tundra (Webber, 1978)

Under the International Biological Programme (IBP), the energy and nutrient budgets for a large number of different world ecosystem types have been examined and quantified. One such study was undertaken for the tundra biome at Point Barrow, in Alaska. The aim was to understand the structure and functioning of a typical tundra ecosystem and the need for such research was accentuated by discovery in the early 1970s of large oil reserves in the region. The exploitation of these reserves would have far-reaching consequences in terms of environmental impact and disturbance. The most important feature of the tundra environment is the climate, with severe cold for much of the year. At Point Barrow, the mean annual temperature is $-12.6°C$, with nine months below freezing. Thus, mean monthly temperatures only exceed freezing for three months of the year, and it is in this very brief period that plant growth and primary production occurs. Consequently, all ecological activity and change is extremely slow, and communities are extremely fragile. Related to the climate is the periglacial geomorphology of the area. Permanently frozen ground or permafrost occurs everywhere at depth, and only a relatively thin surface or active layer thaws in the summer months. As a result, soil-water conditions are highly problematical for plant growth, since when the water is not frozen, drainage is poor and waterlogging is widespread in the summer thaw. Various types of patterned ground occur (Figure 5.20a,b) and these are closely correlated with small-scale variation in the vegetation.

Once the site at Point Barrow, Alaska had been chosen as a typical example of tundra vegetation, one of the first tasks was to describe the temporal

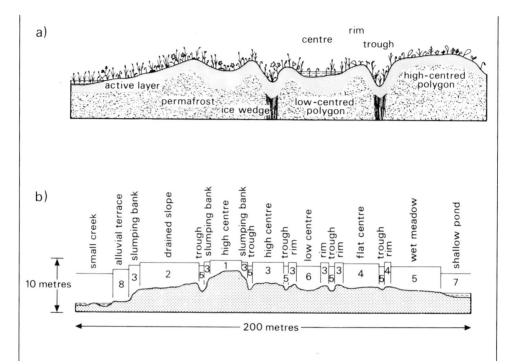

Figure 5.20 (a) Types of patterned ground and (b) associated plant communities at Point Barrow, Alaska (Webber, 1978; redrawn with kind permission of Springer-Verlag Publishers, New York)

and spatial variation of the vegetation, prior to the commencement of detailed work on primary production and nutrient cycling.

This first stage involved both the description and definition of the plant communities present and their underlying environmental gradients. The overall site was approximately 110ha in size, and within this area, in 1972, 43 permanent plots were established, each 1 x 10m in size. These were then monitored over the next four growing seasons. The rectangular shape was used to allow for the patterned ground characteristic of the region.

Percentage cover and frequency of both higher plants and cryptogams were recorded. Detailed site descriptions were made along with the collection and physical and chemical analysis of soil samples, which were collected on a single day in 1973, to a depth of 10cm. The most striking feature of the vegetation of the site was that it changed every few metres in response to variation in microrelief and drainage. Percentages of sands, silts and clays, percentage of organic matter, water-holding capacity and pH were determined for each soil sample. Soluble soil phosphate was also measured for selected soils and the hydrogen sulphide content of the soil was estimated using a four-point subjective scale based on odour. Soil moisture content was measured gravimetrically from two samples at 10cm depth at intervals of two days during two growing seasons. Soil-thaw depth was assessed by pushing a thin graduated rod into the soil every 10 days, and measurements of snow depth, snow cover and date of snowmelt were carried out just prior to and at the start of each growing season. The impact of grazing by lemmings was considered to be the main biotic control and was assessed on a five-point

Table 5.5 The characterising species and microrelief types of the eight vegetation communities/noda identified for the Point Barrow IBP sites in Alaska (Webber, 1978; reproduced with kind permission of Springer-Verlag, New York)

Nodum/community	Characteristic species	Major microrelief types
I Dry *Luzula confusa* heath	*Luzula confusa, Potentilla hyparctica, Alectoria nigricans, Pogonatum alpinum, Psilopilum cavifolium*	High-centred polygons
II Mesic *Salix rotundifolia* heath	*Salix rotundifolia, Arctagrostis latifolia, Saxifraga punctata, Sphaerophorus globosus, Brachythecium salebrosum*	Low-centred polygons and sloping creek banks
III Mesic *Carex aquatilis-Poa arctica* meadow	*Carex aquatilis, Poa arctica, Luzula arctica, Cetraria richardsonii, Pogonatum alpinum*	Hummocky polygon rims and centres and dry, flat polygonised sites
IV Moist *Carex aquatilis-Oncophorus wahlenbergii* meadow	*Carex aquatilis, Oncophorus wahlenbergii, Dupontia fisheri, Peltigera aphthosa, Aulacomnium turgidum*	Moist, flat sites and drained polygon troughs
V Wet *Dupontia fisheri-Eriophorum angustifolium* meadow	*Dupontia fisheri, Eriophorum angustifolium, Cerastium jenisejense, Peltigera canina, Campylium stellatum*	Wet, flat sites and troughs
VI Wet *Carex aquatilis-Eriophorum russeolum* meadow	*Carex aquatilis, Eriophorum russeolum, Saxifraga foliolosa, Calliergon sarmentosum, Drepanocladus brevifolius*	Low-polygon centres and pond margins
VII *Arctophila fulva* pond margin	*Arctophila fulva, Ranunculus pallasii, Ranunculus gmelini, Eriophorum russeolum, Calliergon giganteum*	Pond and stream margins
VIII *Cochlearia officinalis* pioneer meadow	*Cochlearia officinalis, Phippsia algida, Ranunculus pygmaeus, Stellaria humifusa, Saxifraga rivularis*	Snowbeds, creek banks and creek sides

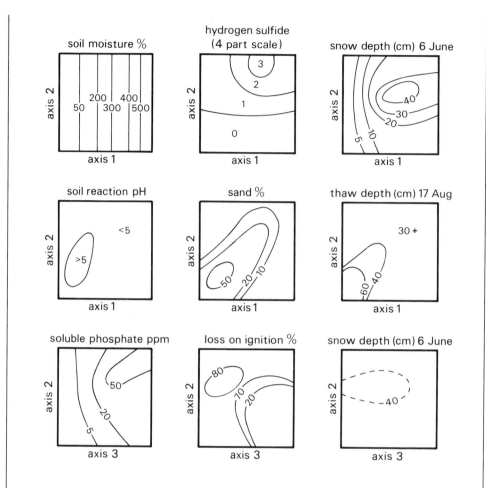

Figure 5.21 Plots of environmental data superimposed on the quadrat polar
ordination plots for tundra communities at Point Barrow, Alaska
(Webber, 1978; redrawn with kind permission of Springer-Verlag, New
York)

scale, where 0 = no activity and 4 = intense grazing.

Data analysis

The recognition of plant communities was achieved by numerical classification, using a method of similarity analysis known as average-linkage cluster
analysis (see Chapter 8). This produced the eight community types or noda
shown in Table 5.5. In order to examine environmental gradients and to relate
the environmental data to the plant community structure, a Bray and Curtis
or polar ordination was performed. This ordination was carried out on 39 of
the 43 sample quadrats. Four stands characteristic of the pioneer vegetation
were so dissimilar that they distorted the initial ordination and were removed.
This immediately demonstrates the point that ordination displays the total
variability within the data, and if several quadrats are very different from all
the rest, a poor initial ordination will result. Removal of these quadrats then

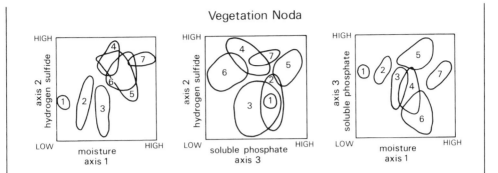

Figure 5.22 The distribution of the seven community types on the quadrat ordination plots from Point Barrow, Alaska (Webber, 1978; redrawn with kind permission of Springer-Verlag, New York)

reduces the overall variability and enables patterns of variation to be observed in the remaining data. For the 39 quadrats, three ordination axes were extracted and correlation coefficients were calculated between the various environmental factors and these axes.

Mean soil moisture content and snow depth were most highly correlated with the first axis, hydrogen sulphide and percentage of organic matter with the second and soluble phosphate with the third. The individual environmental factors, plotted as gradients, are shown in Figure 5.21. Clearly, there is evidence of linear trends with the three axes for those environmental factors that showed significant correlations, but other plots show patterns which imply correlation with more than one axis, for example percentage of sand and snow depth. Still other plots show no evidence of linear relationship at all, for example soil pH.

A further stage of interpretation involved the plotting of quadrat membership of the seven community types recognised by classification on the polar ordination diagrams (Table 5.5; Figure 5.22). The eighth type was the pioneer nodum removed prior to the ordination. The relationships between the noda are quite clearly seen and their extent of overlap can be assessed.

On the basis of these results, it became possible to make inferences about the successional sequences among the noda or community types. These are shown in Figures 5.20 and 5.23. These were related to sequences in the geomorphic and periglacial processes occurring in the area, with one sequence attached to the thaw lakes found in the centres of low-centred polygons and lower areas of patterned ground and another on the alluvial material of high-centred polygons and areas in between, which formed areas of higher ground.

Primary production was assessed by harvesting two small 20 × 50cm quadrats from each of the 43 plots during the period of peak above-ground vascular biomass. Estimates of the net primary production were made from the dry weights of the clipped material, although no allowance was made for grazing and shedding of plant parts. The detailed work on community description and its relationship with local environmental variation meant that quite detailed correlations could be made between variation in primary production, community composition and micro-environmental factors.

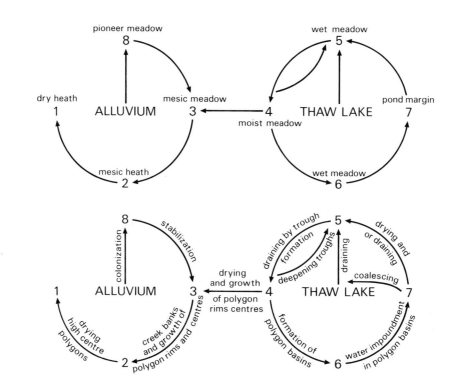

Figure 5.23 Successional sequences in tundra vegetation at Point Barrow, Alaska. Numbers refer to communities/noda in Table 5.5. (Webber, 1978; redrawn with kind permission of Springer-Verlag, New York)

A key point in this study is that for many types of detailed ecological work, the application of ordination methods to data on floristics and environmental controls is a vital prerequisite. In this case, before attempting studies of primary production, the vegetation and its variation needed to be described both as a basis for sampling and to enable estimates of the variability of production to be made.

The vegetation and environmental controls of sea cliffs at South Stack, Anglesey — principal component analysis as an ordination technique (Goldsmith, 1973a,b)

The vegetation of sea cliffs has rarely been studied in detail, principally because of the obvious problems of working in such a difficult environment. Goldsmith (1973a,b) carried out a study of sea-cliff communities at South Stack on Holy Island, Anglesey, Wales, with the aim of generating and testing hypotheses concerning relationships between sea-cliff plant communities, distributions of individual species and causal environmental factors.

The collection of vegetation data on sea cliffs posed a number of problems, particularly over access to sites and sampling, bearing in mind the crenulations of the coastline and the near vertical nature of many cliffs. As a result,

Table 5.6 Environmental factors recorded in each 3×3m sample stand of sea cliff vegetation (Goldsmith, 1973a,b; reproduced with kind permission of *Journal of Ecology*)

pH	Total bases
Conductivity	Phosphate (PO_4)
% soil moisture	Conductivity/soil moisture
Soil water-holding capacity	Height above sea
% organic matter	Distance from sea
Soil depth	General aspect
Sodium (Na)	Specific aspect
Potassium (K)	Slope
Calcium (Ca)	Potassium/calcium (K/Ca)

a stratified sampling system was adopted. The range of exposures, aspects and slopes of the cliffs was determined and then vegetation composition was sampled within each of these types. Quadrats were taken where there was a visually homogeneous stand of vegetation within a certain combination of aspect, slope and geology.

The geology of the cliffs comprised Precambrian rocks derived from grits, sandstones and shales, over which both podsolised and brown-earth soils had formed. A quadrat or sample area with a size of 3×3m was used for vegetation description and data from 65 such stands were collected. In each 3×3m area, percentage cover was assessed using 200 random points and a sampling pin with a diameter of 1.5mm and shoot frequency was measured in 50 randomly placed 10×10cm sub-quadrats within each 3×3m area for all flowering plants. In order to determine the major environmental gradients controlling sea-cliff communities and to enable variation in environmental factors to be synthesised, a total of 18 environmental factors were measured in each quadrat (Table 5.6).

For data analysis, both ordination and numerical classification methods were used. Of primary interest here is the use of principal component analysis on the floristic data to give both species and quadrat ordinations, but also of importance is the further but separate application of PCA to the environmental data. Thus variation in plant response was identified by one ordination, while variation in environmental factors and gradients was characterised by another. This corresponds to both the indirect ordination approach (left pathway) and the synthesis of environmental factors (right pathway of Figure 5.1). The extent to which the two ordinations were related would show how closely variation in floristics was correlated with variation in environmental controls.

Quadrat ordination of floristic data

The ordination diagrams of the quadrat ordination of the 65 quadrats using the frequency data standardised by quadrats are shown in Figures 5.24 and 5.25. The weighted similarity coefficient (Orlóci, 1966) was used as the coefficient of similarity. The gradients of environmental factors on the ordination diagram are shown in the first plot. Note that axes are centred on zero since in PCA, both positive and negative scores occur on each axis.

The first plot (Figure 5.24a) shows the quadrat distribution in relation to the first two axes. In Figure 5.24b, the numbers of species in each quadrat have been plotted as quintiles, while in Figure 5.24c, total plant cover is plotted as

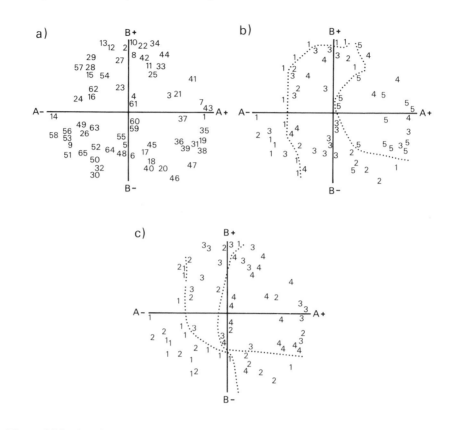

Figure 5.24 Quadrat ordination for principal component analysis of sea cliff data on Anglesey, Wales: (a) quadrat plot; (b) species richness in quadrats plotted as quintiles (1 = low; 5 = high); (c) total plant cover plotted as quartiles (1 = low; 4 = high) (Goldsmith, 1973a; redrawn with kind permission of *Journal of Ecology*)

quartiles. A clear gradient of both species numbers and cover is shown on both diagrams, with low values in quadrats to the left of the first (horizontal) axis and higher values to the right. Quadrats on the extreme left were typical of exposed cliff areas, while those on the extreme right corresponded to cliff-top grassland and heath communities. Further evidence of this was obtained by plotting the abundance of certain species on the quadrat ordination diagram (Figure 5.25). The salt-tolerant species *Armeria maritima* occurs with high frequencies on the left-hand side of the plot, particularly in the sector A^-B^+ and is virtually absent on the right-hand side of the plot. *Festuca rubra* and *Plantago maritima* have distributions centred on the sector A^+B^+, and these quadrats are typical of the sea-cliff grasslands found in more sheltered sites on cliff tops and slightly inland. *Calluna vulgaris*, however, occurs predominantly in sector A^+B and is representative of several dominant heath species, again found in more sheltered sites on the cliff top or inland.

Plots of environmental data superimposed on the quadrat ordination are

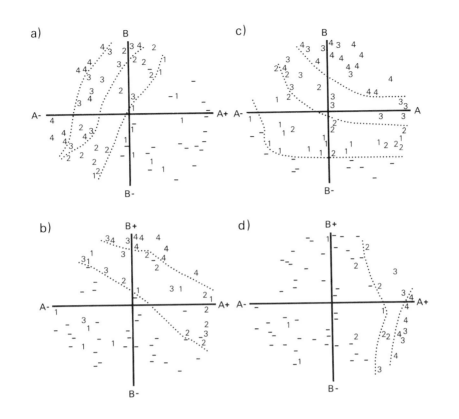

Figure 5.25 Plots of individual species on the quadrat ordination diagram for sea cliff vegetation on Anglesey, Wales. Values are species frequencies in quartiles (1 = low; 4 = high). (a) *Armeria maritima*, (b) *Plantago maritima*, (c) *Festuca rubra*, (d) *Calluna vulgaris* (Goldsmith, 1973a; redrawn with kind permission of *Journal of Ecology*)

shown in Figure 5.26. The plots for conductivity (a) and soil moisture (c) fit in well with the observations on species richness and individual species distributions, with quadrats exposed on high salt-spray concentrations on the left and low-salinity concentrations on the right, where the more sheltered inland grassland and heathland species occur. The aspect plot (b) is more complicated but indicates a visual correlation of quadrats exposed to the prevailing south-west winds with high salinity. Soil moisture shows a reverse trend to salinity/conductivity and is also related to the overall amount of plant and soil cover as well as evaporation and exposure. Last, a plot for bird influence (d) is presented where no definite trend is observable, although it is known that sea birds contribute large amounts of phosphates and nitrates to soils through their droppings and may damage vegetation by trampling and collection of nesting material.

Ordination of environmental factors

The same 65 quadrats were then ordinated using the 18 environmental

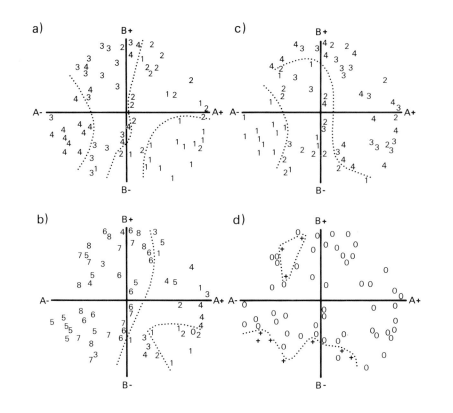

Figure 5.26   Plots of environmental data superimposed on the quadrat ordination diagram for sea cliff vegetation on Anglesey, Wales: (a) soil conductivity/salinity as quartiles (1 = low; 4 = high), (b) aspect (SSW = 9; SW,S = 8; WSW,SSE = 7; W,SE = 6; WNW,ESE = 5; NW,E = 4; NNW,ENE = 3; N,NE = 2; NNE = 1; flat = 0), (c) soil moisture as quartiles (1 = low; 4 = high), (d) bird influence (Goldsmith, 1973a; redrawn with kind permission of *Journal of Ecology*)

variables of Table 5.6. The resulting quadrat ordination diagram for the first two components is presented in Figure 5.27a. Note that when compared to the previous figures, axes A and B are plotted differently. Again, plotting of selected environmental factors shows the first axis to be characterised by salinity (Figure 5.27c) and the second by the distribution and amount of organic matter (Figure 5.27b). Examination of Figures 5.24a and 5.27a shows some similarity of distribution of the quadrats with respect to each other. However, there are also clear differences. One of the reasons for this is that species distribution and response is not solely determined by environmental controls. Other factors, such as competition and plant species strategies, have to be considered as well.

Hypothesis generation and testing

On the basis of this evidence, various hypotheses were erected concerning

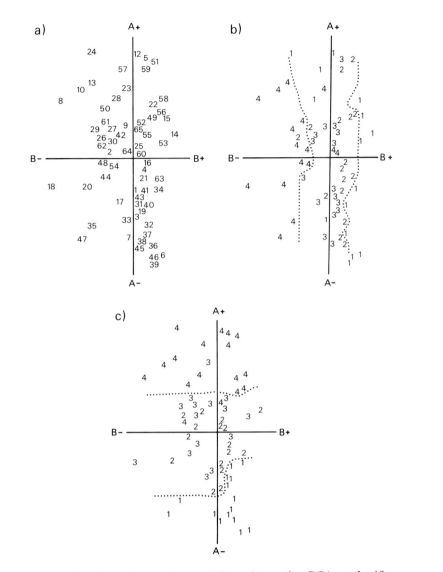

Figure 5.27 Ordination plot of the sea cliff quadrats using PCA on the 18 environmental variables of Table 5.6: (a) quadrat plot, (b) distribution of humus determined by loss on ignition as quartiles, (c) soil conductivity/salinity as quartiles (Goldsmith, 1973a; redrawn with kind permission of *Journal of Ecology*)

plant species distributions on the sea cliffs of Anglesey in relation to environmental controls. Low species cover and numbers and the presence of maritime species such as *Armeria maritima* (Sea pink) were thought to be correlated with high salinity as reflected in the conductivity, exposure and soil moisture data. However, salinity is highest when soils dry out, and the absence of other species was hypothesised as being the result of high total salts plus the dessication factor. The effect of salinity on *Armeria* was thus an indirect one, in that *Armeria* survives not only because of its tolerance of high salinity, but also because other species, which otherwise would compete with it, are precluded by the severe environmental conditions. Thus it is principally the absence of competition which allows *Armeria* to flourish.

These ideas were put forward as hypotheses which were then tested in more detailed experiments. First, the pattern of deposition of salt spray, which determines the salinity levels in the soil, was studied, with the hypothesis that spray deposition would be a function of distance from the sea. Plastic beakers of non-saline soil were placed at carefully chosen locations around the cliffs and left for periods of up to six months. Conductivity analysis of the samples after this time showed that variations in the amount of salt deposited were correlated principally with height and distance from the sea and second with local topographic features which provided increased shelter or exposure. No direct variation in response to a transect up a cliff and away from the sea was observed, however. Instead, variations in aspect leading to differences in evaporation were found to be of significance. This was further reflected in temporal variation between summer and winter, where southerly aspects experienced much higher evaporation rates in summer.

A second experiment investigated the hypothesis that species such as *Armeria* were found in exposed saline sites, not because the salinity was favourable to them but rather that the salinity prevented the establishment of competitive species found further inland. The maritime species *Armeria* and the inland cliff species *Festuca rubra* (Red fescue) were grown separately in pots and half were watered with fresh and half with salt water. Also they were grown together in a replacement series experiment (de Wit, 1960) which was also subject to a fresh and saline watering regime. In such a series, two species are grown together along a gradient of inverse proportion to each other with one species gradually replacing the other, while overall density of plants remains the same. The results showed that when grown alone under either treatment, there was no difference between yields of the two species. However, in the replacement series, *Armeria* had the competitive advantage in the sea-watered pots, whereas *Festuca* had the advantage in the fresh-watered pots. It was concluded that maritime species do not **require** saline conditions but are salt **tolerant** and flourish on exposed cliffs largely because of the absence of other species which would normally out-compete them. This lack of competitive ability is due to their slow growth rate and semi-prostrate form. On exposed sea cliffs, competition from inland species is limited by high soil salinity and damage to leaves by salt spray during gales.

This research demonstrates the use of quadrat ordinations to display data on environmental limiting factors and individual species distributions. More importantly, it also shows the value of ordination as a means of hypothesis generation, with the hypotheses being tested by more detailed experimentation afterwards. A major criticism of the application of ordination methods in plant ecology over the past 20 years is that very few researchers have gone beyond the stage of description of vegetation variation and environmental gradients to generation of hypotheses which they have subsequently tested (Kent and Ballard, 1988). Finally, Goldsmith's work demonstrates the care which is necessary in the interpretation of causal relationships between species and environmental factors. In the case of *Armeria* the simple deterministic explanation that it is found by the sea because of its ability to tolerate salt spray has been shown to be over simplistic, and a more detailed understanding of competitive interaction between *Armeria* and other species is required.

6

Ordination methods II, 1970–92

Developments in ordination techniques since 1970 have principally centred on **correspondence analysis (CA)** also known as **reciprocal averaging (RA)**.

Correspondence analysis and reciprocal averaging

Correspondence analysis (CA) or reciprocal averaging (RA) is related to the method of weighted averaging devised by Whittaker (1967). The technique was first proposed by statisticians as long ago as 1935 (Hirschfeld, 1935; Fisher, 1940) and was initially applied to problems of attitude scaling in the social sciences. In ecology, it was first used by Roux and Roux (1967), Benzécri (1969; 1973), Hatheway (1971) and Guinochet (1973). Hill (1973b; 1974) and Gauch *et al.* (1977) demonstrated its potential application as an ordination method in plant ecology. The method is now widely applied across all the sciences and has been refined and improved (Whittaker, 1978a; Greenacre, 1981; 1984; Digby and Kempton, 1987; Jongman *et al.*, 1987; Legendre and Legendre, 1987; Ludwig and Reynolds, 1988).

As a method, correspondence analysis is very important because it provides the basis for most developments in ordination techniques since 1970. It is also at the heart of two-way indicator species analysis (TWINSPAN), which is probably the most widely used method for numerical classification of vegetation data at the present time (Chapter 8).

One of the attractions of correspondence analysis is that in the handworked version for one axis, it is relatively simple to explain and understand. Another is that the calculation of the quadrat ordination is related to the calculation of the species ordination. The method of weighted averages (p. 169) is applied to a data matrix so that quadrat scores are derived from species scores and weightings and conversely, species scores can similarly be derived from quadrat scores and weightings. These are carried out successively using an iterative procedure and stabilise out to give a set of scores for species which gives axes for a species ordination and a set of scores for quadrats which gives axes for a quadrat ordination. The starting point is to give weights to the species in a matrix or an arbitrary scale from 0 to 100 based on their assumed position along a primary environmental gradient. Equally, weights could be given to each quadrat depending on the level of an important environmental factor. Whether these weights are correct or not is immaterial, since **any** values can be given across the range of the scale 0–100. However, the nearer the values are to the final solution, the less computation is required.

The method is described below in its simplest form as an iterative two-way averaging process; hence the name 'reciprocal averaging' given to it by Hill (1973b). For this reason, this is the name used in the description below. It can also be computed using matrix algebra and eigenanalysis and as such is related closely to the method of principal component analysis (PCA) described in Chapter 5. When computed in this manner it is more usual to refer to the method as correspondence analysis.

Table 6.1 Calculations for the first axis of reciprocal averaging ordination of the New Jersey salt marsh data (Table 3.2)

	Quadrats												Quadrat totals (Q)
	1	2	3	4	5	6	7	8	9	10	11	12	
Species													
1. *Atriplex patula*	1	1	1	1	1	1	1	0	1	0	1	1	10
2. *Distichlis spicata*	0	1	1	1	1	1	1	1	1	1	1	0	10
3. *Iva frutescens*	0	0	0	0	0	0	1	1	1	1	1	1	6
4. *Juncus gerardii*	0	0	1	0	0	1	1	0	0	0	0	0	3
5. *Phragmites communis*	0	0	0	0	0	0	0	1	1	1	1	1	5
6. *Salicornia europaea*	1	1	1	1	1	0	1	0	0	1	0	0	7
7. *Salicornia virginica*	0	0	0	1	1	0	0	0	0	0	0	0	2
8. *Scirpus olneyi*	0	0	0	0	0	1	1	0	0	0	1	0	3
9. *Solidago sempervirens*	0	0	0	0	0	0	0	0	1	1	1	1	4
10. *Spartina alterniflora*	1	1	1	1	1	1	0	1	1	1	0	0	9
11. *Spartina patens*	0	0	0	0	0	0	1	1	1	0	1	1	5
12. *Suaeda maritima*	0	0	0	1	1	0	0	0	0	0	0	0	2
Species totals (S)	3	4	5	6	6	5	7	5	7	6	7	5	
Q1	42.42	34.09	32.73	48.49	48.49	36.36	36.36	47.27	44.16	43.94	41.56	43.64	
Q2	45.75	45.92	36.74	63.95	63.95	32.48	38.26	55.67	54.60	54.52	50.42	56.89	
Q3	49.41	49.34	39.48	66.23	66.23	32.73	38.98	55.64	55.67	56.43	50.58	57.74	
Q14	53.98	51.08	50.02	67.39	67.39	41.92	32.66	20.65	21.29	27.37	17.11	10.85	
Q15	54.22	51.31	50.32	67.54	67.54	42.27	32.94	20.77	21.42	27.50	17.27	10.93	
Q20	54.54	51.62	50.74	67.75	67.75	42.75	33.33	20.95	21.59	27.67	17.49	11.05	
Q20(Sc)	76.70	71.55	70.00	100.00	100.00	55.34	39.29	17.46	18.59	29.31	11.36	0.00	

(Sc); Axis rescaled from 0–100 at the end of each iteration

The method

The method is best explained in relation to an example. For this purpose, the New Jersey salt marsh data coded in presence/absence form are taken (Table 3.2).

(1) First, the row and column totals of the matrix are calculated. With presence/absence data this gives column Q (Table 6.1), which is the number of quadrats in which each species occurs and row S, the number of species in each quadrat.

Derivation of the first axis

(2) The next stage is to allocate weights to the 12 species. Here it is assumed there is no information on species tolerances in relation to a primary gradient. Thus for the 12 species, weights are allocated evenly across the range 0–100 (Table 6.1, column W). Thus species 1 (*Atriplex patula var. hastata*) is given a score of 0, species 2 (*Distichlis spicata*) 9.09, species 3 (*Iva frutescens*) 18.18, species

W	S1	S1 (Sc)	S2	S2 (Sc)	S3	S3 .... S15	S15 (Sc)	S16	S16 .... S21	S21 (Sc)
0.00	40.83	42.58	48.90	46.47	50.64	46.57.... 41.37	46.05	41.57	46.20.... 41.86	46.40
9.09	41.35	46.44	49.65	49.16	51.13	48.22.... 39.69	42.57	39.89	42.71.... 40.16	42.89
18.18	42.86	57.52	51.73	56.54	52.50	52.94.... 21.65	5.18	21.80	5.23.... 22.01	5.30
27.27	35.15	0.00	35.83	0.00	37.06	0.00.... 41.53	46.40	41.84	46.76.... 42.28	47.26
36.36	44.12	67.21	54.42	66.12	55.21	62.22.... 19.45	0.62	19.56	0.62.... 19.75	0.62
45.45	40.93	43.34	49.87	49.93	52.30	52.24.... 49.99	63.92	50.19	64.06.... 50.49	64.26
54.55	48.49	100.0	63.95	100.00	66.23	100.00.... 67.39	100.00	67.54	100.00.... 67.75	100.00
63.64	38.10	22.08	40.39	16.22	40.76	12.69.... 30.56	23.66	30.82	23.93.... 31.19	24.31
72.73	43.32	61.28	54.11	65.01	55.10	61.84.... 19.15	0.00	19.27	0.00.... 19.45	0.00
81.82	42.00	51.32	50.40	51.81	52.35	52.41.... 44.57	52.69	44.76	52.81.... 45.03	52.98
90.91	42.60	55.84	51.17	54.55	51.72	50.25.... 20.51	2.81	20.66	2.88.... 20.88	2.96
100.00	48.49	100.00	63.95	100.00	66.23	100.00.... 67.39	100.00	67.54	100.00.... 67.75	100.00

4 (*Juncus gerardii*) 27.27 and so on up to species 12 (*Suaeda maritima*) with a score of 100.0.

(3) The process of reciprocal averaging then commences. The procedure is iterative in that the averaging process is repeated numerous times for quadrats and then species. On the first iteration, the aim is to derive row Q1 which is the first set of **quadrat scores**. The value for the first quadrat (42.42) is calculated as follows. Taking the species scores for the first quadrat, each is taken in turn and multiplied by its weighting in column W. Those values are then summed and averaged. Thus:

$$(1 \times 0.00) + (0 \times 9.09) + (0 \times 18.18) + (0 \times 27.27) + (0 \times 36.36)$$
$$+ (1 \times 45.45) + (0 \times 54.55) + (0 \times 63.64) + (0 \times 72.73) + (1 \times 81.82)$$
$$+ (0 \times 90.91) + (0 \times 100.0) = 127.27$$

This value is then divided by the species total for quadrat 1 (3) = 127.27/3 = 42.42. The remaining values in row Q1 of Table 6.1 are calculated in exactly the same way.

(4) The averaging process is then applied in reverse to give a **new set of scores for the species** using the values just calculated for row Q1. In Table 6.1, the revised value for *Atriplex patula* is calculated by multiplying the species score by the new quadrat scores, summing these results and then averaging them. Thus for *Atriplex patula (var. hastata)*:

$$(1 \times 42.42) + (1 \times 34.09) + (1 \times 32.73) + (1 \times 48.49) + (1 \times 48.49)$$
$$+ (1 \times 36.36) + (1 \times 36.36) + (0 \times 47.27) + (1 \times 44.16) + (0 \times 43.94)$$
$$+ (1 \times 41.56) + (1 \times 43.64) = 408.30.$$

This score is then divided by the quadrat total for species 1 (10) = 408.3 = 40.83.

Scores for the other species in column S1 are calculated in the same manner.

(5) For convenience in computation and to prevent working with very small numbers, the species scores are then rescaled to the range 0–100. The highest value in column S1, 48.49 for *Suaeda maritima*, is rescaled to 100, while the lowest, 35.15 for *Juncus gerardii*, is given a value of 0.0. The range of values (48.49 − 35.15 = 13.34) is made equal to 100, and the remaining scores are rescaled accordingly:

$$\text{S1(Sc)} = \frac{(\text{Species value} - \text{lowest species value})}{\text{Range of species values}} \times 100.0$$

where S1(Sc) = rescaled species value

Thus for *Atriplex patula (var. hastata)*:

$$\text{S1(Sc)} = \frac{(40.83 - 35.15)}{13.34} \times 100.0 = 42.58$$

The other values are similarly rescaled to give column S1(Sc) of Table 6.1.

(6) The process of reciprocal averaging from stages 3 to 5 above is then repeated to give row Q2 for the second estimate of quadrat scores and column 2 for the second estimate of species scores, which are rescaled from 0 to 100 in column S2(Sc) (Table 6.1).

(7) Reciprocal averaging is then repeated until the amount of change which occurs in the species and quadrat scores is minimal. By the 15th iteration the scores are relatively stable (Table 6.1). The first species *Atriplex patula* has a scaled score of 46.05 in column S15(Sc). By the 21st iteration this is only changed to 46.40 in column S21(Sc). The decision as to when to stop the iterations depends on the accuracy required and whether or not the analysis is being done by computer. Clearly the process of averaging is extremely tedious, and the only reason for carrying out a small analysis by hand would be to demonstrate the workings of the method. In computer programs, the average change in species scores between each iteration is calculated and when it falls below a critical value, iteration of the averages is terminated. Obviously, many more iterations can be calculated by computer.

(8) An estimate of the eigenvalue for the first axis is obtained by taking the range of the unscaled scores on the final iteration (column S21) (Max = 67.75; Min = 19.45; Range = 48.3) and expressing this as a proportion of the range of the scaled values for the previous iteration (S20) which obviously had a range of 100. This estimate is thus 0.48 and is an approximation to the eigenvalue

Table 6.2 Species and quadrat scores for the second axis of reciprocal averaging ordination of the New Jersey salt marsh data

	Species			Quadrats
1.	*Atriplex patula*	46.85	1.	43.31
2.	*Distichlis spicata*	44.70	2.	43.65
3.	*Iva frutescens*	32.52	3.	54.92
4.	*Juncus gerardii*	100.00	4.	29.11
5.	*Phragmites communis*	20.84	5.	29.11
6.	*Salicornia europaea*	42.52	6.	63.12
7.	*Salicornia virginica*	0.00	7.	55.13
8.	*Scirpus olneyi*	83.49	8.	34.89
9.	*Soligado sempervirens*	21.00	9.	34.62
10.	*Spartina alterniflora*	40.56	10.	33.69
11.	*Spartina patens*	35.82	11.	40.75
12.	*Suaeda maritima*	0.00	12.	31.41

for the first axis of the ordination. It may be seen as a measure of the proportion of the total variation in the data explained by the axis in the same way as for principal component analysis.

(9) The scores in column S21(Sc) (Table 6.1) represent the first axis of the species ordination, while the rescaled values of row Q20, row Q20(Sc) represent the first axis of the quadrat ordination. These are plotted as single axis ordinations in Figure 6.1. and may be interpreted in the standard way described in Chapter 5.

Derivation of the second axis

(10) Hill (1973b) in an Appendix to his paper gives a simple example of how a second axis can be extracted. The same reciprocal averaging process is repeated but with a new set of initial scores. He recommends a set of starting values for species weights near to the end of the iterations for the first axis, for example column S16(Sc) (Table 6.1). Also, at the end of each iteration, when the new set of species scores has been calculated, a multiple of the first axis has to be subtracted from the species scores. If this correction is not made, the first axis will slowly re-establish itself.

In the above example, the second axis scores for the 12 species and the 12 quadrats, calculated using the computer program of Kent (1977) and after 18 iterations, are shown in Table 6.2. The eigenvalue estimate for this second axis is 0.29.

(11) A third axis and subsequent axes may be extracted in a similar manner.

(12) The two axis species and quadrat ordination plots are presented in Figure 6.1. These plots are interpreted in the usual manner as described in Chapter 5.

Most computer programs for CA use matrix algebra to provide a more efficient and accurate solution, and the matrix algebra approach is described in the Appendix to Hill's (1973b) paper as well as in Jongman *et al.* (1987) and Ludwig and Reynolds (1988). The analysis is very similar to principal components analysis (PCA) and produces eigenvalues and eigenvectors. As with PCA, the eigenvalues decrease in size and importance with successive axes but the cumulative sum of the eigenvalues is not the same as the sum of the variances of either species or quadrats.

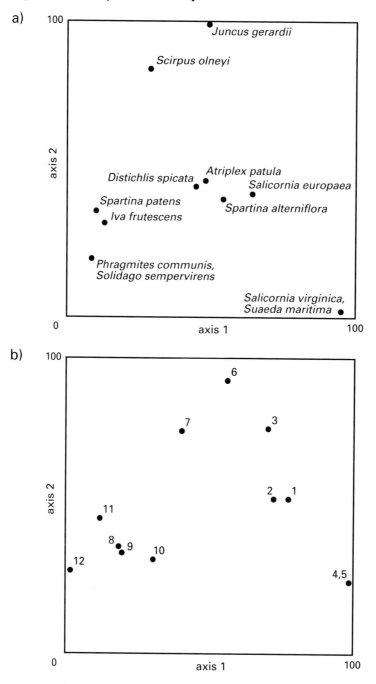

Figure 6.1 Two axis ordination plots produced by reciprocal averaging of the New Jersey salt marsh data (Table 3.2). (a) the species ordination, (b) the quadrat ordination

Hence the eigenvalues in CA are less important than they are in PCA. The individual elements of the eigenvectors represent the axis scores that are used for the construction of ordination plots. Hill (1973b) points out that CA/RA is very similar to non-centred PCA standardised by species.

As with PCA, CA/RA can be described in geometric terms. Like PCA, CA/RA aims to find new axes that summarise the variation within a multidimensional cloud of points representing distances between quadrats or species, reducing the dimensionality in the process. RA/CA, however, uses chi-square distances, an origin at the centroid (the centre of gravity of the swarm of points making up the cloud) and has both quadrat and species weights proportional to the quadrat or species totals (double transformation by totals). Thus the methods are similar but differ in the way that the points are projected on to the axes.

The great advantage of CA/RA is the simultaneous analysis of species and quadrats and the relationship between the species ordination plot and the quadrat ordination plot. Plotting the distribution of scores of individual species in each quadrat (Table 3.2) on the quadrat ordination plot (Figure 6.1b) and comparison of the resulting quadrat plot with the position of that species on the species ordination plot (Figure 6.1a) will show this relationship. The position of the species on the species ordination plot will be at the centre of gravity or the 'average' position of the distribution of the same species on the quadrat ordination plot.

Gauch (1982b) states that CA/RA is usually best for analysing long community/environmental gradients and since a majority of analyses are of this type, it was an optimal ordination method. However, for relatively homogeneous data sets with short gradients, he accepts that other methods such as PCA may be better.

CA/RA has been introduced here using presence/absence data. It is equally applicable to quantitative data with the species or quadrat weightings being multiplied by the abundance data rather than the 1/0 of the qualitative data.

Problems with correspondence analysis and reciprocal averaging

Although Hill (1973b) and Gauch *et al.* (1977) claimed that CA/RA was generally superior to all other ordination techniques which existed at that time, particularly for data collected across long environmental gradients, two major problems soon emerged in its application:

The 'arch effect'

This is well demonstrated in the ordination plot of the New Jersey salt marsh data (Figure 6.1) and also in the case study of the Narrator Catchment, Dartmoor described at the end of this chapter (Figure 6.11). The arch is produced when the first two axes are plotted as in Figure 6.1 and reflects the fact that the second axis may be simply a quadratic distortion of the first axis. This problem continues into higher dimensions — the third axis may have a cubic distortion, the fourth a quartic and so on. However, because most ordination plots use only the first two axes, it is the quadratic distortion of the second axis that causes most difficulty. As a result, important secondary gradients in the data may not emerge until higher axes, because the eigenvalue of the quadratic distortion may be greater than that of the actual secondary gradient, particularly when the secondary gradient is less than about half the extent of the primary gradient (Gauch, 1982b). Interpretation thus becomes problematic, and it is difficult to predict which axes carry ecologically meaningful information. Where a clear arch occurs, it can be beneficial to interpret

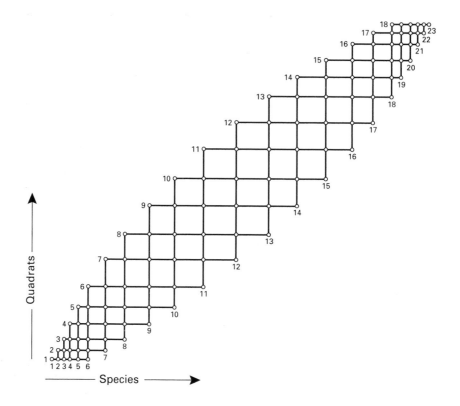

Figure 6.2 The arrangement of an artificial and regular data structure containing 18 quadrats in rows and 23 species in columns spaced according to the first axis of a reciprocal averaging ordination. The presence of a species in a sample is indicated by a dot. In theory, these quadrats and species should be spaced evenly, but the effect of reciprocal averaging is to compress the ends of the axes relative to the middle (Hill and Gauch, 1980; reproduced with kind permission of *Vegetatio* and Kluwer Academic Publishers)

the ordination by examining the distribution of points along the length of the arch. This will usually clarify the primary gradient of the first axis. However, the second axis should then be ignored, and it may then be useful to examine the third and fourth axes for meaningful environmental relationships.

The axis compression effect

This second problem is demonstrated in Figures 6.2, 6.3 and 6.11. Figure 6.2 shows the results of a one axis CA/RA ordination of a set of artificial data made up of 18 quadrats and 23 species (Hill and Gauch, 1980). Both the quadrat and species single axis ordinations are plotted together and in theory, because of the properties of the artificial data, the points in the grid should all be equally spaced in relation to both the quadrat and the species axes. In practice, this does not occur and points nearer to the ends of the axes are compressed, while those nearer the centre are more spread out. This is related to the arch effect, and the combined arch and

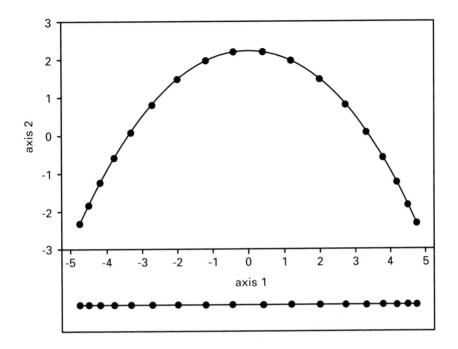

Figure 6.3 The two major faults of reciprocal averaging: (a) the arch effect and distortion of the second axis and (b) the compression of the ends of the first axis relative to the middle. These are the same data as for the quadrat ordination in Figure 6.2, with 18 equally spaced points in a straight line (Hill and Gauch, 1980; reproduced with kind permission of *Vegetatio* and Kluwer Academic Publishers)

compression effects for the quadrat ordination of the artificial data are shown in Figure 6.3.

Detrended correspondence analysis

Detrended correspondence analysis (DCA) and the associated FORTRAN computer program DECORANA was devised by Hill (1979a) and Hill and Gauch (1980) in order to attempt to solve the two problems of CA/RA: the 'arch effect' and compression of points at the ends of the first axis (Figures 6.2 and 6.3). The 'arch effect' occurs because, although the second and higher axes of CA/RA are uncorrelated, they are not independent of each other. The first two axes are uncorrelated, or orthogonal, because the positive correlation on one side of the arch is matched by the negative correlation of the other side of the arch as in Figure 6.3. The arch, however, is a result of the quadratic relationship between the first and second axes and is thus rarely a reflection of the ecological content of the data but rather is a consequence of the mathematics behind the method.

Removal of the 'arch effect' by DCA involves what is known as **detrending**. The first axis is divided into a number of segments and within each segment, the second axis scores are recalculated so that they have an average of zero, as in Figure 6.4. When this is done for all segments, it means that all second axis scores are expressed as deviations from a mean of zero. In the computer program

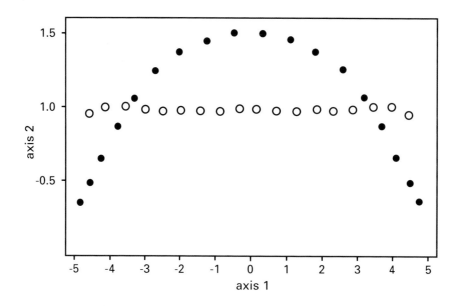

Figure 6.4 A simplified representation of the method for detrending used in detrended correspondence analysis. For explanation, see text. Quadrat scores before detrending are shown as ●, after detrending as ○. (Redrawn from Hill and Gauch, 1980; with kind permission of *Vegetatio* and Kluwer Academic Publishers)

DECORANA, the first axis is divided into many segments and the averaging is achieved through a sophisticated running-averages procedure. Figure 6.4 shows how this reduces the arch effect and straightens out the arch trend in relation to the first and second axes.

Detrending is applied to the quadrat scores at the end of each iteration of the correspondence analysis, except that once convergence is achieved, the final quadrat scores are derived from weighted averages of the species scores without detrending. For the third axis, quadrat scores are detrended in relation to both first and second axes and the pattern is similar for higher axes.

The second problem of CA/RA is the compression of points at the ends of the first axis relative to the middle (Figure 6.3). This difficulty is overcome in principle again by segmenting the axis and expanding those segments at the end and contracting those in the middle, so that the species turnover (the arrival and departure of species along the first axis gradient) occurs at as uniform a rate as possible. Hence equal distances on the quadrat ordination axes correspond to equal differences in species composition. The rescaling of axis segments is achieved by expanding or contracting small segments of the **species ordination**, while trying to equalize the average within-quadrat dispersion of the species scores at all points along the quadrat ordination axis (Figure 6.5). Thus the **species ordinations** are adjusted so that the quadrat scores are the weighted mean values of the scores of the species that occur within them (Hill and Gauch, 1980; Gauch, 1982b).

Another important feature of DCA is that the axes are scaled in units of the **average standard deviation of species turnover (SD)** (Gauch, 1982b). Along a

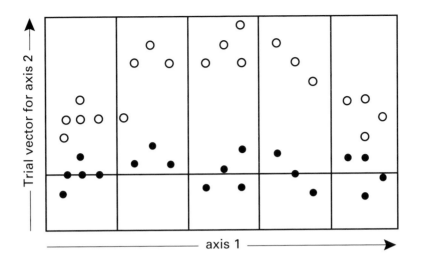

Figure 6.5 Within-quadrat standard deviation of species scores in relation to position along the first (quadrat) axis in detrended correspondence analysis (●), compared with reciprocal averaging/correspondence analysis (○) (Redrawn from Hill and Gauch, 1980; with kind permission of *Vegetatio* and Kluwer Academic Publishers)

gradient, a species appears, rises to its mode and then disappears over a distance of almost 4SD and similarly a complete turnover in the species composition of samples occurs in about 4SD. A change of 50 per cent in the composition of a quadrat (called a half-change) occurs in about 1 SD or slightly more. DCA scales axes in these SD units. Thus in DCA, axes can be of variable length and unlike CA/RA are not scaled into an arbitrary range of 0–100 according to the size of the eigenvalues.

Gauch (1982b) concluded that DCA gives 'results at least as good as and usually superior to other ordination techniques' (p. 159). However, Hill and Gauch (1980) did acknowledge that there were still problems, even in DCA ordination, although these have tended to be forgotten, given the enthusiasm with which DCA was received during the 1980s. Outliers (individual quadrats or species that are well separated from the rest of the points on an ordination diagram and hence are very different in species composition or distribution) and discontinuities (gaps in species or quadrat distribution along axes) both present problems. Outliers are best dealt with by removal from the data set and the computer program DECORANA has an option for this. Large discontinuities mean that the width of gaps in the gradient have to be estimated and this can lead to inaccuracies.

In the 1980s, with the widespread availability of the computer program DECORANA, DCA achieved considerable prominence and became widely used (Kent and Ballard, 1988). Its application has not been without criticism.

Dargie (1986) noted that a certain amount of distortion in DCA ordination is attributable to variations in species richness (alpha diversity) along gradients and axes in addition to the arch and compression effects. DCA does not necessarily correct for this. He concluded that the literature has emphasised beta diversity (between-habitat diversity) as a significant source of distortion, but species richness is probably also important because it varies systematically along most successional and environmental gradients.

Wartenberg *et al.* (1987) comment on the manner in which DCA has been adopted rather uncritically by plant ecologists and they argue that there is 'no empirical justification for the method, since the DCA model is not consistent with the structure of the data . . . and . . . there is no theoretical justification for the method since DCA is, as Hill and Gauch (1980) point out, an *ad hoc* adjustment of CA/RA' (p. 438). In relation to the 'arch effect', the position of points on the first axis is not changed, and all DCA does is to flatten the arch on the original plot. They continue, 'this deception does not enhance our understanding of the data or help to identify the cause of the observed distortion' (p. 435). Concerning the end of the axis-compression problem, they also question the assumption that rates of species turnover are constant or even along gradients and axes and that all species can be treated equally. Also Wartenberg *et al.* (1987) and Minchin (1987a) point out that the detrending and flattening of the arch may result in loss of ecological information if some of the arch form represents a real pattern in the original data.

Peet *et al.* (1988) replied that although they agreed with many of the basic criticisms of Wartenberg *et al.* (1987), it would be a mistake to reject DCA as an ordination method for those reasons. Despite its limitations, DCA still remains one of the most powerful methods for indirect ordination and is computationally very efficient.

The method of detrending has also been criticised and refined (Greenacre, 1984; Oksanen, 1988; ter Braak, 1988a,b and Knox, 1989). The original detrending using segments and subtracting of a moving average has been shown to be unstable in some situations and with certain types of data. Ter Braak (1988a,b) has provided an alternative approach called **polynomial detrending**, in which axis scores are replaced by residuals from a multiple regression on polynomial functions (up to a specified degree) of the axes already obtained. This provides an alternative for the reduction of the arch effect but does not assist with rescaling to reduce end of axis compression. Knox (1989), however, has questioned whether polynomial detrending represents any real improvement and Jackson and Somers (1991) have shown that the number of segments selected for the detrending procedure gives varying results on second and higher axes. They state (p. 711) 'Although detrended solutions may be interpretative (an attribute that encourages the use of DCA), the arrived at solution may be only one of many possible results . . . we caution against the acceptance of DCA as an ecological panacea.'

In summary, DCA still remains as a widely used and effective indirect ordination technique. Despite the various problems, the method is probably still as good as any other in most situations and better than most in many. The key point in these complex discussions is that interpretation of results from DCA is best carried out with some knowledge of its limitations and in comparison with other ordinations of the same data (see below).

For comparative purposes, the plots for the first two axes of the DCA of the New Jersey salt marsh data are presented in Figure 6.6. When compared to the original CA/RA plots (Figure 6.1), the effects of detrending and rescaling of the first axis to avoid compression are clearly seen.

Non-metric multidimensional scaling

Non-metric multidimensional scaling (NMDS) was first used as an ordination method in plant ecology by Anderson (1971), and developed further by Austin

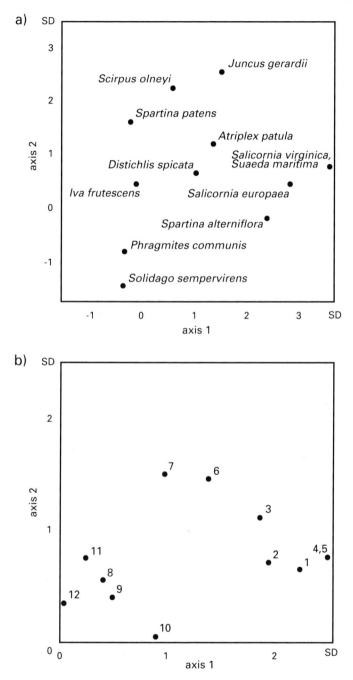

Figure 6.6 Ordination plots for the first two axes of detrended correspondence analysis of the New Jersey salt marsh data (Table 3.2)

(1976), Fasham (1977), Prentice (1977; 1980) and Kenkel and Orlóci (1986). The method and programs have been reviewed more recently by Carroll (1987). NMDS is really a set of related techniques that use the rank order information in a matrix of dissimilarities between species or quadrats. The earliest and most frequently used method is that of Shepard (1962) and Kruskal (1964a,b) known just as **multidimensional scaling** (MDS). In MDS, quadrats or species are positioned within a few dimensions or axes, so that the distances between the points representing quadrats or species on the ordination diagram have the same rank order as the interpoint dissimilarities in the dissimilarity matrix calculated between all pairs of quadrats or all pairs of species.

In most NMDS analyses, the number of ordination axes must be specified at the outset and depending on whether a quadrat or a species ordination is being computed, an ordination produced by another method must be supplied as input, along with a matrix of dissimilarity coefficients between the species or the quadrats. NMDS then modifies this ordination, so that the rank order of the interpoint distances between the quadrats or the species are as close as possible to the rank order of the equivalent values between quadrats or species in the dissimilarity matrix. The measure of how good a fit or match occurs between the two is called a **stress function** and can be expressed as a single value $\geqslant 0$. If there is a perfect match then stress = 0. The aim of the method is to reduce stress as much as possible. The stress function is normally drawn as a diagram, known as a **Shepard diagram**, with the dissimilarity values in the floristic data plotted in rank order against the distances on the ordination plot. If there is a good match and stress is low, the points will lie on a steadily increasing curve. The more the deviations from this smooth curve, the higher the stress.

The other approach of Shepard and Carroll (1966) is known as **continuity analysis** and it has the aim of defining a low-dimensional ordination within which species response curves are as smooth or as continuous as possible.

Various authors have tested MDS and NMDS and their variants against other ordination methods such as polar ordination, principal components analysis, reciprocal averaging and detrended correspondence analysis (Austin, 1976; Fasham, 1977; Prentice, 1977; 1980; Oksanen, 1983; Kenkel and Orlóci (1986) and Minchin, 1987a). Some (for example Oksanen and Minchin) have claimed superior results to other methods, while others have found little advantage in using NMDS (Gauch *et al.*, 1981). This seems to imply that in certain cases, NMDS will give better results, but this is not universal. NMDS is also good at recovering gradients of high beta (between habitat) diversity.

A major problem with NMDS is that the computational procedures are very complex. Computation time increases with the square, cube or greater powers of the number of samples or species and, even with modern computers, analysis of large data sets is time-consuming. Also in all methods, quadrat and species ordinations must be analysed separately. The calculation process is iterative and convergence to the best solution does not always occur, with several different solutions being possible for some data sets, depending on the initial ordination input. Results have also been shown to be prone to the same 'horseshoe' or 'arch' effect as PCA and RA/CA. Finally, calculated dissimilarity between species or quadrats may vary along a gradient, particularly if there is a trend in species richness. This problem can be overcome by constructing separate Shepard diagrams for each quadrat or species in turn, where all the dissimilarities and ordination plot distances are plotted for each quadrat or species separately against all other quadrats or species. This is known as 'local' stress, and a 'local' stress value can be calculated in each case. If all the individual stress values are summed, then this gives an indication of 'global' stress.

Despite these various problems, the method has its proponents. A good recent example of its application to the analysis of Mediterranean vegetation is provided in Tong (1989).

Canonical correspondence analysis

The most recent development in ordination techniques is canonical correspondence analysis (CCA) developed by ter Braak (1985; 1986a; 1987a,b; 1988a,b,c). The application of the technique has been greatly aided by the availability of the CANOCO computer program (ter Braak, 1988a,c). CCA is different from all the ordination methods so far discussed. Ultimately, ordination is an exercise in examining relationships between species distributions and the distribution of associated environmental factors and gradients. Of particular relevance in this is the application or methods of **correlation and regression** (ter Braak and Prentice, 1988). All the ordination methods discussed in this chapter, along with polar ordination and principal component analysis in the previous chapter, have this goal but they are all **indirect** in that the analysis is performed on the **species data alone** first, and then environmental interpretation is made by superimposing environmental data on the ordination plots and looking for patterns and correlations. Some analyses may go as far as correlation and regression of quadrat axis scores with environmental factors but for various reasons, this is not always satisfactory.

CCA differs from this classical indirect approach because it incorporates the correlation and regression between floristic data and environmental factors within the ordination analysis itself. Thus the input to CCA consists of not just a data matrix of species × quadrats but also a second data matrix of environmental factors × quadrats. Using multivariate analysis and particularly techniques of multiple regression (Chapter 4, case studies), together with various forms of correspondence analysis or reciprocal averaging (CA/RA), an integrated ordination of species together with associated environmental data is obtained. In view of this, CCA is best defined as a method of **direct** ordination with the resulting ordination being a product of the variability of the environmental data as well as the variability of the species data. It also follows from this that CCA may only be performed effectively if a good set of environmental data has been collected for the samples or quadrats in the analysis.

This approach of using both species and environmental data in the actual ordination process is known as a form of **canonical analysis**. The resulting ordination diagram thus expresses not only patterns of variation in floristic composition but also demonstrates the principal relationships between the species and each of the environmental variables.

The exact process by which CCA works is very complex and is explained in detail in ter Braak (1986a; 1987b). The method uses multiple regression to select the linear combination of environmental variables that explains most of the variation in the species scores on each axis. Using the iterative approach of correspondence analysis or reciprocal averaging (CA/RA), within each iteration or averaging cycle, a multiple regression is carried out between the quadrat ordination scores for an axis (dependent variable) and various combinations of the environmental variables (the independent variables). The calculated best-fit values for quadrats for the combination of environmental variables which gives the highest explained variance in the original axis scores are then taken as an improved estimate of those quadrat ordination axis scores. CCA iteration then continues with another multiple regression being performed to improve fit on the next iteration and so on. The scores eventually stabilise in the same manner as for CA/RA.

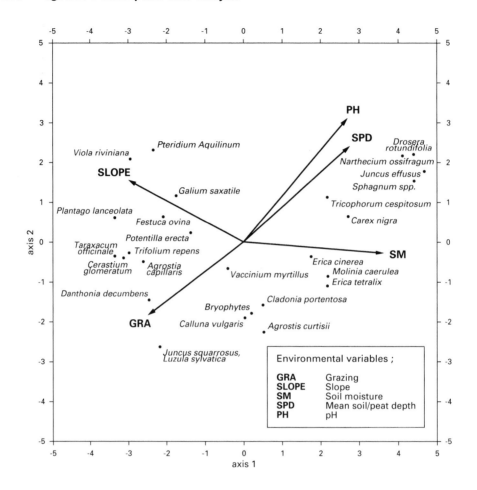

Figure 6.7 Species/environment biplot from canonical correspondence analysis (CANOCO) of the Dartmoor data (Table 3.4). Associated correlations, eigenvalues and percentage variances explained are given in Table 6.3. For explanation see text

CCA thus becomes a restricted correspondence analysis. Ter Braak and Prentice, 1988) call this a 'constrained ordination': the quadrat axis scores are constrained by the environmental variables. These restrictions are reduced as more environmental variables are included in the analysis, since more of the variation in the quadrat scores is likely to be accounted for. The 'arch effect' described for CA/RA can also occur in CCA (ter Braak, 1986a; 1987b), and the method of detrending used in detrended correspondence analysis (DCA) (Hill and Gauch, 1980) can be applied to remove it (giving DCCA). In the computer program CANOCO (ter Braak, 1988a,c), options are available for detrending by segments, as in the original DCA program DECORANA (Hill, 1979a) or by polynomials (ter Braak, 1988a,c). However, ter Braak (1987b) points out that the arch can be removed 'more elegantly' (p. 139) by dropping superfluous environmental variables. The environmental variables most likely to be superfluous are those most highly correlated with the arch axis (second axis). CCA with such variables removed should not need detrending.

An important part of the output from (D)CCA, in addition to standard plots for species and quadrat ordinations is the use of **biplots** of the species ordination diagram together with the environmental factors (Gabriel, 1971; ter Braak, 1986a; 1987b). The principles of biplots were introduced in Chapter 5. They greatly enhance the interpretation of environmental gradients and in particular allow individual species to be related to all major environmental factors.

As an example, the Gutter Tor data from Dartmoor, south-west England (Table 3.4) have been analysed by CCA using the CANOCO program. In addition to the species data, environmental data were available on soil/peat pH, soil moisture content, soil/peat depth and animal grazing intensity as measured by faecal units.

In Figure 6.7, the species-environment biplot is presented. The points represent individual species, and an arrow representing each environmental variable is plotted pointing in the direction of maximum change of the environmental variable across the diagram. The length of the arrow is proportional to the magnitude of change in that direction; and for interpretation purposes, each arrow can also be extended backwards through the central origin. Those environmental factors that have long arrows are more closely correlated in the ordination than those with short arrows and are much more important in influencing community variation. A point corresponding to an individual species can be related to each arrow representing an environmental factor by drawing a perpendicular from the line of the arrow up to the point representing the species. The order in which the points project on to the arrow from the tip of the arrow downwards through the origin is an indication of the position of the species in relation to the environmental factor. Species with their perpendicular projections near to or beyond the tip of the arrow will be strongly positively correlated with and influenced by the arrow. Those at the opposite end will be less strongly affected (ter Braak, 1987b).

The biplot for the Dartmoor data (Figure 6.7) shows the species and community distribution very clearly, with various groupings of species emerging. Species of the wet bogs and flushes (*Drosera rotundifolia*, *Narthecium ossifragum*, *Juncus effusus*, *Sphagnum spp.*, *Trichophorum cespitosum* and *Carex nigra*) are shown to the right of the plot. Heath species are found in the lower centre (*Calluna vulgaris*, *Erica tetralix*, *Vaccinium myrtillus*, *Cladonia portentosa*, *Agrostis curtisii*). Finally, the species of the improved acidic grasslands (*Agrostis capillaris*, *Festuca ovina*, *Trifolium repens*, *Danthonia decumbens*) are located towards the left of the diagram.

These groupings and the individual species are clearly shown in relation to the arrows representing environmental factors and gradients. The wet bogs and flush species are shown to have higher pH, deeper soil/peat depths and high soil moisture with low slope angles. Equally, the grazing axis is most strongly influencing the improved acidic pastures to the lower left. The effects of slope are again well illustrated with *Pteridium aquilinum*, *Galium saxatile* and *Viola riviniana*, all found in well-drained soils with steeper slopes, which is well known to correspond to their environmental preference.

Finally, the position of each environmental arrow with respect to each axis indicates how closely correlated the axis is with that factor. Thus soil moisture is most highly correlated with the first axis, and the CANOCO results confirm this correlation as 0.96 (Table 6.3). The other factors are correlated with both first and second axes, although the highest correlations are with the first axis in each case. The eigenvalue for the first axis of the species/environment biplot is 0.61 and the second 0.12, representing 45.1 per cent and 26.6 per cent of the total variance,

Table 6.3   Canonical correspondence analysis of the Dartmoor data (Table 3.4).
Correlations of species ordination axes with environmental factors, eigenvalues
and percentage variances explained

Factor	Axis I	Axis II
Soil/peat depth	0.74	0.41
Slope angle	− 0.80	0.26
pH	0.69	0.51
Soil moisture	0.96	− 0.05
Grazing	− 0.68	− 0.32
Eigenvalue	0.61	0.12
% variance explained	45.10	26.60

Cumulative percentage variance explained by first two axes of the joint species/environment biplot = 71.7.

respectively. Thus the first two axes account for 71.7 per cent of the variance in the species/environment data.

In addition to the species–environment biplot, the standard quadrat or sample ordination plot can be produced in the usual way. The arrows for the environmental data can also be superimposed on this plot, or the environmental data may be plotted as centroids (averages) of their distribution in the quadrats in relation to the ordination axes.

Partial canonical correspondence analysis

CCA also has the facility to carry out 'partial' analyses of species–environmental relationships, where certain environmental 'covariables' can be controlled for and eliminated from the ordination. Such an analysis examines the residual variation in the species data and its relationship with any specific environmental variables which may be of interest. It is also suitable for analysing successional or permanent plot data. However, such analyses are very advanced and lie beyond the scope of this book.

Hypothesis testing using canonical correspondence analysis

Canonical correspondence analysis and the CANOCO program can also be used in experimental community ecology for hypothesis testing — **deductive rather than inductive analysis** (Chapter 1). A Monte Carlo significance test is available to test for the effects of specific environmental variables after the influence of other variables has been removed or 'controlled'. In this way, the method can be used to analyse data from randomised block experiments or data from a number of different locations. Also it is possible to restrict the permutations to those among samples-within-blocks or samples-within-locations. Again, this represents more complex and advanced analysis. Good examples of this approach are described in Wassen *et al.* (1990) and Pyšek and Lepš (1991).

Problems in the application of ordination methods

The nature of the species-response model and conflict with the assumptions of linear relationships between species

The prevailing model of species response to an environmental factor is the Gaussian or bell-shaped response curve (Figures 1.6 and 1.7) and most researchers using ordination accept this as being the most realistic model (Whittaker, 1953; 1975; Gauch and Whittaker, 1972a,b). Clear understanding of the idea of bell-shaped species-response curves has been confused because sometimes position on an environmental gradient has been equated with actual physical location on a transect. It is very important to appreciate that environmental gradients are an abstract concept representing a theoretical distribution of a species and do not necessarily imply an underlying spatial relationship (Austin, 1985).

Testing for the existence of these bell-shaped response curves in the real world has been carried out by Austin (1980; 1985; 1987), Austin and Austin (1980), Austin *et al.* (1984) and Austin and Smith (1989). The results show that although bell-shaped species response curves are found, they are not universal, they may be bimodal and are often positively skewed. Patterns of species response curves along environmental gradients are seen to be highly variable and are greatly influenced by species richness (beta diversity). They conclude that an adequate general model of species response does not yet exist.

Most methods of ordination make some assumptions about the nature of the data, one of the commonest being that species are **linearly** related to each other. This problem was discussed briefly in Chapter 5 (Figure 5.5a–d). The idealised linear relationship between two species is shown in Figure 5.5a. This is the model underlying the mathematics of many ordination methods. If, however, the bell-shaped species response model is accepted, then the joint relationship between two species along an environmental gradient produces the highly distorted curve of Figure 5.5d. Clearly most species are not linearly related, and Austin's work indicates that many species do not have perfect bell-shaped response curves either. Thus there is a serious problem regarding the inadequacy of the existing underlying model of species response to environment and the way in which ordination methods treat data from the real world.

The consequence of this is that, when applied to real-world data, all methods of ordination cause distortion of the true relationships between species and samples and associated environmental data because of the disparities between the ecological and mathematical models. The key question is how serious is this distortion and does it matter?

The range and choice of ordination techniques and the search for a 'best' method

The range and choice of methods for ordination is a substantial problem for both researchers and students (Kent and Ballard, 1988). Even now, there is no absolute agreement over which is the 'best' method which gives the least distortion to the data. A clear consensus over which method should be recommended for general use has never emerged. Rather, as each technique or group of techniques has evolved, it has usually been assumed to represent the best available. Unfortunately, subsequent evaluation and testing through either real world application or using simulated data has virtually always led to a reappraisal of methods. Thus the exact reason why a particular researcher chooses a certain method or combination of methods for the analysis of a given set of data is often not very clear. Choice is

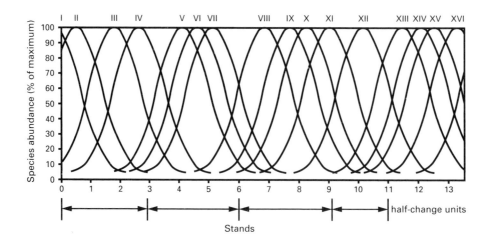

Figure 6.8   A simulated coenocline with 17 species and 13 quadrats or samples. Single half-changes are shown along the coenocline, calculated on the basis of Czekanowski's similarity coefficient (Chapter 3) (Redrawn from Causton, 1988; with kind permission of Chapman and Hall)

often based on the availability of computer programs or the current feeling on the 'goodness' of certain methods based on recent articles or textbooks.

Numerous authors have compared different ordination methods on both simulated data and on real-world data sets and in the process have tried to evaluate the degree of distortion inherent in different methods. Gauch and Whittaker (1972a,b; 1976) introduced the idea of simulation of species response to environmental gradients and the construction of artificial data sets based on the **coenocline** or simulated gradient of community composition. From such an artificial gradient, data sets of known properties can be produced. Figure 6.8 shows a simulated coenocline with 17 species and 13 quadrats (Causton, 1988). The 17 species are each shown with a bell-shaped response curve, and the 13 quadrats have been sampled at 13 equally spaced points along the coenocline.

As species come and go along the coenocline, **species turnover** occurs. The amount of species turnover along the coenocline is the same as the **beta diversity**. Gauch and Whittaker (1972a,b; 1976) devised a measure of this beta diversity by defining units of **'half change'** along the gradient. A half change is defined as the distance or separation along the coenocline at which the similarity between samples or quadrats is 50 per cent. In Figure 6.8, the number of half changes from left to right is shown, based on the Czekanowski coefficient (Chapter 3). A single gradient coenocline can be extended to two gradients producing a **coenoplane**. Again these can be used to test for distortion in different ordination methods.

Using the coenocline principle, Gauch and Whittaker (1972a,b) and Gauch *et al.* (1977) compared the distortion attributable to various ordination methods. The results for correspondence analysis/reciprocal averaging and two variants of principal component analysis are shown in Figure 6.9. In all cases, the line representing quadrats or samples along the coenocline should be straight and parallel to the first axis, but it becomes distorted into the now familiar 'arch' of CA/RA and 'horseshoe' of PCA. These results of these and other comparisons (Gauch *et al.*,

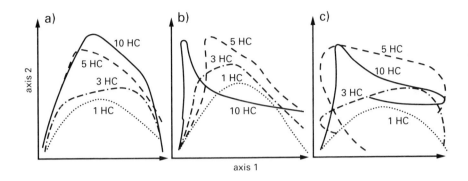

Figure 6.9 The 'arch' and 'horseshoe' effects demonstrated by analysis of simulated coenocline data: (a) correspondence analysis/reciprocal averaging; (b) centred and species-standardised principal components analysis; (c) centred and non-standardised principal components analysis. Sample ordinations for four levels of beta diversity (1, 3, 5, 10 half-changes) are shown. In all cases, the result should be a horizontal straight line. (After Gauch *et al.*, 1977; with kind permission of *Journal of Ecology*)

1981) have indicated that detrended correspondence analysis (DCA) is best, followed by polar ordination (PO), then CA/RA and finally PCA. The results for PO were a surprise and as an ordination method, it still has its supporters (Beals, 1984; Causton, 1988). More recent refinements for simulation of community data have been described by Minchin (1987b).

However, a good performance using simulated data does not automatically mean that a method always works best with different real-world data sets. Numerous other researchers have compared different combinations of ordination methods on their own real-world data (Austin, 1976; Fasham, 1977; Whittaker and Gauch, 1978; Whittaker, 1978a; Gauch *et al.*, 1977; Gauch *et al.* 1981; Oksanen, 1983; Brown *et al.*, 1984; Ezcurra, 1987; Minchin, 1987a; Podani, 1989). No completely consistent results have emerged and recommendations have depended very much on the properties of the particular data sets analysed and the combinations of methods applied. Thus no consensus over a 'best' method has emerged. At the present time detrended correspondence analysis (DCA) and canonical correspondence analysis (CCA) are widely accepted as 'best' methods, but they each have their problems. The problems of DCA have been discussed earlier and (D)CCA is a substantial improvement if a good set of environmental data can be supplied with the species data. However, a great deal of effort in environmental measurement is necessary in order to obtain such data. Soil variables are clearly often very important but such measures, particularly of soil chemistry, can be very time consuming for large numbers of samples. Biotic factors such as grazing, burning and human impact are often important, but obtaining reliable and consistent data on these variables is difficult. In summary, where a good set of environmental data are available in addition to the species data, then CCA is most suitable. Where such data are not available, DCA probably still remains the best and most appropriate choice.

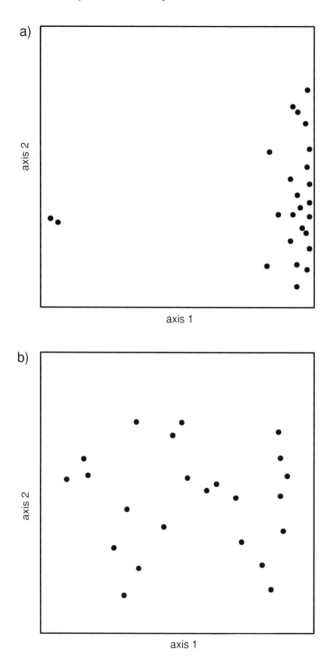

Figure 6.10 (a) The effect of 'outliers' on a two-dimensional ordination; (b) The same plot with the outlying quadrats removed from the data set. The remaining points are spread out and the relationships between them are demonstrated much more clearly.

The problem of outliers

Outliers will affect an ordination regardless of which technique is applied. An outlier is a quadrat which is very different from the others in a data set in terms of species composition or a species which is very different in its distribution in a set of quadrats. Thus when a quadrat ordination is performed, the resulting ordination plot shows the outlying quadrat or quadrats at one extreme and the rest of the quadrats grouped tightly together at the other extreme (Figure 6.10a), since the majority of quadrats are all much more similar to each other than to the outlying quadrat or quadrats. By definition, an outlying quadrat usually contains several species which only occur in the outlying quadrat and nowhere else in the data. A similar situation can occur in a species ordination.

The solution is to remove the offending quadrat(s) and to reanalyse the data. This has the effect of spreading out the remaining quadrats and displaying the similarity among them much more clearly (Figure 6.10b). In the process, unusual and outlying species are generally also removed. Computer programs such as DECORANA and CANOCO have options which allow for the deletion of quadrats while the program is running, so that data files do not have to be edited to remove quadrats and/or species. Removal of a quadrat does not mean that it can then be conveniently forgotten. In interpretation, it should be explained why the quadrat or quadrats were outliers, and the particular ecological conditions that caused the problem to occur should be investigated.

The problems of hypothesis generation and over-interpretation

At the start of Chapter 5, ordination techniques were described as a group of methods for **hypothesis generation**. Once an ordination has been run, study of ordination plots and species/quadrat — environment relationships should enable patterns in the data to be seen and new ideas to be generated. However, Kent and Ballard (1988) showed that very few examples of published research clearly progress from the description of the results to the generation of hypotheses leading to further new research involving the actual testing of the hypothesis. The most notable exception to this is the work of Goldsmith (1973a,b) described in Chapter 5. There is a definite need for plant ecologists and vegetation scientists to address this problem. Ordination has tended to become an end in itself, whereas it should really just be a beginning — a method of **exploratory data analysis** leading to the formulation of new ideas and new research based on hypothesis testing, experimentation and modelling.

A related problem is the tendency to over-interpret ordination results and to try to read too much into them. A student or researcher seeing ordination as an end in itself will be tempted to take those results as final and conclusive, rather than accepting that further data collection and experimentation centred on a newly formulated hypothesis is probably necessary.

Ordination and multivariate analysis as a panacea

In recent years, various authors have raised questions about the application and teaching of multivariate analysis generally and ordination in particular (Gower, 1987; Gittins *et al.*, 1987; James and McCulloch, 1990). The problems identified in the previous section are again highlighted. James and McCulloch strike a cautionary note when they state:

Ecologists and systematists need multivariate analysis to study the joint relationships of variables. That the methods are primarily descriptive in nature is not necessarily a disadvantage . . . we are forced to agree in part with the criticism that multivariate methods have opened a Pandora's box. The problem is at least partly attributable to a history of cavalier applications and interpretations. We do not think that the methods are a panacea for data analysis, but we believe that sensitive applications combined with focus on natural biological units, modelling and an experimental approach to the analysis of causes would be a step forward. [pp. 158–9]

This seems an appropriate caution with which to conclude this chapter. Few vegetation scientists who have used ordination methods, and certainly not the authors, could say that this criticism, at least in part, did not apply to them and some of their work!

Case studies

Reciprocal averaging in the analysis of the forest vegetation of the Lower Alabama Piedmont, USA — Chapter 1 (Golden, 1979)

This project was introduced in Chapter 1 (p. 3). The aim was to examine the forest vegetation at the southern end of the Alabama Piedmont and its relationships to environmental factors. All the forests had been cleared at some point in their history and there were numerous different areas of regrowth and successional stages. Details of sampling are given in Chapter 1. The reason for including the study here is that reciprocal averaging ordination was used to analyse the 84 samples that were described.

The input data were unusual. Since this was a forest study, basal areas for trees with stems over 2 cm diameter breast high were taken. A total of 49 tree species occurred in the 84 samples. Another interesting feature was that in the first ordination, two sample stands were clearly outliers, giving a poor ordination. They were thus removed and the remaining 82 stands were reanalysed, again using reciprocal averaging. When the ordination plots were examined for environmental gradients, the first axis was readily interpretable, but a large undifferentiated cluster of stands and species was found in the upper middle of the diagram. Higher axes provided no further separation within this cluster. Thus to obtain additional resolution within this group, Golden partitioned the data and the 49 sample stands in this central area of the ordination plot were extracted and analysed separately.

Having analysed the tree component of the forests, a further and separate reciprocal averaging ordination of the ground flora data (common shrubs, vines, pteridophytes and herbs) was performed. In order to focus on the main species only those species present in over 25% of the stands were included. Thus rarer species were deleted.

Interpretation of results was extensive and wide-ranging and it is impossible to present them all here. However, the primary gradient was related to topography and drainage characteristics with three categories being recognised: streambottom communities, mesic upland sites and xeric (dry) upland sites. A secondary gradient correlated with management history and time since abandonment following clearance of the original forest was crucial. It was particularly interesting to compare the ordination of the tree data with that of the ground flora and to study the relationships between the two. A good measure of agreement was found.

Suggestions for management were also made since there was a tendency for these oak-pine forests to move to entirely hardwood forests following selective logging of the pine. This tendency to hardwood dominance was most rapid on the stream bottomland sites. Pine is also a successional stage to eventual hardwood oak/hickory forest and if the diversity of forest types was to be maintained, then management would be required to maintain sufficient examples of all successional stages.

This study illustrates a number of interesting variations in the application of ordination methods, as well as being an example of how a primarily academic study nevertheless provided information which was of relevance to management practice and biological conservation.

Correspondence analysis/reciprocal averaging (CA/RA) and detrended correspondence analysis (DCA) as ordination methods used to analyse species environment relationships in the Narrator Catchment, Dartmoor (Kent and Wathern, 1980)

The Narrator Catchment and the experimental hydrological and ecological work which has been completed there were introduced in the case studies of Chapter 3. χ^2 and 2×2 contingency tables were used to define species assemblages in this upland catchment in order to provide background information for more detailed research on the movement of water through the land phase of the hydrological cycle and to assess the effects of vegetation type on water chemistry.

In addition to carrying out the χ^2 analysis, ordination methods were also used to show the ecological relationships between the 162 quadrats and the 82 species. In the original paper (Kent and Wathern, 1980), correspondence analysis/reciprocal averaging (CA/RA) was applied to the data, using the computer program of Kent (1977) and giving the quadrat ordination diagram shown in Figure 6.11. This diagram is an excellent example of the 'arch effect' attributable to CA/RA. Nevertheless, it was possible to make a very clear interpretation of the data, particularly by following the trend of the arch. In addition, rank correlation was used to relate quadrat ordination axis scores to environmental data. Significant correlations emerged between the first axis and soil moisture content, peat/soil depth, slope angle and peat/soil pH. Thus the primary gradient was inferred as being drainage quality and related variables. The second axis showed significant correlations with altitude, *Calluna vulgaris* (Heather) age and pH. Taken together, these were interpreted as being a reflection of biotic factors and land use management practices such as burning and grazing.

At the time of the original data analysis (1979), detrended correspondence analysis (DCA) was not yet available. When the computer program DECORANA was released, the data were reanalysed in order to see the effects of detrending on the ordination. The quadrat ordination diagram for DCA is presented in Figure 6.12. Comparison of Figures 6.11 and 6.12 clearly shows how the detrending process flattens the arch. However, even after detrending, a form of arch structure still remains, suggesting that some of the original 'arch effect' was attributable to the inherent properties of the data, as well as to the use of CA/RA.

In terms of interpretation, the DCA ordination probably makes very little difference, since the relative positions of points on the first axis are very

a)

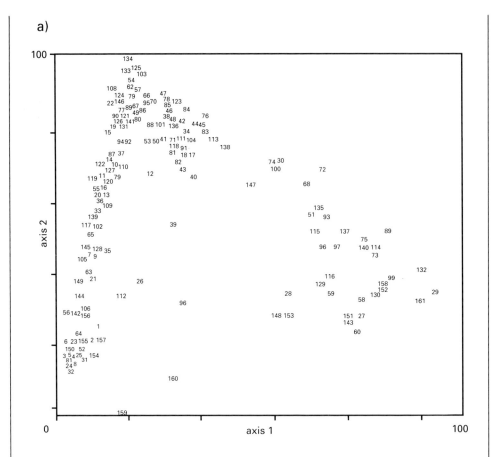

Figure 6.11 Quadrat ordination of vegetation data from the Narrator Catchment, Dartmoor using correspondence analysis/reciprocal averaging (Kent and Wathern, 1980; reproduced with kind permission of *Vegetatio* and Kluwer Academic Publishers)

similar, although there is a significant change in the distribution of points on the second axis. Thus although DCA clearly gives an 'improved' ordination, the extent of that improvement is hard to assess accurately. However, significantly higher rank correlations were found between the second axis and the associated environmental variables of altitude, *Calluna vulgaris*, age and pH. Thus the interpretation of both ordinations remains much the same.

Use of canonical correspondence analysis (CCA) to examine fen vegetation gradients, groundwater flow and flooding in an undrained valley mire at Biebrza, Poland (Wassen et al., 1990)

The River Biebrza valley mires are one of the few remaining extensive examples of this habitat in Central Europe. The valley has a catchment area of around 7000 km² and lies between 100–130m. Mean rainfall is 583mm with 244mm falling in the wet summers. The valley has been glaciated with some extensive moraines. Some of the valley is grazed by cattle and the rest

b)

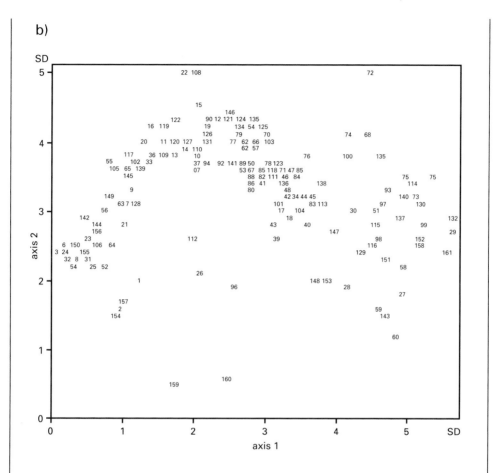

Figure 6.12 The same data as in Figure 6.11 analysed by detrended correspondence analysis

is grazed by elk. The vegetation of the whole valley has been described as rheophilous mire. The basin is divided into three sections referred to as upper, middle and lower. The upper basin is not flooded by the river and the surface rises and falls as it absorbs and loses water in alternate wet and dry periods. The middle Biebrza is much more extensively modified by drainage than the upper and lower sections.

In this study, a total of 58 stands or samples of fen vegetation were described from the Biebrza mire complex. Both rich (minerotrophic) and poor (ombro-trophic) mires were examined. The dominant factors controlling floristic composition were believed to be direction and rate of water flow, supply of calcium and pH. Water flow velocity is closely related to nutrient, mineral and oxygen supply in the top layer of a mire. Calcium and pH are also significantly intercorrelated, being high in fens that are supplied by groundwater but low in bogs that are flushed principally by rainwater. The authors state that the role of calcium is largely that of a conditioning factor controlling other determinants of plant growth such as pH, solubilities of other elements, cation-saturation of exchange sites and microbiological processes affecting nutrient availability.

A distinction was also made between highly productive and less productive rich fen, the differences in productivity being related to phosphorus and potassium concentrations. The aim of the research was to assess the effects of groundwater quantity and quality on floristic composition and productivity of the fens and to generate hypotheses about the influence of water supply on nutrient availability in the various peat-forming communities.

Sampling

Sampling was carried out using transects located on the basis of existing vegetation maps (Palczynski, 1984). Three transects were laid out, one in the upper and two in the lower basins. The middle basin was not sampled because of the highly modified environment. Vegetation was recorded using a decimal scale for cover abundance (van der Maarel, 1979) in $10m^2$ plots at around 20 sites at regular intervals along each transect. Water levels were recorded at 10 sites on each transect using piezometers, and the ground-water was sampled at 2.5m and 4m depth. Also at each of the 58 sites, groundwater was sampled at 5–15 cm below the surface along with peat at the same depth at 30 of the sites. Waters were analysed for conductivity, pH, SO_4^{2-}, Cl^-, NO_3, NH_4^+ and H_2PO^{4-} (orthophosphate). Ca^{2+}, Mg^{2+}, Na^+, K^+, total $Fe(Fe_t)$, total $Al(Al_t)$, total $Mn(Mn_t)$ and Si content was also determined. Further details of measurement techniques are given in the original paper.

Peat was analysed for water content, pH (using calcium chloride), carbon, total nitrogen, easily extractable PO_4-P (per cent of dry weight). C–N ratios were calculated along with the lime potential (p_{lime}).

Ordination analysis

Canonical correspondence analysis (CCA) (ter Braak, 1988a,b,c) was applied to the data using the computer program CANOCO. The principal aim of this particular set of analyses was to define environmental gradients within the floristic data of the mires and to assess the relative importance of the numerous environmental variables. This case study illustrates immediately the necessity for a full set of environmental data to have been collected at each sample site so that CCA can be successfully applied. The data set consisted of the cover abundance values for the plant species plus a matrix of environmental variables on water quality and peat, the height of the water table and the frequency of river flooding. Several analyses of different subsets of the 58 sites were performed. Figure 6.13 shows the CCA species-environment biplot for the 30 sites where a full set of environmental data were available. Three distinct assemblages of species were identified (see Chapter 7 for explanation of species terminology):

(a) *Glycerietum maximae* association — tall productive swamp vegetation, relatively species-poor. Characterising species are shown by circles in Figure 6.13.

(b) *Scheuchzerio-Caricetea fuscae* association — various combinations of *Carex* sedge community — this could be subdivided into further sub-types. Characterising species are indicated by triangles in Figure 6.13.

(c) *Lolio-Cynosuretum* grassland association found at the edge of the river and grazed by cattle. Characterising species are shown as squares in Figure 6.13.

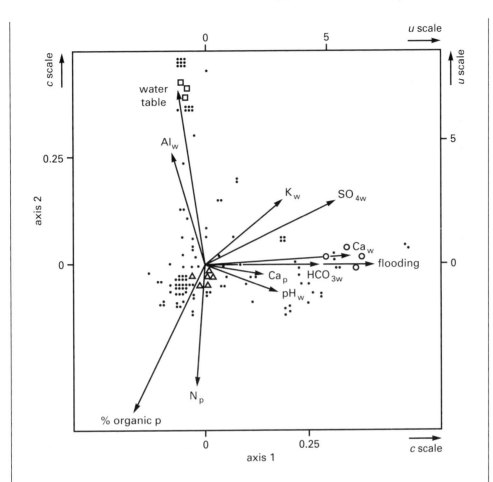

Figure 6.13   Canonical correspondence ordination of the fen vegetation at Biebrza mire, Poland, showing correlation between species and environmental variables. Peat and mire water were sampled in July 1987 at 5–15cm depth. p = peat; w = mire water. The *c* scale applies to environmental variables, represented by arrows; the *u* scale to species, represented by points or symbols. Each point and each symbol represents the weighted average of one species. Unfilled symbols represent characteristic species of the communities: (○) *Glycerietum maximae*; (△) *Scheuchzerio-Caricetea fuscae*; (□) *Lolio-Cynosuretum* (Wassen *et al.*, 1990; reproduced with kind permission of *Journal of Ecology*)

The environmental gradients and the relative importance and intercorrelation of the environmental variables is shown by the arrows in Figure 6.13. The length of an arrow is proportional to its importance and the angles between the arrows reflect the intercorrelations between the variables. The angle between an arrow and each axis is a representation of its degree of correlation with the axis. Thus on Axis 1 flooding and the calcium content of the water are the most significant variables determining variation in species composition, followed by SO_4 and HCO_3 content of the waters. On the second axis, the water table and the organic content of the peat were most important, followed by the Al content of the waters and the N content of the peat. Interestingly, the calcium content of the peat and the pH of the waters are shown to be of lesser significance in explaining variation in species composition.

The three different associations emerge clearly in three separate parts of the diagram. Species of community (a) *Glycerium maximae* occur to the right of the diagram, and are species from flooded wet sites with high concentrations of major ions. Community (b) *Scheuchzerio-Caricetea* species are most abundant on non-flooded undrained sites with a high organic matter content and occur in the centre and towards the bottom of the plot. Species of community (c) *Lolio-Cynosuretum* correlate with mineral soils with deeper water tables and higher aluminium concentrations in the groundwater (that is valley margins) and emerge as a group near the top centre of Figure 6.13. Although they are not identified, points representing individual species could be projected on to each environmental factor to show their relationships to each environmental control.

Further more detailed analyses were then performed on other subsets of the data. Overall the results indicate that differences in vegetation floristics and production within the mire are largely controlled by floods, which are responsible for increasing potassium supply and by groundwater input, which is poor in phosphate but rich in calcium which is a sink for potassium. In those locations where rainwater is an important control, this potassium sink can be mobilised. The authors suggest that further hypotheses should be generated and tested, centred on seasonal changes of water and peat chemistry, the uptake of nutrients by vegetation and experimental fertilization trials using N, P and K. However, more detailed hypotheses and actual hypothesis testing are still in progress. Once again, it would seem that the application of ordination methods has, as yet, not gone far beyond the generation of basic hypotheses and much more detailed research is required on this particular topic. Nevertheless, this is an excellent example of the type of relatively detailed situation where CCA can be used very effectively to explore plant community-environment relationships and to define environmental gradients.

7

Phytosociology and the Zurich–Montpellier (Braun-Blanquet) school of subjective classification

Introduction

The concept of the plant community was discussed at length in Chapter 1. The next two chapters are concerned with methods for **recognising** and **defining plant communities**. This process is known as **phytosociology** — phyto means plant and sociology means assemblages or groupings (of plant species). The theory and practice of phytosociology implies that workers must agree to a large extent with the ideas of Clements (1916) and his belief that distinct assemblages of plant species which repeat themselves over space can be identified and called plant communities. In contrast, the individualistic view of Gleason (1917; 1926) did not believe in identifiable communities; instead plant species are seen as being distributed as a series of continua, responding to different environmental gradients. Supporters of the Gleasonian view therefore reject the whole basis of phytosociology. Most researchers, however, tend towards the Clementsian view and the climax pattern ideas of Whittaker (1953).

Classification

All methods for recognising and defining plant communities are methods of **classification**. The aim of classification is to group together a set of **individuals** (quadrats or vegetation samples) on the basis of their attributes (floristic composition). The end product of a classification should be a set of groups derived from the individuals where, ideally, every individual within each group is more similar to the other individuals in that group than to any individuals in any other group. In practice, this ideal is rarely achieved, particularly in phytosociology. Nevertheless, this remains the ultimate maxim of the process of classification. The groups derived from a set of individual quadrats through classification on the basis of their floristic content are usually taken as the plant communities of the area under study.

The methods for carrying out classification are many and varied. Earliest methods, such as those of the Braun-Blanquet school, were based on sorting of floristic data tables by hand and have often been described as 'subjective'. Since the advent of computers in the late 1950s and early 1960s, various numerical methods based on mathematics and statistics have been devised, and these are described as 'objective', although the word 'objective' needs to be defined and used with care. This point is discussed further at the start of the next chapter. Although the distinction between 'subjective' and 'objective' methods is made, the difference relates more to the whole approach, including both purpose and methods of data collection as well as the methods of classification. Thus the 'subjective' methods are dealt with separately in this chapter, and the 'objective' techniques are covered in the next.

A final introductory point is that all the methods of classification in the next two chapters are applied to floristic data, that is data collected on species presence/absence and/or abundance.

Phytosociology using subjective classification

There is a very long history and tradition for the methods of subjective classification. This history is described in detail in Whittaker (1962), Shimwell (1971), Werger (1974a), van der Maarel (1975) and Westhoff and van der Maarel (1978). The methods have been developed primarily in Europe, and Shimwell identifies four major groups or schools of method which evolved during the period 1900–1960.

(a) **The Zurich–Montpellier School** — this group was established by Professor Braun-Blanquet in 1928, when he published his book *Pflanzensoziologie*, which was translated into English in 1932. This text includes a classification of the vegetation of the French Mediterranean and the central Alps and shows the regional nature of the whole approach. Subsequently Professor Braun-Blanquet's ideas were developed further by Professor Rheinhold Tüxen in Germany, and together Braun-Blanquet and Tüxen carried on the development of the classification system well into the 1960s.

(b) **The Uppsala School** — was based in Scandinavia and its origins can be traced back to the work of von Post (1862). However, the most significant developments of the school came in the 1920s with Professor du Rietz's paper on method in 1921. The centre of research in Uppsala produced many students who together produced a text on *The Plant Communities of Sweden*.

(c) **The Raunkaier (Danish) School** — Raunkaier is most famous for his work on vegetation life-form (1934; 1937), but he also developed a method for describing vegetation based on floristics and for tabulating vegetation samples into community types (Raunkaier, 1928). Subsequently, however, his methods have been little used.

(d) **'Hybrid Schools'** — several researchers developed their own subjective methodologies based primarily on the methodology of the Zurich–Montpellier and Uppsala Schools. Most notable among these was the 'British' School of Poore (1955a,b,c; 1956) and the work of Poore and McVean (1957) and McVean and Ratcliffe (1962) on the vegetation of the Highlands of Scotland. Rieley and Page (1990) present a recent phytosociological account of British vegetation.

Two points need to be made about the traditions of subjective classification. First, the methods have been devised and applied largely in Europe, although some attempts have been made to apply them in North America (see case study). Outside of these areas and particularly in the tropics, they have been very limited in their application. Second, although the various schools did have differences in initial approach, most methodology has now converged on the technique of the Zurich–Montpellier School and Braun-Blanquet. For this reason, the method which is usually described is that of the Zurich–Montpellier School, and it is this that is included here. However, there are problems in that the method is not very well covered in the literature and a number of aspects of the method remain unclear, even to experienced workers.

The method of the Zurich–Montpellier School (Braun-Blanquet)

The following description of the method is derived from several sources, notably Braun-Blanquet (1932/1951), Poore (1955a,b,c; 1956), Pawlowski (1966), Becking (1957), Shimwell (1971), Werger (1974b) and Mueller-Dombois and Ellenberg (1974).

The purpose of the whole methodology of Braun-Blanquet is to construct a global classification of plant communities. The method is based on several fundamental concepts and assumptions.

The relevé (or aufnahme)

This is a **vegetation sample** or **stand**, equivalent in terms of vegetation description to a quadrat. The most important point is that the location of all relevés is entirely **non-random**. The site for vegetation description is thus deliberately and carefully selected as **a representative area of a particular vegetation type**. This presupposes that the worker already has a very thorough knowledge of the vegetation in the region under study and already has a clear, if subjective, impression of the major vegetation types. The samples are thus selected to be representative of those types. This means that the methodology is really only capable of being used by those who have had a long and intimate experience of the vegetation.

Homogeneity

A further complication is that the relevé or sample should be **uniform** and **homogeneous**. This means that the particular assemblage of species which are believed to be representative of the community type being described should exist over a sizeable local area without any detailed variations within it. Thus local micro-environmental and micro-habitat variations should be either avoided or ignored. The existence of such uniform or homogeneous plots in all vegetation types is sometimes questionable, particularly if there are mosaics within the vegetation.

Minimal area

The method also states that the relevé or quadrat must be large enough for a representative sample of uniform vegetation to be taken. This will vary according to the life-form and physiognomy of the dominant vegetation type and the number of species that are found in the relevé as the size of the plot increases. To determine the relevé size, the method of **minimal area** is used (Chapter 2). This is a graph of species numbers against increase in relevé or quadrat size, also known as the **species/area curve**. The curve is derived by taking a small quadrat size and counting the number of species. Quadrat size is then doubled and the number of species found is recorded again. The process is repeated with the quadrat size being progressively doubled and species numbers being counted. At some point, depending on the diversity and physiognomy of the vegetation, the graph will level off. The point at which this occurs is known as the **minimal area**, and the quadrat size at that point is the smallest area that will adequately describe the vegetation. An example of the graph is shown in Figure 2.7. In practice, considerable difficulty often occurs in determining the exact point on the graph at which the 'break' occurs and the line for species numbers against area levels off. Cain (1932; 1934b) described a means of 'recognising' this break point, but later he admitted that it depended on the ratios of the axes used for the graph. The subject of minimal area curves has been reviewed by Hopkins (1955; 1957) and Dietvorst *et al.* (1982). They stress the difficulties of defining the break point, and Dietvorst *et al.* devised a

Table 7.1 An association table from McVean and Ratcliffe's classification of Scottish Highland vegetation (1962) (reproduced with kind permission of English Nature)

Relevé reference number	1	2	3	4	5	6	7	8	9
	M55	M55	R36	M58	M57	M58	M58	M58	M58
Map reference	8652	8645	8645	8312	8898	8075	8078	8075	8042
	1999	2006	2004	2225	2463	2899	2984	2930	2883
Altitude (feet)	150	150	150	600	700	850	1300	1000	1300
Aspect (degrees)	325	360	360	360	360	—	—	—	—
Slope (degrees)	5	3	5	20	8	0	0	0	0
Cover (%)	100	100	100	100	100	100	100	100	100
Height (cm)	60	60	65	60	30	—	—	—	—
Plot area (cm²)	16	16	4	8	16	4	4	4	4
Trees, shrubs and dwarf shrubs									
Betula pubescens	+	+							
Calluna vulgaris	6	7	8	7	7	4	5	7	8
Empetrum nigrum		3		1	2			+	3
Erica cinera					+				
Erica tetralix	1	3		1		+			
Ilex aquifolium		+							
Pinus sylvestris	1	+		1				1	
Sorbus aucuparia	2	1			1				
Vaccinium myrtillus	6	5	4	4	6	7	7	5	6
Vaccinium vitis-idaea	3	3	3	3	3	3	4	6	3
Ferns and fern allies									
Blechnum spicant				2	1		1		1
Pteridium aquilinum	1	2	1	2			2		
Grasses, sedges and other monocotyledons									
Deschampsia flexuosa	2	1	1	3	1	3	3	3	3
Listera cordata	+			2				1	1
Luzula multiflora						+			
Dicotyledon forbs									
Melampyrum pratense	2	1							
Oxalis acetosella							1		
Bryophytes									
Aulacomnium palustre				1	3				2
Dicranum majus	3	3	1	3	2		1		+
Dicranum scoparium		1	1			2		3	
Hylocomium splendens	3	5	7	4	3	4	8	9	5
Hylocomium umbratum			2						
Hypnum cupressiforme			2	1			2	1	
Plagiothecium undulatum	2	2	2	3	2	3	2	+	2
Pleurozium schreberi			3	2		3	2	4	2
Polytrichum commune							+		
Polytrichum formosum									3
Ptilium crista-castrensis	1	2	3	4	+	9	3	4	4
Rhytidiadelphus loreus	3	1	3		2		4	2	2

Table 7.1 contd

	1	2	3	4	5	6	7	8	9
Sphagnum girgensohnii		+			5	+			+
Sphagnum nemoreum	7	8	7		8		2		2
Sphagnum palustre	+								
Sphagnum quinquefarium	7		7				2	+	7
Sphagnum russowii				8					
Thuidium tamariscinum	3	+	2		2				
Liverworts									
Anastrepta orcadensis				3					
Calypogeia trichomanis				3	3		2		3
Cephalozia bicuspidata					3				
Cephaloziella sp.				3					
Frullania tamarisci	+		2						
Herberta hutchinsiae		+							
Leptoscyphus taylori	1			4					
Lophocolea bidentata				1		2			
Lophozia floerkii				2	3				
Lophozia obtusa							3		
Lophozia ventricosa				2				3	2
Mastigophora woodsii	+								
Plagiochila asplenoides							2		
Scapania gracilis			2						
Lichens									
Cladonia carneola							1		
Cladonia coccifera								2	
Cladonia cornuta							1	2	
Cladonia floerkeana								1	
Cladonia furcata				1					
Cladonia gracilis				1					
Cladonia impexa				1				1	
Cladonia pyxidata		+				1	1		
Cladonia rangiferina agg.								1	
Cladonia squamosa							1	2	
Peltigera horizontalis					1				
Number of species (60)	23	23	19	29	20	15	23	22	20

Localities 1–3 Coille na Glas Leitre, Loch Maree, Wester Ross
4 Mullardoch, Glen Cannich, Inverness
5 Amat Wood, Bonar Bridge, Wester Ross
6 Loch an Eilean, Rothiemurchus, Inverness
7 Glenmore, Inverness
8 Iron Bridge, Rothiemurchus, Inverness
9 Invereshie, Inverness

method based on the calculation of similarity between plots in a series of nested samples of increasing size. The method is, however, described as tedious, although it does overcome some of the problems of subjectivity.

Concepts of minimal area are also related to **homogeneity**. A smooth minimal-area curve, levelling off beyond a break point, will only occur if the vegetation is homogeneous. If the doubling of quadrat size brings the relevé into an adjacent local area of different vegetation, the curve may level off but then start to rise again. This indicates that the sample is not homogeneous. Also, in highly diverse tropical environments, particularly rain forests, minimal areas may be impossible to define or could be represented by areas of several square kilometres.

The association

This is the basic unit of the classification system, corresponding to the level of the plant community. An association is a plant community type, found by grouping together various sample relevés that have a number of species in common. An example of such a set of relevés constituting an association from McVean and Ratcliffe's survey of the vegetation of the Scottish Highlands (1962) is presented in Table 7.1

Abstract and concrete communities

An important feature of Braun-Blanquet associations is that they are 'abstract' in type. Thus in the McVean and Ratcliffe classification of the plant communities of the Scottish Highlands, the various communities found in the region are presented with lists of typical relevés (Table 7.1). The complete set of such tables, showing the various communities and associations, represents an inventory of the range of vegetation types in the Scottish Highlands. Apart from the few example relevés listed in each table, there is, however, no indication of the detailed location and distribution of the various community types. The associations are thus 'abstract' and are simply described as occurring 'somewhere' at various or numerous locations in the Scottish Highlands. Researchers who had described the plant communities and vegetation types at a particular location in Scotland could then look up McVean and Ratcliffe's abstract classification to see where their results 'fitted' within the overall structure of Highland communities.

The alternative to the 'abstract' community is the 'concrete' community. If the vegetation of one area of the Scottish Highlands were described in detail and perhaps mapped, then the communities would be termed 'concrete' because **they are precisely shown in terms of location**.

Tabular comparison and sorting of relevés

The final associations, which represent groups of similar relevés, are derived by a subjective process of tabular sorting and rearrangement of both relevés and species. The exact method varies and is not well described. Thus difficulties often occur even when the approach is being applied by experienced workers. Generally, sorting involves the following stages and a diagrammatic summary of the whole process is shown in Figure 7.1:

(1) Compilation of the **raw data table**. This comprises a set of relevés from the region under study. The data for these relevés will have been collected following minimal-area analysis in 'representative' samples of homogeneous vegetation. The data will usually be based on the Braun-Blanquet or Domin cover abundance scales (Table 2.5). To illustrate the application of the method, the

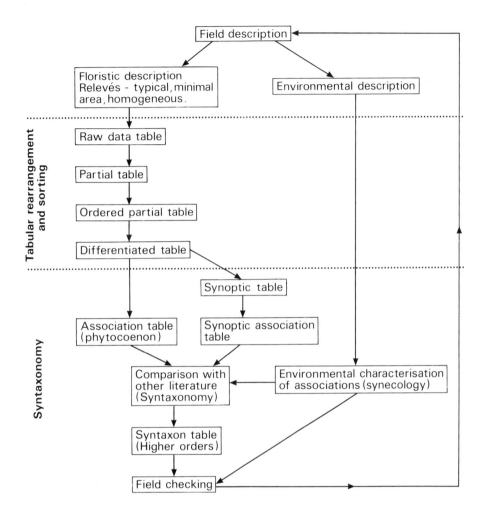

Figure 7.1 Flowchart of stages in the subjective classification of relevés using the Braun-Blanquet method (adapted from Westhoff and van der Maarel, 1978)

Gutter Tor data, from Dartmoor, (Table 3.4), transformed into Braun-Blanquet cover values, is presented in Table 7.2. It should be remembered that this is a small and simplified data set. In practice, all plant groups are normally included.

(2) Calculation of the **constancy** of each species. This is simply the number of relevés within which each species occurs. The species in the raw data table are then rearranged on the basis of constancy, from high to low (Table 7.3). The purpose of this is to assist in the identification of **differential species** at the next stage.

(3) Finding good **differential species**. These will be species of medium to low constancy which tend to occur together in a series of quadrats and can thus be used to characterise groups. In Table 7.3, examples of sets of good differential species have been underlined using different symbols for different groups.

Table 7.2 Quadrat data from Gutter Tor (Table 3.4) transformed to Braun-Blanquet cover values (Table 2.5) to provide a raw table (C = constancy)

Species	|										Quadrats																C
	1	2	3	4	5	6	7	8	9	10	11	12	13	14	15	16	17	18	19	20	21	22	23	24	25	
Agrostis capillaris	5		5										5	5				5		5				5		7
Agrostis curtisii				5		5														2						4
Bryophytes				2		1	3	3							1							1				7
Calluna vulgaris				2	3	1	2				2			1	5	1					5	1	5	1		6
Carex nigra	2	2							1	2	1	1				+		1				1			2	10
Cerastium glomeratum																		1								2
Cladonia portentosa				1					2	2	1		+	2		2					2	2				8
Danthonia decumbens													1							1						2
Drosera rotundifolia	1			1						1														1		3
Erica cinerea																										1
Erica tetralix											2	2					2			2	1	2				5
Festuca ovina			2	3	3	1	3	3	1		2		3	2	3	1	2	2	2	2	1		3	3		14
Galium saxatile			2	1	1	1	3	3	1	2	1	1	2		2	1	1	2	2				3	2		15
Juncus effusus	1	2						1	1																2	5
Juncus squarrosus																		1		1						1
Luzula sylvatica																		1	1							1
Molinia caerulea	3	3		3		2				2	4						5				5			2		9
Narthecium ossifragum	2											2												1		3
Plantago lanceolata																		1			1					2
Potentilla erecta			2	1		1	2	2	2	2	1		2	1	1	2	2	1	1					2		14
Pteridium aquilinum								5							4				5				5			4
Sphagnum sp.	5	5							4			4													5	5
Taraxacum officinale																		+								1
Trichophorum cespitosum		1					1		1		2							1						1		4
Trifolium repens			1																							2
Vaccinium myrtillus					3		2	2		2	2		2	2	1	1	1					2				10
Viola riviniana										1									1			1		1		3

Table 7.3 Raw table with species rearranged by constancy (C) and with groups of matching species underlined

Species	1	2	3	4	5	6	7	8	9	10	11	12	13	14	15	16	17	18	19	20	21	22	23	24	25	C
Galium saxatile	2		2		1	1	3	3	1	2		1	2	2	2	1	1	2	2	2			3	2		15
Festuca ovina			2		3		3	3					3	3	3			2	2	2	1		3	3		14
Potentilla erecta			2	1		1		2	1	2	1		2	1	1	2	2	1	1			1		2		14
Carex nigra	2	2							1	1	1	1				+	1		1			1			2	10
Vaccinium myrtillus				3	3		2	2	1	2	2			2		1	1				2	1				10
Molinia caerulea	3	3		3		2				2	4						5				5	5			2	9
Cladonia portentosa				1					2	2	1			2		2					2	2				8
Agrostis capillaris			5						1				5	5				5		5			5			7
Bryophytes														1		1					1	1				7
Calluna vulgaris				2	1	1				2						5					5			1		6
Erica tetralix				1	3		2			2	2	2					2					2				5
Juncus effusus	1	2							1			1													2	5
Sphagnum sp.	5	5							4			4													5	5
Agrostis curtisii				5		5		5		2										2						4
Pteridium aquilinum															4				5				5			4
Trichophorum cespitosum		1							1			2													1	4
Drosera rotundifolia	1											1													1	3
Narthecium ossifragum	2											2													1	3
Viola riviniana													1						1	1				1		3
Danthonia decumbens													1					1								2
Cerastium glomeratum													+					1								2
Plantago lanceolata																		1								2
Trifolium repens										1								1								2
Erica cinerea																				1						1
Juncus squarrosus																				1						1
Luzula sylvatica			1																							1
Taraxacum officinale																			+							1

Table 7.4 The rearrangement of quadrats as part of construction of a partial table from the raw table of the Gutter Tor, Dartmoor data (Table 7.3)

Species	\-	\-	\-	\-	\-	\-	\-	\-	Quadrats																
	1	2	9	12	25	11	17	22	8	15	19	23	3	13	14	18	20	24	16	21	4	6	5	7	10
Galium saxatile			1	1					3	2	2	3	2	2	2	2	2	2	1	1	1	1	1	3	2
Festuca ovina									3	3	2	3	2	3	2	2	2	3	1				3	3	
Potentilla erecta			1	1		1	2		2	1	1		2	2	1	1		2	1		1	1			
Carex nigra	2	2	1	1		1	1												+						2
Vaccinium myrtillus	3	3				2	1		2		1				2				1				3	2	2
Molinia caerulea			2	2		4	5	5													3	2			2
Cladonia portentosa			1		2	1	2	2											2		1				2
Agrostis capillaris													5	5	2	5	5	5	2						
Bryophytes													5	5	1	5	5	5	1		2	1		2	
Calluna vulgaris			2	1	1	2									1				5	5	2	1	3		
Erica tetralix			1	2		2	2	2													1				
Juncus effusus	1	2	1	1	2																				
Sphagnum sp.	5	5	4	4	5																				
Agrostis curtisii																	2				5	5			2
Pteridium aquilinum									5	4	5	5													
Trichophorum cespitosum		1	1	2	1																				
Drosera rotundifolia			1	1	1																				
Narthecium ossifragum	2		2	2	1																				
Viola riviniana											1							1							
Danthonia decumbens														1			1								
Cerastium glomeratum														1		1									
Plantago lanceolata														+		1									
Trifolium repens																1		1							
Erica cinerea													1												
Juncus squarrosus																	1								
Luzula sylvatica																	1								1
Taraxacum officinale																+									

Table 7.5 Further rearrangement of species to give a partial table of the Gutter Tor vegetation data

Species

					Quadrats																				
													Groups												
	I					II			III				IV						V		VI		VII		
Differential species	1	2	9	12	25	11	17	22	8	15	19	23	3	13	14	18	20	24	16	21	4	6	5	7	10
Carex nigra	2	2	1	1	2														+						2
Juncus effusus	1	2	1	1	2																				
Sphagnum sp.	5	5	4	4	5																				
Trichophorum cespitosum	1	1	1	1	1	1	1	1			1														
Drosera rotundifolia	1		1																						
Narthecium ossifragum	2		2	2	2																				
Molinia caerulea	3	3			2	4	5	5	2						2										
Vaccinium myrtillus						2	1	1							2				1	2	3	2	3		2
Erica tetralix				2		2	2	2																2	2
Pteridium aquilinum																									
Festuca ovina									5	4	5	5	2	3	2	2	2	3					3		2
Galium saxatile			1			1	1		3	3	2	2	2	2		2		2	1	1	1	1	1	3	2
Potentilla erecta			1				2		3	3	2	3	2	2	1	1		2			1	1		3	
Agrostis capillaris			1						2	1	2	1	5	5	5	5	5	5	5	5	2	1	3		
Calluna vulgaris						2		2							2		2		2	2	1				2
Cladonia portentosa			2	1		1															5	5			2
Agrostis curtisii															1				1		2	1			
Bryophytes								1																2	
Companion species																									
Viola riviniana											1		1	1				1							
Danthonia decumbens														1		1	1								
Cerastium glomeratum														+											
Plantago lanceolata														1		1	1								
Trifolium repens														1		1		1							
Erica cinerea													1												
Juncus squarrosus																									
Luzula sylvatica																	1	1							
Taraxacum officinale																+									1

There may be as few as two and as many as six or eight differential species characterising a group.

(4)   Drawing up of **partial or extract tables**. These show groups of quadrats characterised by sets of differential species. For the Dartmoor data, this is completed in two stages. First, in Table 7.4, the quadrats of Table 7.3 have been rearranged to place relevés containing similar differential species into groups. Second, the species are then also rearranged, so that the differential species which characterise each group are placed next to each other (Table 7.5). The result is an **ordered partial table**. This last stage has the effect of concentrating entries in the data matrix down the diagonal. The top of the table containing the differential species is also separated from the rest.

(5)   Secondary sorting of relevés within the groups of the partial table, so that the most similar quadrats are placed next to each other. This has been done in Table 7.6. The process of secondary sorting will involve species that are not differential species and known as the **companion species**. The resulting table is known as the **differentiated table**. The various groupings of relevés then emerge. Occasionally, a relevé or sample may not fit in any of the established groupings. Quadrat 10 of the Dartmoor data is a good example. This relevé must be discarded.

(6)   Each of these groups would then be characterised as an **association** or **plant community**. In the Zurich–Montpellier method, there is a system of nomenclature using the names of the characterising species and suffixes to denote the association. Thus in the first group of Table 7.6, the characterising species names would be run together to give the title Sphagnetno–Junceto–Caricetum.

Fidelity

These characterising species are also described as having **degrees of fidelity**. Thus Braun-Blanquet (1951) defines five degrees of fidelity:

Fidelity 5 — **exclusive species**, completely or almost completely confined to one community;

Fidelity 4 — **selective species**, found most frequently in a certain community but also rarely in other communities;

Fidelity 3 — **preferential species**, present in several communities more or less abundantly but predominantly in one certain community and there with a greater deal of vigour;

Fidelity 2 — **indifferent species**, without a definite affinity for any particular community;

Fidelity 1 — **accidentals**, species which are rare and accidental intruders from another community or relics of a preceding community.

The concept of fidelity has been much criticised and misunderstood. Poore (1955a) points out that the degree of fidelity of a particular species can only be fully assessed when **all** the vegetation of a region has been described. It is thus very much a geographically defined concept. However, the concept must also depend on the size of the geographical region used to define fidelity. Also fidelity has been confused with **constancy of species within groups or associations**. A species which has a high constancy in an association, that is occurs in every or almost every relevé of the association group, may not necessarily have the highest degree of fidelity. As a small example, in the Dartmoor data, *Juncus effusus* is entirely faithful to group I and is 100 per cent constant because it occurs in all five quadrats of the group.

Table 7.6 The final differentiated table following secondary sorting within groups

Species

Quadrats

	Associations																								
	I					II			III				IV						V		VI		VII		
Differential species	12	25	1	2	9	11	22	17	8	15	19	23	3	13	18	24	14	20	16	21	4	6	5	7	10
Carex nigra	1	1	2	2		1		1			1								+						2
Juncus effusus	1	2	2	2	1																				
Sphagnum sp.	4	5	5	5	4																				
Trichophorum cespitosum	2	1	1	1	1																				
Drosera rotundifolia	1	1	1																						
Narthecium ossifragum	2	1	2	2																					
Molinia caerulea		2	3	3					2								2								
Vaccinium myrtillus	2					2	1	1											1	2	3	2	3	2	
Erica tetralix	2					2	2	2												1					
Pteridium aquilinum	1				1	1		1											1	1	1	1	3	3	2
Festuca ovina									5	4	5	5	2	3	2	3	2	2	1						2
Galium saxatile					1	1		1	3	3	2	3	2	2	2	2	2		1	1	1	1	3	3	2
Potentilla erecta					1			2	3	2	2	2	2	2	1	2	1				1	1	1	1	
Agrostis capillaris					1				2	1	1	1	5	2	5	5	5	5	5	2	2		3		2
Calluna vulgaris		1				2													5	5	2	1			2
Cladonia portentosa					2	1	2									2		2	2	2	5	5			2
Agrostis curtisii							1									1	1	1	1		5	1	1	2	
Bryophytes																		2			2	1			
Companion species																									
Viola riviniana											1		1	1	1			1					1		
Danthonia decumbens													1	1											
Cerastium glomeratum													+		1										
Plantago lanceolata													1	1	1			1							
Trifolium repens															1										
Erica cinerea													1												
Juncus squarrosus																		1							
Luzula sylvatica																		1							
Taraxacum officinale																+									

Table 7.7 A synoptic table of common species in the 22 rich-fen communities of England and Wales (Wheeler, 1980); the digits are constancy values: 1 = 5–20%; 2 = 21–40%; 3 = 41–60%; 4 = 61–80%; 5 = 81–100%; the occurrence of species with very low constancy is not shown (reproduced with kind permission of *Journal of Ecology* and Dr B.D. Wheeler)

	1	2	3	4	5	6	7	8	9	10	11	12	13	14	15	16	17	18	19	20	21	22
Phragmites communis	5	5	5	4	3	5	5	5	5	3	5	1	2	3	2	2	4	5	3	2	4	4
Galium palustre	5	5	2	5	5	5	3	5	2	5	1	1	4	2	2	1		3	2	5	3	4
Mentha aquatica	2	3	1	4	3	4	2	5	3	5	5	1	4	2	2	1			1	3	3	4
Acrocladium cuspidatum	1	1	1	2	4	5	3	5	2	5	5	4	5	5	5	4		5	1	5	5	5
Angelica sylvestris		2			4	3	5		3	5	5	1	3	3	5	5			1	3	1	3
Eupatorium cannabinum	1	3		4	2	4	4	4	5	2	5		2	4		2	1	5	2	1	1	3
Lythrum salicaria		3		4	3	3		4	3	2	4		3	1			1	3			3	3
Juncus subnodulosus	2	2		4	2	1	1	4	5	2	5		4	5		1	1	4			3	3
Carex rostrata				2	5					5	2	1	1							4		2
Eriophorum angustifolium	1				5					5	3	3	1							2		
Equisetum fluviatile	1				4					5	3		2		1	1				4		2
Menyanthes trifoliata	3			5	5	2		2		5	4	2	2		1					4		1
Potentilla palustris				5	5	2		3		5			2		1		1	3	2	3		3
Cladium mariscus		5			1		1	4	5	1	2		1	1			2	2			1	1
Ranunculus lingua	2	1	5		3	3		1		2			1						1	1		
Typha angustifolia	3	1	5			2		1										2				
Sium latifolium	1	1	5					1														
Cicuta virosa	1	1	5					1	1													
Carex pseudocyperus	1	1	5		1			2				1	1									2
Carex lasiocarpa	2							2		4	1		1							2		
Carex diandra								1		5	1		1									
Acrocladium giganteum								1		5	2		1							2		
Carex elata				5				4	1	1	2		1					1	2	2	2	2
Carex paniculata						5	3	3	5	3	2		2	3		1					2	2
Campylium stellatum						2	1	3	5	4	5	4	3	3	2							5

Table 7.7 cont.

	1	2	3	4	5	6	7	8	9	10	11	12	13	14	15	16	17	18	19	20	21	22
Valeriana officinalis						2	4	3	2	2	4		2	1	2	1		3		4	3	4
Filipendula ulmaria						4	2	3	2	5	3		5	2	4	5	1	3	1	5	3	2
Lysimachia vulgaris						4	2	4	1				2				1	3	1	2	2	3
Peucedanum palustre						4		5				1	1				2	5	3		1	1
Thelypteris palustris						4	1	4					1				1	5	3		2	2
Calamagrostis canescens						1	1	4					1				1	5	2			2
Epilobium hirsutum						1	4	1					2			5						1
Thalictrum flavum							2	1					2			1						
Carex panicea								3	2	5	5	5	4	5	4					1		
Carex nigra								1	1	5	5	5	2	3	5					2		
Carex lepidocarpa								1		3	5	5	1									
Pedicularis palustris								2		3	5	4	2	1								
Schoenus nigricans								2	3	1	5	4	2	1								
Epipactis palustris								2	1	1	5		2	2								
Anagallis tenella										1	4		2	1								
Eriophorum latifolium										1	5	4	1									
Drepanocladus revolvens										3	5	4	1									
Pinguicula vulgaris											5	5	1									
Carex dioica										1	3	4										
Molinia caerulea								2	5	3	5	5	3	5	5	3			3	3		1
Cirsium dissectum								1	4	3	3		1	5		3						
Potentilla erecta								1	5	2	5	4	3	5	4				2	2		
Succisa pratensis								1	3	3	5	5	4	5	4				2	2		
Cirsium palustre						1	2	3	2	3	5	3	5	4	3	3		3		4	3	5
Vicia cracca							1	1	3	1	4		5	4						2		
Galium uliginosum							1		2	3	3	1	4	5	2	1				2	3	1
Valeriana dioica							1	2	2	4	3	5	4	4	4	1		1		4	2	3

Table 7.7 cont.

	1	2	3	4	5	6	7	8	9	10	11	12	13	14	15	16	17	18	19	20	21	22
Caltha palustris		2		1		2	2	3		5	3	2	4		3	1				5	1	4
Iris pseudacorus						1	3	3					3					2	1		4	4
Holcus lanatus										2	3		5	3		1				3	2	1
Ranunculus acris										2	2	1	5	2	3					4	2	2
Poa trivialis											1	1	5			1				5	2	3
Carex acutiformis											1		3			2				1	3	4
Scrophularia aquatica						2					1		3									2
Epilobium parviflorum											1		3									
Hypericum tetrapterum											2		3	1		1						
Cerastium holosteoides												1	4	2								
Rumex acetosa													4	4		1						
Sanguisorba officinalis										2			1		5					2		
Geum rivale										1					3					3		
Climacium dendroides										2		1	1		5					4		
Crepis paludosa										1					4					5		
Myrica gale											1						5	3	5	2	2	1
Alnus glutinosa						4	2	3			2		1	1			4	3	1	2	2	5
Salix cinerea						4	3	3			2		2	1		1		3	3	5	5	5
Betula pubescens							1	2			3		1	1	2		1	5	5	4	3	3
Mnium undulatum								1			2		3		1					3	3	3

Key to community-types: 1, *Scirpo-Phragmitetum*; 2, *Cladietum marisci*; 3, *Cicuto-Phragmitetum*; 4, *Caricetum elatae*; 5, *Potentillo-Caricetum*; 6, *Caricetum-paniculatae*; 7, *Angelico-Phragmitetum*; 8, *Peucedano-Phragmitetum*; 9, *Cladio-Molinietum*; 10, *Acrocladio-Caricetum*; 11, *Schoeno-Juncetum*; 12, *Pinguiculo-Caricetum*; 13, Fen meadow communities; 14, *Cirsio-Molinietum*; 15, *Carex nigra-Sanguisorba officinalis* community; 16, *Epilobium hirsutum-Filipendula ulmaria* community; 17, *Myricetum gale*; 18, *Betulo-Dryopteridetum cristatae*; 19, *Betulo-Myricetum*; 20, *Crepido-Salicetum*; 21, *Salix cinerea* carr; 22, *Osmundo-Alnetum*

Table 7.8 The hierarchical classification units of the Braun-Blanquet system

Rank	Suffix	Example
Class	-etea	Molinio-Arrhenatheretea
Order	-etalia	Molinietalia
Alliance	-ion	Junco-Molinion
Association	-etum	Cirsio-Molinietum
Sub-association	-etosum	Cirsio-Molinietum caricetosum
Variant	—	Specific names used
Facies	—	Specific names used

However, *Galium saxatile* is a constant species of associations III, IV and V but is not faithful to any one of those groups. Even in the Dartmoor example, the apparent high fidelity of *Juncus effusus* to group I of the data set does not mean that it automatically has high fidelity when compared to other community types from other studies that may have been completed on Dartmoor and surrounding regions. The same principle applies to much larger-scale studies. Again, the key point is that fidelity is entirely dependent on the size of the vegetation region that is being described and the extent to which that description is complete.

Further discussions of the problems of fidelity are presented in Poore (1955a), Becking (1957), Moore (1962), Shimwell (1971), Werger (1974b), Mueller-Dombois and Ellenberg (1974), Westhoff and van der Maarel (1978) and Kershaw and Looney (1985).

Synoptic tables

Once associations have been defined and recognised, a synoptic table can be produced summarising the data for each association. Each community type is represented by a column in which each characterising species of each association is indicated as a percentage or class value. An example of such a table is shown in Table 7.7, which contains data from a survey of all the rich-fen systems of England and Wales (Wheeler, 1980). In the first of several papers, a total of 22 communities or associations were identified, and the synoptic table shows percentage constancy values on a five-point scale.

Higher orders of classification

In the Braun-Blanquet system, the level of **the association** is fundamental and represents the basic unit of vegetation description, equivalent to the plant community. However, under the system, higher and lower orders can be recognised within an overall **floristic association system**, as shown in Table 7.8.

A grouping of two or more associations which have their major species in common and which differ only in detail can be combined to give **an alliance**. Alliances can similarly be grouped at a higher level into **orders** and orders into **classes**. Equally, the association can be subdivided into **sub-associations**; sub-associations can be divided into **variants** and within these various **facies** can be recognised. An example of how this hierarchical structure works is shown in Table 7.9, which is again taken from Wheeler's work on rich-fen systems in England and Wales (1980). In this manner, the whole hierarchy of vegetation units in a region can be described and their relationships to each other displayed. To assist with this process over large areas, an international code of botanical nomenclature has been established, based on the idea of **syntaxonomy**, which is a set of rules governing

Table 7.9 A synopsis of the plant communities of the rich-fen systems of England and Wales (Wheeler, 1980) (reproduced with kind permission of *Journal of Ecology* and Dr B.D. Wheeler)

PHRAGMITETEA
  PHRAGMITETALIA
    PHRAGMITION (COMMUNIS)
      Scirpo-Phragmitetum W. Koch 1926
        Phragmites-dominated swamp of pools and wet places in fens.
      Cladietum marisci Zobrist 1935 em. Pfeiffer 1961
        Cladium-dominated swamp of pools and wet places in fens.
      Cicuto-Phragmitetum Wheeler 1978
        Semi-floating fen dominated by *Phragmites communis*, with much *Carex pseudocyperus, Cicuta virosa, Ranunculus lingua* and *Sium latifolium.*
  MAGNOCARICETALIA
    MAGNOCARICION
      Caricetum elatae Koch 1926
        Carex elata-dominated swamp of pools and wet fens.
      Caricetum paniculatae Wangerin 1916
        Carex paniculata-dominated swamp of pools and wet fens.
      Potentillo-Caricetum rostratae ass. nov. prov.
        Wet sedge fen, often dominated by *Carex rostrata* or *Juncus effusus* (sometimes *Carex nigra* or *Phragmites communis*), usually with much *Eriophorum angustifolium* and *Potentilla palustris.*
      Angelico-Phragmitetum ass. nov. prov.
        Tall fen dominated usually by *Phragmites communis* or sometimes *Carex paniculata*, with a range of tall herbaceous dicotyledons. Widespread.
      Peucedano-Phragmitetum Wheeler 1978
        Tall fen dominated by *Phragmites communis, Cladium mariscus* or *Calamagrostis canescens*, with *Carex elata, Juncus subnodulosus, Lysimachia vulgaris, Peucedanum palustre, Thelypteris palustris*, etc. Mainly in the Norfolk Broads.
      Phragmites communis-dominated communities
        Species-poor vegetation characterized by dominant *Phragmites.*
      Cladium mariscus-dominated communities
        Species-poor vegetation characterized by dominant *Cladium.*
      Glyceria maxima-dominated communities
        Species-poor vegetation characterized by dominant *Glyceria maxima.*
      Cladio-Molinietum ass. nov.
        A mixed-sedge community, generally species-poor, with much *Cladium mariscus* and *Molinia caerulea*. East Anglia and Anglesey.
PARVOCARICETEA
  TOFIELDIETALIA
    CARICION DAVALLIANAE
      Schoeno-Juncetum subnodulosi Allorge 1922
        Sedge communities of calcareous fens. *Schoenus nigricans* and/or *Juncus subnodulosus* dominate, with a wide range of associates including *Anagallis tenella, Caraex lepidocarpa, Epipactis palustris, Eriophorum latifolium*, etc. Mainly in southern Britain.
      Pinguiculo-Caricetum dioicae Jones 1973 em. Wheeler 1975
        Communities of calcareous soligenous mires on peat or mineral gleys, often displaying a sedge sward of *Carex dioica, C. lepidocarpa, C. nigra, C. panicea*, and *Eriophorum latifolium*. Mainly, but not exclusively, in N. England, both lowland and upland.

Table 7.9 cont.

Acrocladio-Caricetum diandrae (Koch 1926) nom. nov.

Communities of very wet calcareous fens, usually topogenous. *Carex diandra, C. lasiocarpa* and *Acrocladium giganteum* are usually abundant and characteristic, often with much *Carex rostrata, Eriophorum angustifolium, Menyanthes trifoliata* and *Potentilla palustris*.

MOLINIO-ARRHENATHERETEA

MOLINIETALIA

CALTHION PALUSTRIS

Fen-meadow communities

Rush- or sedge-dominated communities, usually in spring fens or wet grassland. *Juncus subnodulosus, J. articulatus* and *J. acutiflorus* may all be important. Important sedges include *Carex acutiformis* and *C. disticha*. A very variable group.

JUNCO (SUBULIFLORI)-MOLINION

Cirsio-Molinietum Sissingh et De Vries 1942

Grassland dominated usually by *Molinia caerulea*, with *Carex hostiana, C. panicea, C. pulicaris, Cirsium dissectum, Gymnadenia conopsea, Potentilla erecta, Succisa pratensis* (*Juncus subnodulosus*).

Carex nigra-Sanguisorba officinalis community Proctor 1974

Herb-rich vegetation normally dominated by *Molinia caerulea* or *Carex nigra*, with *Crepis paludosa, Epilobium palustre, Geum rivale, Sanguisorba officinalis, Climacium dendroides*.

Molinia-Myrica community

Species-poor vegetation dominated by *Molinia caerulea* and *Myrica gale*.

Molinia sociation

FILIPENDULION

Epilobium hirsutum-Filipendula ulmaria community

Vegetation dominated by tall herbs, normally *Epilobium hirsutum* and/or *Filipendula ulmaria*. Often species-poor.

Phragmites-Urtica dioica community

Species-poor *Phragmites*-dominated vegetation with much *Urtica dioica*.

FRANGULETEA

SALICETALIA AURITAE

SALICION CINEREAE

Myricetum gale (Gadeceau 1909) Jonas 1935

Low scrub dominated by *Myrica gale*

Betulo-Dryopteridetum cristatae Wheeler 1975

Acidophilous open birch scrub with much *Sphagnum*, and with *Dryopteris cristata, D. carthusiana, Calamagrostis canescens, Cladium mariscus, Peucedanum palustre* (*Pyrola rotundifolio*). Confined to Broadland.

Betulo-Myricetum ass. nov.

Birch scrub with much *Myrica gale*.

Crepido-Salicetum pentandrae Wheeler 1975

Salix pentandra- or *Salix cinerea-*dominated carr, with *Betula pubescens, Equisetum fluviatile, Geum rivale* and *Crepis paludosa*. N. Britain.

Salix cinerea carr.

Species-poor *Salix cinerea*-dominated vegetation.

Frangula alnus carr

Low dense scrub dominated by *Frangula alnus*.

Table 7.9 cont.

Rhamnus catharticus carr
　　Species-poor carr dominated by *Rhamnus catharticus.*
ALNETEA GLUTINOSAE
　ALNETALIA GLUTINOSAE
　　ALNION GLUTINOSAE
　　Osmundo-Alnetum Klötzli 1970
　　　Alder carr vegetation, typically with *Carex acutiformis, C. paniculata, Eupatorium
　　　cannabinum, Iris pseudacorus, Lythrum salicaria, Solanum dulcamara* and *Urtica
　　　dioica.* Widespread in lowland fens.

---

the naming of communities and hierarchies under the Braun-Blanquet system
(Moravec, 1971). Table 7.9 is an example of a syntaxon table. The higher order
classification of syntaxonomy for the British Isles is presented in Shimwell (1971),
Appendix III.

Computerised methods of tabular rearrangement

The subjective nature of the process of tabular rearrangement has been reduced
over the past 20 years by the writing of various computer programs to carry out
tabular rearrangement, for example Benninghoff and Southworth (1964), Moore,
G.W. *et al.* (1967), Moore, J.J. *et al.* (1970), van der Maarel *et al.* (1978), Westfall
et al. (1982), Wildi and Orlóci (1983; 1990). Some of these programs are based on
simple matching coefficients, while others are related to the more objective methods
of numerical classification, particularly similarity analysis, described in the next
chapter. It is important to realise that even when these more 'objective' methods
are used, there is still a high degree of subjectivity in the overall approach,
particularly with regard to sampling and the selection of typical or representative
relevés by the Braun-Blanquet method and also in the selection and designation of
final groupings or associations. Numerical approaches have also been applied to the
process of syntaxonomy and are reviewed by Mucina and van der Maarel (1989) and
Fischer and Bemmerlein (1989).

Environmental data

Virtually all tabular rearrangement and interpretation of association and group
structure are carried out on the basis of floristic data. However, it is usual to collect
a certain amount of environmental data on each relevé, for example height, aspect,
soil type, drainage quality. Although these are not used in the actual process of
classification, it is clear that the associations which result are also characterised by
certain environmental conditions. Braun-Blanquet himself introduced the concept of
the **synecosystem**, whereby environmental factors are reflected in the vegetation and
vice versa. Final description and discussion of associations will thus often refer to
their site character and the local environment.

Discussion of the Zurich–Montpellier system

The techniques of the Zurich–Montpellier School and Braun-Blanquet are very
important in vegetation science. The main reason for this is that they provide the

methodology for the major classification system of European vegetation types as well as some other parts of the world. The approach is, however, not without its problems. Egler (1954) presented one of the most eloquent criticisms, claiming that the method was over-simplified and represented the forcing of a weak methodology on to a much more complex real world. A major reason for Egler's comments was that he was a follower of the Gleasonian individualistic view of the plant community. Even he, however, admitted that much valuable work had been completed by Braun-Blanquet and his colleagues in Europe. Nevertheless, valid criticism of the method exists and centres on the following points:

(a) The subjectivity of the whole methodology, particularly the methods of field sampling. The selection of 'typical' or 'representative' relevés is often said to be highly biased and does assume a substantial knowledge of the vegetation prior to any attempt at description. Also, non-homogeneous and transitional or ecotone areas between typical and representative samples are not recorded under this method, yet are still clearly plant assemblages. Questions may thus be asked as to when does a transitional community between two major community types become a new community in its own right? Part of the answer to this problem was provided by Poore (1955a,b,c; 1956) with his concept of the **nodum**. Within any region, there are a set of major plant communities or associations that are distinctive and characteristic and can be recognised consistently in the field. However, between these major types, there are zones of transition and ecotones. Thus each of the major plant communities can be described as **a nodum** and every region has a set of **noda**. However, certain transitional types do exist between the noda and should be recognised as such.

(b) The concept of 'abstract' communities. This has proved a confusing concept particularly for students and inexperienced researchers.

(c) The process of tabular rearrangement. The exact methodology for carrying this out varies from one worker to another, although the principles are generally agreed. Development of computerised methods has helped with the practical aspects of relevé sorting.

(d) The discarding of relevés which do not fit any of the associations which have been defined from a set of data has also been questioned. The reason for doing so lies under (a) above, in that such a relevé must have been badly chosen at the stage of field description. However, this inevitably raises questions of circular arguments along the lines that the communities must already have been defined in the worker's mind before data collection started and data which do not fit the preconceived ideas of community structure may be discarded, when in reality they represent an association not previously recognised and thus not sampled. The answer to this point lies in the necessity of the researcher having a **very** thorough knowledge of the vegetation before work commences.

(e) Terminology — much strange and perhaps unnecessary terminology is used at all stages. Also different schools have invented various new terms, each with slightly different meanings, a process which has not been helped by the use of various European languages.

(f) Perhaps most important of all is the criticism that the whole methodology is not well described in the literature, has many variations and has an air of mystique about it which is difficult to dispel. This is not really surprising when the whole approach can only be applied by very experienced workers in the field. This is unfortunate and makes the topic a difficult one for students and young ecologists.

On the positive side, the system undoubtedly works, and a very detailed and accurate classification of most of the vegetation of Europe has been achieved. In addition to the work of McVean and Ratcliffe (1962) and Wheeler (1980), good examples of the application of the method can be found in Shimwell (1971) and Pawlowski (1966) and numerous further examples in the journals *Vegetatio* and *Phytocoenologia*.

Case studies

The phytosociology of the heathlands of Newfoundland (Meades, 1983)

As part of a symposium volume on the biogeography and ecology of Newfoundland (South, 1983), Meades presented a full phytosociological account of the heathland vegetation of the island. The greater part of the island is made up of forests of balsam fir (*Abies balsamea*) and black spruce (*Picea mariana*) (56 per cent) and wetland (24 per cent). The remaining 20 per cent is heathland distributed within five general groups as shown in Figure 7.2. Meades aimed to define these heath communities in much greater detail using the methods of subjective classification. Given the typical vegetation types of Newfoundland, he used terminology which was closer to that of the Uppsala School and Du Rietz (1942a,b). Although field methods are not described in detail, apart from a short Appendix, the methods are very similar to those of the Zurich–Montpellier School described in this chapter, including the use of Braun-Blanquet cover scales for recording.

   At the start of the survey, the major community types were already known, and Meades was able to relate them to three basic phytosociological alliances of heathlands proposed by Du Rietz in 1942:

1. Empetrion — representing exposed alpine and sub-alpine heaths dominated by *Empetrum hermaphroditum*, a very close relative of *Empetrum nigrum* (crowberry). On Newfoundland, this is represented by the very similar species *Empetrum eamesii*. This alliance includes the alpine heath, *Empetrum* heathland and moss heath in Figure 7.2.
2. Myrtillion — these are the less exposed heaths below the treeline, where there is adequate snow accumulation in winter. The heaths are dominated by *Vaccinium myrtillus* (bilberry), a species restricted to the northwest part of the American continent. This is the Kalmia heath in Figure 7.2.
3. Dryadion — this alliance is the species-rich vegetation of neutral and basic soils dominated by *Dryas octopetala*. These are the limestone and serpentine heaths in Figure 7.2.

   As an example, part of the differentiated table for the limestone and serpentine heaths is presented in Table 7.10. The first point to note is the double score for each species. The first value is for abundance using a slightly modified scale:

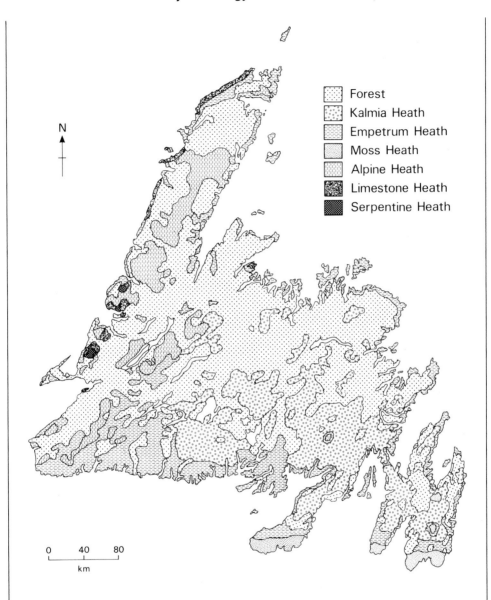

N

Forest
Kalmia Heath
Empetrum Heath
Moss Heath
Alpine Heath
Limestone Heath
Serpentine Heath

0   40   80
km

Figure 7.2 The distribution of major forest and heath types in Newfoundland —
Meades (1983) (redrawn with kind permission of Junk, The Hague)

Table 7.10   Part of the differentiated table of acidic dwarf shrub heaths in Newfoundland (Meades, 1983) (reproduced with kind permission of Junk, The Hague)

Altitude	70	70	240	240	150	120	30	75	30	120	170	30	240	120	30	150	30	30	90	170	90	75	120	60	180	180	230	180	60	75	90	150	150	150	15	150
Slope	0	0	20	20	5	0	5	10	30	30	30	10	25	30	0	0	5	0	20	30	30	30	5	20	10	10	30	40	0	5	0	50	40	0	0	0
Aspect	–	–	N	N	S	–	SE	NE	S	NW	W	W	N	N	S	–	E	E	N	SE	E	NW	NE	W	–	W	–	W	S	–	E	S	E	–	E	–
Site	71	72	71	71	71	71	71	71	71	71	71	71	71	71	71	71	71	71	71	71	71	71	71	71	71	71	71	77	71	71	71	71	71	71	71	71
Number	137	138	74	75	20	42	18A	5B	18B	1A	42P	45B	90	49J	55	15	8D	24C	46C	47B	5C	52B	41C	1A	65B	69C	5	51C	26C	37A	38G	39A	36C			

Differential Species of the Association

Species	A	Re	E	K	V
Arctostaphylos alpina	.1 .1 1.2 1.2				
Loiseleuria procumbens	.1 1.2 1.2 1.2				
Diapensia lapponica	.1 2.2 1.2				
Cetraria nivalis	.1 1.2 1.2	.1			
Ochrolechia frigida	.2 .2 .2	.1			
Rhacomitrium lanuginosum	.2 1.2 2.2 2.2	4.4 4.4 4.4 4.4	2.2		
Cladonia boryii	1.2 1.2 1.2 1.2	4.4 1.2 1.1 2.2	.2 .2		
Sphaerophorus globosus	1.1 1.1 2.2 2.2	2.2 .1 .1 1.2	.2		
Cetraria islandica	.1 .1 1.2 1.2	1.2 .1 .1 1.2			
Vaccinium uliginosum	.1 .2 2.1 2.1	.2 2.1 2.1	2.1	1.1 1.1	
Empetrum eamesii	2.2 2.2 .1 .1	2.2 1.2 2.2 1.2			1.2
Potentilla tridentata	2.1 1.1 1.1 1.1	1.1 1.1 1.1	1.1 1.1 1.1 1.1 .1	2.1	1.1 1.1 1.1
Deschampsia flexuosa	1.2 1.2 1.2	1.2 1.2 1.2	1.2 1.2 1.2 1.2	2.1	2.2 2.2 1.2
Calamagrostis pickeringii		1.2 .1 1.1 1.1	2.1 1.1 2.1 1.1 1.2	1.1	1.2 1.1
Juniperus communis	1.2 .2	2.2	1.2 1.2		2.2 1.2
Prenanthes trifoliolata	1.1 1.1 .1 .1	.1 1.1	2.2 .1		.1 1.1
Solidago uliginosa	.1 .1	.1 .1	1.1		
Rhododendron canadense			1.1	2.1 1.1 3.1 .1 2.1 1.1	2.1
Viburnum cassinoides			.1	.1 2.2 2.1 1.1 1.1 1.1	
Nemopanthus mucronata			1.1	1.1 1.1 .1 .1	1.1
Amelanchier bartramiana			.1	.1 .1 2.1 .1 1.1	1.1
Luzula campestris				.1	1.2 1.2 1.2 1.2
Lycopodium obscurum					.1 .1 .1 1.1
Taraxacum officinale					.1 .1 .1 .1
Achillea millefolium					.1
Hieracium murorum					.1
Dicranum fulvum					2.2 2.2
Anthoxanthum odoratum					1.1 1.1
Solidago rugosa					1.1 1.1 .1

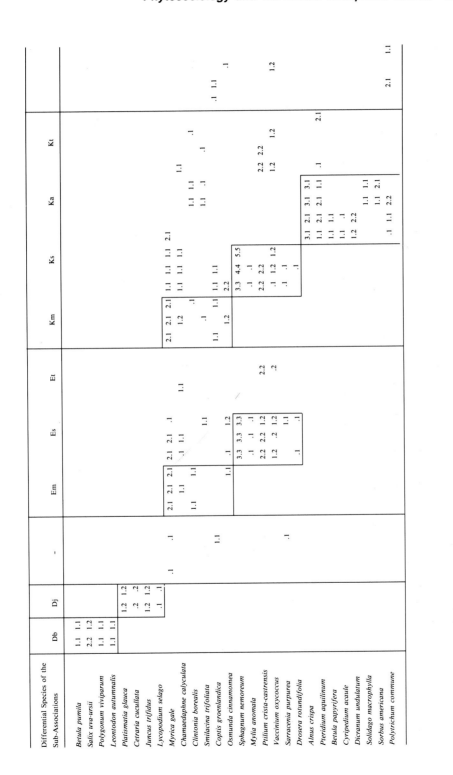

Differential Species of the Sub-Associations	Db	Dj	-	Em	Es	Et	Km	Ks	Ka	Kt	
Betula pumila	1.1 1.1										
Salix uva-ursii	2.2 1.2										
Polygonum viviparum	1.1 1.1										
Leontodon autumnalis	1.1 1.1										
Platismatia glauca		1.2 1.2									
Cetraria cucullata		.2 .2									
Juncus trifidus		1.2 1.2									
Lycopodium selago		.1 .1									
Myrica gale			.1	2.1 2.1 2.1	2.1 2.1 .1		2.1 2.1 2.1	1.1 1.1 1.1			
Chamaedaphne calyculata				1.1	.1 .1		1.2	1.1 1.1			
Clintonia borealis				1.1				1.1 1.1	1.1 1.1	1.1	
Smilacina trifoliata				1.1	1.1	1.1	.1	.1	.1 .1	.1	.1 1.1
Coptis groenlandica			1.1	1.1 .1			1.1	1.1 1.1			
Osmunda cinnamomea				1.2	1.2		1.2	2.2			.1
Sphagnum nemoreum				3.3 3.3	3.3 3.3			3.3 4.4 5.5			
Mylia anomala				.1 .1	.1 .1 .1			.1 .1			
Ptilium crista-castrensis				2.2 2.2	2.2 2.2 1.2	2.2		2.2 2.2	2.2 2.2	1.2
Vaccinium oxycoccus				1.2 1.2	1.2 .2 1.2	.2		.1 1.2 1.2	1.2	.1.2
Sarracenia purpurea			.1	1.1	1.1			.1 .1			
Drosera rotundifolia				.1 .1	.1 .1			.1			
Alnus crispa								3.1	3.1 2.1 3.1 3.1		
Pteridium aquilinum								1.1	1.1 2.1 3.1 1.1	.1	
Betula papyrifera								1.1	1.1 1.1		2.1
Cyripedium acaule								1.1	1.1 .1		
Dicranum undulatum								1.2	2.2		
Solidago macrophylla									1.1 1.1		
Sorbus americana									1.1 2.1		
Polytrichum commune								.1	1.1 2.2		2.1 1.1

	A				Re				E							K										V			
Altitude	70	70	240	150	120	30	75	30	30	75	120	170	30	240	120	30	90	170	90	60	180	180	230	180	60	75	90	150	150	150	150	15	150
Slope	0	0	20	5	0	30	30	30	30	30	30	10	25	30	20	0	0	30	0	20	10	10	20	5	20	0	0	40	0	0	40	15	0
Aspect	–	–	N	S	–	S	NW	W	S	N	N	W	N	N	W	–	E	E	E	NW	W	W	W	SE	E	N	–	W	S	–	S	E	–
Site	71	72	71	71	71	71	71	71	71	71	71	71	71	71	71	71	71	71	71	71	77	71	5	71	71	71	71	71	71	71	71	71	71
Number	137	138	74	75	20	18A	18B	5B	42P	45B	49J	19	55	15	8D	24C	46C	47B	1A	65B	69C	41D	35D	51C	26C	37A	38G	39A	36C				

Companion Species

Vaccinium angustifolium
Kalmia angustifolia
Ledum groenlandicum
Cladonia arbuscula
Cladonia mitis
Cladonia rangiferina
Cladonia alpestris
Cornus canadensis
Empetrum nigrum
Vaccinium vitis-idaea
Pleurozium schreberi
Hylocomium splendens
Dicranum scoparium
Maianthemum canadense
Trientalis borealis
Linnaea borealis
Pyrus floribunda
Dicranum fuscescens
Cornicularia aculeata
Cladonia uncialis
Cladonia terrae-novae
Cladonia elongata
Kalmia polifolia
Hypogymnia physodes
Spiraea latifolia
Polytrichum juniperinum
Aster novi-belgii
Gaultheria hispidula
Larix laricina
Hypnum imponens
Sanguisorba canadensis
Lycopodium clavatum

Additional Species:

72137:   Campanula rotundifolia (.1)
71074:   Picea glauca f. parva (.1); Stereocaulon paschale (1.2); Cladonia coccifera (.1)
71018A: Pinus sylvestris (.1)
71041D: Abies balsamea (.1)
71035D: Rubus pubescens (.1); Lycopodium annotinum (.1)
71052B: Geocaulon lividum (.1)

71069C: Prunus pennsylvanica (1.1)
71036C: Anaphalis margaretacea (.1)
71052B: Sphagnum papillosum (1.2); Aulacomnium palustre (.)
71037A: Fragaria virginiana (.1); Lonicera villosa (1.1); Rosa nitida (.1)
71036:   Cladonia cristatella (.1); Cladonia pyxidata (.1)
71046:   Cladonia squamosa (.2)
71055:   Cladonia coccifera (.1)

1 — cover less than 5%
2 — cover 5–25%
3 — cover 25–50%
4 — cover 51–75%
5 — cover 76–100%

The second digit in the table is the Braun-Blanquet index of sociability:

1 — growing singly
2 — in small tufts or tussocks
3 — in large patches
4 — carpet-forming, carpet covers at least half plot
5 — forming an almost continuous carpet

Note in Table 7.10, that the differential species are clearly delimited at the top of each group and the companion species are listed below.
   Overall, the following list of communities was recognised:

Heath communities of acid soils

1. DIAPENSIO-ARCTOSTAPHYLETUM ALPINAE (alpine heath association)
    Named after *Diapensia lapponica* and *Arctostaphylos alpina*

    Sub associations: (i)   JUNCETOSUM (*Juncus trifidus*)
                       (ii) BETULETUM (*Betula pumila*)

2. EMPETRO-RHACOMITRIETUM LANUGINOSAE (moss heath association)
    Named after *Empetrum eamesii* and *Rhacomitreum lanuginosum*

3. EMPETRETUM (*Empetrum* heaths)
    Named after *Empetrum eamesii*

    Sub-associations (i)   EMPETRETUM TYPICUM (*Empetrum eamesii*)
                     (ii)   EMPETRETUM MYRICETOSUM (*Myrica gale*)
                     (iii) EMPETRETUM SPHAGNETOSUM (*Sphagnum nemoerum*)

4. KALMIETUM (*Kalmia* heaths)
    Named after *Kalmia angustifolia*

    Sub-associations (i)   KALMIETUM TYPICUM (*Kalmia angustifolia*)
                     (ii)   KALMIETUM MYRICETOSUM (*Myrica gale*)
                     (iii) KALMIETUM SPHAGNETOSUM (*Sphagnum nemoerum*)
                     (iv)   KALMIETUM ALNETOSUM (*Alnus crispa*)

5. VACCINIETUM ANGUSTIFOLII (blueberry heaths)
    Named after *Vaccinium angustifolium*

Heath communities of limestone soils

1. EMPETRETUM (*Empetrum* heaths on limestone)
    Named after *Empetrum eamesii*

    Sub-associations (i)   EMPETRETUM-SALICETOSUM CORDIFOLIAE
    (*Salix cordifolia*)
    (ii)   EMPETRETUM-SALICETOSUM RETICIULATAE
    (*Salix reticulata*)

2. HERACULETUM-SANGUISORBETUM CANADENSE (forb-dominated
    association)
    Named after *Heracleum maximum* and *Sanguisorba canadensis*

3. POTENTILLETUM (shrubby cinquefoil association)
    Named after *Potentilla fruticosa*

    Sub-associations (i)   POTENTILLETUM-DRYADETOSUM
    INTEGRIFOLIAE (*Dryas integrifoliae*)
    (ii)   POTENTILLETUM-JUNCETOSUM ALPINAE
    (*Juncus alpinus*)

Heath communities of serpentine soils

Sub-associations (i)   LYCHNETUM TYPICUM (*Lychnis alpina van.
americana*)
(ii)   LYCHNETUM ADIANTETOSUM (*Adiantum
pedatum*)

Following this categorisation of the associations, the origins and the successional and climax status of the heathlands are discussed. This study is an
excellent example of the application of the methods of subjective classification
and is unusual in being from North America. Finally, it is worth noting that no
higher classification was attempted by Meades. In the early 1980s, when this
was written, he states 'the existing Zurich–Montpellier classification of heath . . .
will have to undergo considerable revision before the North American heaths
can be incorporated into it'. This again demonstrates the emphasis of the whole
approach on European vegetation.

Phytosociological studies in the Hoyfjellet, Southern Norway (Coker, 1988)

The purpose of this research was to carry out phytosociological studies of an
area across the tree line in Southern Norway and to map and investigate the
distribution of the major plant noda (syntaxa). A further aim was to evaluate
the efficiency of various methods of multivariate analysis when applied to
phytosociological work. The multivariate methods described in Chapters 5, 6
and 8 can be used to assist with tabular sorting and the classification of
quadrats or relevé samples leading to an interpretation in the conventional
phytosociological terms described earlier in this chapter. Finally, the associations and phytosociological groups that were identified were correlated with
the major environmental and temporal features of the montane environment
using ordination methods.

Method

In order to describe the spatial variation in floristics within the 8km^2 area across the tree line, a stratified random sampling design was adopted. For each relevé or quadrat, a 4m^2 area was taken, with species abundance measured using subjective estimates of percentage cover. All groups of plants apart from algae and fungi were recorded in a total of 500 relevés. Within these, 425 species were found.

Eighty relevés selected from the original 500 were also investigated for environmental variables. Soil samples were taken for pH, organic matter content, total nitrogen and exchangeable phosphate and calcium. Slope, aspect and an estimate of the length of snow lie were also recorded. Aspect data were not included owing to problems of transformation. It was realised subsequently that angular transformation would have been possible (see Batschelet, 1981).

Analysis

Hand-sorting of the 500 relevés was regarded as totally impracticable and thus numerical methods were applied. A number of different techniques for both classification (Chapter 8) and ordination (Chapters 5 and 6) were applied and evaluated. An early version of the MULVA phytosociological analysis program (Wildi and Orlóci, 1983; 1990) and two-way indicator species analysis (TWINSPAN) (Hill, 1979b), were used for classification, while detrended correspondence analysis (Hill, 1979a) and canonical correspondence analysis (ter Braak, 1988a,b) were used for ordination.

Classification

The MULVA program produced interesting results and showed the promise of this numerical development of the traditional approach to multivariate analysis. Unfortunately, however, at that stage (1987) the program was difficult to run on a standard personal computer and the program was not well documented. The more recent version (Wildi and Orlóci, 1990) is much improved. Better results were obtained from TWINSPAN, particularly when rare species (those occurring in only one relevé) were removed from the data set. This reduced the species total from 425 to 336.

The TWINSPAN groups were therefore interpreted giving the following syntaxa (alliances and associations) using nomenclature provided in Dahl (1956; 1985) and Økland and Bendiksen (1985). Alliance names are in bold:

Arctostaphyleto-Cetrarion nivalis
  Cetrarietum nivalis typicum trifidetosum
  Alectorieto-Arctostaphyletum uvae-ursi
  Potentilleto-Festucetum ovinae

Phyllodoco-Vaccinion myrtilli
  Phyllodoco-Vaccinetum myrtilli dicranetosum

Oxycocco-Empetrion hermaphroditi
  Chamaemoreto-Sphagnetum acutifolii
  Betuleto-Sphagnetum fusci

Lactucion
Geranieto-Betuletum
Corno-Betuletum
Rumiceto-Salicetum lapponae

Aconition septentrionalis
Aconitum septentrionalis ass.

Nardeto-Caricion bigelowii
Hylocomieto-Betuletum nanae juniperetosum

Kobresio-Dryadion
Dryas-Antennaria alpina (nodum)

Cratoneureto-Saxifragion aizoidis
Cratoneureto-Saxifragion aizoidis
Cariceto-Saxifragetum aizoidis (?)

Cassiopeto-Salicion herbaceae
Dicranetum starkei
Luzulo-Cesietum
Lophozieto-Salicetum herbacae typicum conostometosum

Ranunculeto-Oxyrion digynae
Alchemilletum alpinae
Polygoneto-Salicetum herbaceae

Sphagneto-Tomenthypnion
Carex atrofusca ass.
Carex dioica-Tomenthypnum nitens ass.
Filipendula-Mnium ass.
Salix glauca-Paludella-Sphagnum warnstorfianum ass.

Stygio-Caricion limosae
Scorpidieto-Caricetum limosae

Mniobryo-Epilobion hornemannii
Mniobryo-Epilobietum hornemanni
Philonoto-Saxifragetum stellaris

Ordination

These groupings were further confirmed by the ordination analyses, with CANOCO giving much better results than DECORANA. The major use of ordination, however, was on the subset of 80 quadrats for which environmental data had been collected. These variables were standardised to zero mean and unit variance (Chapter 5) and the CANOCO analyses showed that duration of snow-lie was the most significant environmental factor related to the first axis and calcium concentration and pH were of significance on the second. Surprisingly, phosphate levels, slope and organic content appeared to be of little significance.

Examination of biplots for both species and quadrat ordinations revealed species groups that were strongly correlated with snow patches and with calcium-rich soils. There were also distinctive groups for calcareous woodland and mire communities. Although soil moisture values were not determined, there appeared to be evidence of a wet-dry gradient corresponding to the second axis. At one end, mire species and those typical of irrigated sites formed a well-defined group, while at the other there was a clear group of lichen-heath species characterising dry sites. The relationship of this potential moisture gradient to the pH and calcium gradients was investigated further and it was concluded that the wetter sites were also the most calcareous while the drier were the least calcareous.

This research is important in demonstrating the way in which methods of numerical analysis can be used to assist with classical phytosociology, particularly where there is a very large set of data. The relationships between classification and ordination methods are also shown, with classification assisting primarily with the phytosociology but ordination proving invaluable in the exploration of environmental characteristics of the phytosociological groups and in determining the positions with respect to the primary environmental gradients. A final point is that the associations recognised above were mapped within the study area to show their biogeographical and spatial relationships.

8

Numerical classification and phytosociology

Introduction

The principles of classification and their relevance to phytosociology using tabular rearrangement were introduced in Chapter 7. This chapter is concerned with objective or numerical methods for classification. The goals of classification using numerical methods are the same as for tabular rearrangement: the grouping of a set of **individuals (quadrats or vegetation samples)** into classes on the basis of their **attributes (floristic composition)**. Ideally each group should contain quadrats with very similar species composition. These groups or classes are then interpreted and used to define a set of plant communities for the area under study.

The field of numerical classification is a very broad one and the techniques described here have not simply been used for phytosociology within plant ecology but have been applied across the whole range of sciences from biology (taxonomy), geology, geography, chemistry, medicine and astronomy to psychology, sociology, archaeology and history (Cormack, 1971; Williams, 1971; Sneath and Sokal, 1973; Sokal, 1974; Clifford and Stephenson, 1975; Whittaker, 1978b; Everitt, 1980; Greig-Smith, 1980; Gillison and Anderson, 1981; Dunn and Everitt, 1982; Gordon, 1987; van Tongeren, 1987; Gower, 1988). The methods are also widely described as techniques of **cluster analysis**, based on the concept of grouping together points representing individuals with similar characteristics in mathematical space.

Methods of numerical classification, like methods of ordination, are techniques for **data reduction and data exploration**. They are used to look for pattern and order in a set of data. The search for order is frequently an end in itself with the use of classification to production of a set of groups as the only concern. However, as with ordination, the methods can be used for **hypothesis generation**, leading to further more detailed research. The development of such methods has been closely linked with the advent of computers and their increasing power and sophistication. Classification is a tedious process, and the subjective methods of tabular arrangement described in Chapter 7 are only suitable for relatively small sets of data. The numerical techniques now available can handle hundreds and even thousands of samples or species in one analysis.

Numerical methods of classification are defined as **objective** only in the sense of **repeatability**. When used for phytosociology, a method for numerical classification represents a set of rules governing the process of grouping individuals or quadrats together. Thus for one set of data, any researcher using the same numerical method should obtain the same result. The element of subjectivity in the classification process is removed. However, as will be seen later, different numerical methods give varying results on the same set of data and these variations are dependent on the mathematical properties of each technique. Thus although any one numerical method is objective in this sense of repeatability on one set of data, there is no unique solution or single classification of a set of data. As with ordination methods, the 'best' classification is one which enables a clear ecological interpretation to be made. The idea of user satisfaction is very important and although the classification

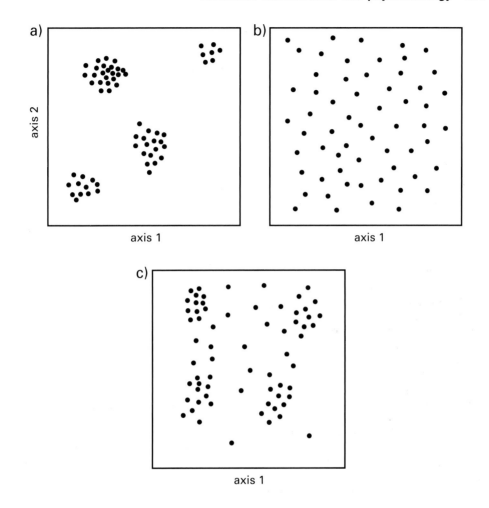

Figure 8.1 Principles of grouping or clustering shown on a two-dimensional ordination diagram: (a) clear group structure with tight clusters — numerical classification should find these easily; (b) a continuum of points with no group structure — numerical classification would arbitrarily partition the continuum; (c) the more common situation between (a) and (b) where some group structure exists but there are transitional points between groups — if numerical classification is applied, it will find the 'cores' of the groups and transitional points between groups will be allocated to the nearest group, depending on which method of numerical classification is applied

process can be described as objective, the interpretation still remains subjective. This point is discussed very effectively by Goodall (1978).

Natural and forced classifications of vegetation data

When classification is applied to a set of vegetation data, ideally the 'natural' group structure within the data should be found. Different data sets will contain individuals or quadrats that have varying intensities of group structure. Ordination diagrams (Chapters 5 and 6) are a good way of illustrating this point. In Figure 8.1a, the points representing quadrats on the ordination diagram show a distinct clustering. Remembering the fundamental principle of ordination diagrams, that increase in distance between any two points is a reflection of decrease in their degree of similarity of species composition, then a set of points close together, separated from another set, as in Figure 8.1a represents a clear group structure. Classification of these data would produce well-defined groups and these natural groupings would be be found relatively easily.

However, if classification is applied to a set of data represented on an ordination diagram such as Figure 8.1b, then it is unlikely that any meaningful groupings will be found, since the vegetation data are distributed as a **continuum** (Chapter 1). Nevertheless, it is still possible to apply methods of numerical classification to these data. If this is done, the points will be forced into groups and quite arbitrary boundaries would be drawn between the points. It is for this reason that some ecologists, particularly in America, believe that phytosociology and classification of vegetation samples is inherently wrong, since if vegetation is distributed as a continuum, then forcing a classification upon it will produce misleading results. McIntosh (1967b) discusses this at length.

In reality, most vegetation data produce ordination diagrams that lie between these two extremes, as in Figure 8.1c. Here there are clusters of points which act as cores for groups and that definitely indicate some form of group structure, which may be taken as representing the major plant communities. However, there are also some points which are transitional between groups and have properties (floristic composition) common to more than one group. There is a debate as to how to interpret this situation. Some subjective classifiers (Chapter 7) would tend to reject or ignore the transitional points, thus artificially strengthening the group structure. Other ecologists would regard these points as interesting and possibly representing samples from **ecotones** or **transitional** areas between major plant communities. However, numerical classification of the data portrayed in Figure 8.1c would cause **all** points to be allocated to their nearest group. At the interpretation stage, the decision would have to be made as to whether to disregard intermediate points between groups or to identify them as transitional types worthy of particular attention.

The important conclusion to this discussion is that when numerical classification is applied to a set of vegetation data, the researcher must believe that some group structure is present in the data and that reasonably distinct community types exist. Otherwise, the classification will only arbitrarily partition a continuum. The differences of viewpoint between Clements and Gleason on the nature of the plant community, described in Chapter 1, are worth remembering here.

This section also serves as an introduction to the relationships between classification and ordination. Increasingly, the two groups of methods are being used together on the same set of data in order to examine patterns and to search for group structure (Kent and Ballard, 1988). This is known as **complementary analysis**.

Characteristics of methods of numerical classification

A set of general principles which apply to the various methods of numerical classification has evolved over the past 25 years.

(a) Hierarchical or non-hierarchical techniques
   Most techniques devised over the past 25 years have been **hierarchical** in nature. This means that the results can be portrayed as a **dendrogram (tree or**

linkage diagram). The reason why hierarchical methods are more common is that such a dendrogram shows different levels of similarity or dissimilarity very clearly, and the different levels displayed on the dendrogram are often very helpful when it comes to making ecological interpretations. A recent alternative for the presentation of hierarchical data is **icicle plots**, but these have not yet been widely used in plant ecology. The principles of their construction are described in Kruskal and Landwehr (1983). However, some methods are **non-hierarchical** and present results in the form of a **plexus** or **constellation diagram** (Figure 3.9). Individuals or points are merely assigned to clusters and their hierarchical relationships are not shown.

(b) Divisive or agglomerative

Divisive methods start with the total population of individuals and progressively divide them into smaller and smaller groups. Division ceases either when each group is represented by a single individual or when some form of predetermined stopping rule is applied to halt division. **Agglomerative methods** start with each individual and join individuals and then individuals and groups together into larger and larger groups until all the individuals are in one big group. Again decisions are usually made to halt the process of agglomeration before this point when an interpretable set of groups has emerged.

(c) Monothetic or polythetic

With **polythetic** methods, the process of classification and allocation of individuals to groups is based on **all** the data. **Monothetic** methods allocate individuals to groups on the basis of the presence or absence of one variable or species. In practice, this terminology is now redundant, since most methods are polythetic in nature. The only major monothetic method is **association analysis**, which is described below.

(d) Quantitative or qualitative data

Methods of classification may be applied to either **quantitative** or **qualitative** data. Most methods will accept either type of data (except association analysis), and the decision on whether to use quantitative data depends on the type of problem being analysed. More recent methods, notably **two way indicator species analysis (TWINSPAN)**, employ the idea of the **pseudospecies**, whereby the presence of a species at different predetermined levels of abundance is used (Hill *et al.*, 1975; Hill, 1977; 1979b). In TWINSPAN the percentage cover scale is often divided into six using five cut levels. Thus the first pseudospecies may be 1–2 per cent cover of the species, 3–5 per cent the second pseudospecies, 6–10 per cent the third, 11–20 per cent the fourth, 21–50 per cent the fifth and over 50 per cent the sixth. These six levels of abundance of a species are then used in presence/absence form to make the classification.

(e) Equal emphasis of species

Most analyses assume that all species present are given equal importance in the analysis. It is possible, however, to downweight rarer species or to increase the importance of common or dominant species in a classification.

(f) Normal and inverse analysis

A **normal** analysis is where samples or quadrats are classified into groups on the basis of species composition. **Inverse** analysis is when groupings of species are produced on the basis of their distribution in a series of samples or quadrats.

(g) Single or joint analysis

Until recently, most methods analysed quadrats or samples separately from species (separate normal and inverse transposed analyses). Thus there was no direct mathematical relationship between the two. More recent methods and TWINSPAN in particular, carry out a **joint classification** of quadrats and species simultaneously. Considerable advantages are claimed for this approach (Hill, 1979b; Gauch, 1982b).

(h) Robustness

The idea of **robustness** is that the effectiveness of a method of classification should not be dependent on properties of a particular set of data, and a technique should perform well in most applications. Methods vary greatly in their robustness and thus the extent to which they have found widespread use.

For purposes of explanation, it is convenient to discuss hierarchical methods first under the subheadings of agglomerative and divisive.

Hierarchical classification

The following discussion assumes that a normal analysis is being performed (grouping of quadrats or samples).

Agglomerative techniques

Agglomerative methods proceed from individual samples or quadrats and progressively combine them in terms of their similarity until all the quadrats are in one group. The use of terms such as similarity and dissimilarity appears confusing at first (Chapter 3). Similarity coefficients measure how **alike** any two quadrats are in terms of species composition. Dissimilarity coefficients assess how **unlike** any two quadrats are in species composition. Dissimilarity is thus the complement of similarity. Most hierarchical agglomerative techniques are grouped under the general heading **similarity analysis**.

Similarity analysis

As shown in Figure 8.2, most methods contain four stages:

The raw data matrix
Data are first arranged in a raw data table as described in Chapter 3.

Standardization of raw data
In numerical classification, data are sometimes standardized before analysis. The following standardizations are possible:

(a) **Standardization to sample/quadrat total** Species scores within a quadrat are summed and each abundance is divided by the total. The effect of this is to correct for the overall abundance or 'size' of the sample (total cover or number of individuals). This method is sensitive to species richness.

(b) **Standardization to species total** In this case, abundances for one species are totalled over all the quadrats and then individual scores are divided by that total. This tends to overweight rare species and downweight common species. It is not widely used.

(c) **Standardization to sample/quadrat maximum** Here, all species scores are divided by the maximum attained by any species in the sample or quadrat.

(d) **Standardization to species maximum** In a similar way to (c), all quadrat scores

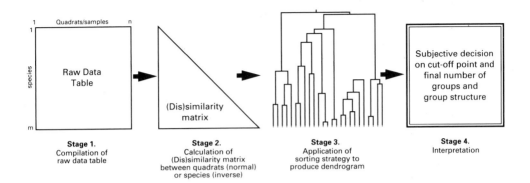

Figure 8.2 Stages in agglomerative classification using similarity analysis

for a species are divided by the maximum attained by the species across all the quadrats. This tends to weight less abundant species more equally. The data are also less dependent on the type of abundance measure used (percentage cover, frequency, density or biomass).

(e) **Standardization to unit variance and vector length** Species abundances in a quadrat are divided by the square root of the sum of the squares of the abundances. In this way, all samples are scaled equally to a range of 1.0.

The calculation of a similarity or dissimilarity matrix

The degree of matching between each pair of quadrats is then calculated on the basis of a **similarity** or **dissimilarity** coefficient. Various (dis)similarity coefficients were introduced in Chapter 3, notably the coefficient of squared Euclidean distance. Many others also exist and new and improved ones are still being devised (Crawford and Wishart, 1967; Noest and Van der Maarel, 1989). In the CLUSTAN computer package for numerical classification (Wishart, 1987), for example, over a dozen coefficients for continuous variables are presented. Nevertheless, squared Euclidean distance, which is a **dissimilarity measure**, is the most widely used. When calculated between all pairs of samples or quadrats, **a (dis)similarity matrix is produced.** An important point about dissimilarity coefficients is that they are often equated with **distances between points in mathematical space.** Thus with squared Euclidean distance, the higher the value, the greater the distance between two points or between points and groups. Further descriptions of other (dis)similarity coefficients are given in Clifford and Stephenson (1975), Everitt (1980), Orlóci (1978) and Wildi and Orlóci (1990).

The application of a sorting strategy

A sorting strategy is a set of rules bý which a set of samples or quadrats is progressively allocated to groups on the basis of the information in the (dis)similarity matrix. An important feature of such methods is the ideas of **iteration** and **fusion**. Most sorting strategies are **iterative** in that a series of passes is made through the similarity matrix, looking for the pair of individuals, the individual and group or pair of groups that are most similar to each other. This will be the pair with either the highest similarity value or the lowest dissimilarity value. When the most similar pair are found, they are **fused** into the same group. Then the next most similar pair is found and so on. This iterative process of successive grouping is

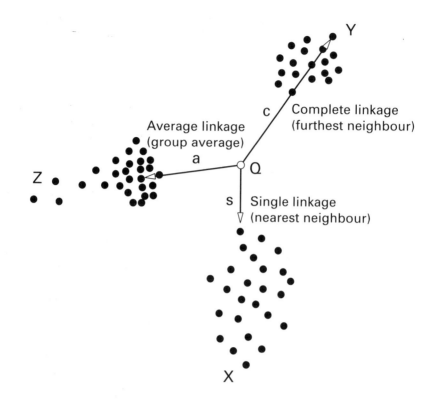

Figure 8.3 Graphical illustration of three different sorting strategies or clustering methods. Q represents a quadrat to be assigned to one of three groups, X, Y and Z. The shortest distance from Q to the nearest neighbour (single linkage) = s; furthest neighbour (complete linkage) = c; and average linkage (group average) = a

known as **fusion**. Each fusion that occurs, either between individuals, individuals and group or between groups decreases the number of groups by one.

A number of different sorting strategies have been developed under the general heading of similarity analysis. Those most frequently used in vegetation analysis have been:

Single-linkage clustering (minimum or nearest neighbour method) (Sneath and Sokal, 1973)

The method looks for the most similar pair of quadrats in the (dis)similarity matrix in order to form groups. In geometrical terms, this means that quadrats or groups are joined or fused by looking for the nearest individual or member of another group (Figure 8.3). Single linkage clustering tends to produce 'straggly' clusters and is not usually that effective for phytosociological work. Nevertheless it is more frequently used in taxonomic research.

Complete-linkage clustering (maximum or furthest neighbour method) (Sneath and Sokal, 1973)

This is the opposite of single-linkage clustering. The distances between individuals and groups is now defined as the distance between their most remote pair of individuals (Figure 8.3). The method starts in the same way as single linkage by looking for the most similar pair but further fusions depend on finding the minimum distance between the furthest points in existing groups (not the nearest, as with single linkage).

Average-linkage (group average) (Sokal and Michener, 1958)

In this case, the process of fusion is based on the minimum **average** distance between individuals and groups (Figure 8.3). The most commonly used is the unweighted pair-groups method using arithmetic averages (UPGMA).

Centroid sorting (Sokal and Michener, 1958)

The method works by finding the most similar pair of quadrats in the (dis)similarity matrix and then fusing them by averaging their attributes. The (dis)similarity between these fused quadrats and the rest of the data is then recalculated. The process is repeated with the most similar pair of individuals, individual and group or group and group, the fusion on each iteration being carried out in the same manner by averaging their attributes. The method is different from average linkage because in the latter case it is the (dis)similarity values of all individuals in a group which are averaged, rather than the attributes on fusion of individuals and groups. Centroid sorting has a number of problems attached to it, notably **reversals**, where after one dendrogram fusion, the next takes place at a lower level than the original when normally it should be higher (Clifford and Stephenson, 1975).

Minimum variance or error sums of squares clustering (Ward's method) (Ward, 1963)

This is similar to average-linkage clustering and is based on the idea that at each stage of the analysis, the loss of information on fusing quadrats into groups can be measured by the sum of squared deviations of every quadrat from the mean of the group to which it belongs. On each iteration, every possible pair of quadrats and groups is taken and those two individuals or groups whose fusion results in the lowest increase in the error sum of squares (or the variance) are combined.

Although these are the five most commonly used sorting strategies for similarity analysis (Williams *et al.*, 1966), others exist, notably **median-cluster analysis** and **McQuitty's method**. Several of these methods have been been shown to be related (Lance and Williams, 1966; 1967) and can be very efficiently programmed as computer algorithms.

Information Analysis (Williams et al., 1966)

This could be described as another type of similarity analysis but deserves specific mention because it is based on the idea of **information content** as an approach to similarity and was devised and used specifically for phytosociological purposes. The method is based on the **Information Statistic (I)**. A group of quadrats contains a certain amount of information, and the greater the differences between the quadrats making up a group, the greater the information content and the higher the value of I. The Information Statistic (I) for a group is calculated as:

$$I = \sum_{j}^{1} [n \log n - a_j \log a_j - (n - a_j)\log(n - a_j)] \qquad (8:1)$$

where n = the number of quadrats
j = the species
a_j = the number of quadrats in the group containing the jth species

The data are in presence/absence form.

Individual quadrats have an information content of zero. When two quadrats are combined, the more closely the species composition of the two quadrats matches and the lower the I values of the combined quadrats. When two groups of quadrats are combined, the information content of the new group is equal to the information content of the first group plus the information content of the second plus an increment in information content caused by the differences in species composition between the two groups. Thus:

$$I_{group\ 1} + I_{group\ 2} + I_{heterogeneity} = I_{new\ group}$$

Those two groups resulting in the least increase in information content (that is which are most similar) are combined. At the start, those two quadrats forming a group with the lowest value of I are found and combined. Worked examples of Information Analysis are given in Poole (1974) and Causton (1988). Orlóci (1972) presents alternative information functions.

In this way, fusions are progressively made and a dendrogram is constructed. The method was particularly favoured by Australian ecologists (the Canberra School) in the 1960s and 1970s. Elsewhere, it has never been widely used, but it nevertheless provides an interesting alternative to similarity analysis.

Interpretation

In agglomerative polythetic methods of similarity analysis and information analysis, a complete dendrogram is usually drawn. The problem of how to choose the most meaningful number of final groups, which are then taken to represent plant communities, still remains. This is a **subjective decision**, which relies on the **ecological knowledge and experience of the user**. One commonly applied method is to draw a line through the higher levels of the dendrogram at a suitable point and to interpret the groups that result. While this is a sensible starting point for interpretation, there is no reason why all groups should be taken from this one arbitrary level. If, on interpretation, a large group looks as though it can be sub-divided on phytosociological and ecological grounds, then it should be.   Equally, two groups which do not appear to be clearly distinct could be recombined. The key point is that at this interpretive stage, the knowledge of the researcher is paramount and phytosociological and ecological sense should prevail.

Various authors have attempted to devise more rigorous and objective methods for determining an optimal set of groups or clusters (Mojena, 1977; Gates and Hansell, 1983; Ratkowsky, 1984; Bock, 1985). Although the idea is attractive, none of these provide easily applicable answers, and a good discussion of the whole issue is provided in Dale (1988).

Similarity and information analysis of the Gutter Tor data (Table 3.4)

In Figure 8.4, the Gutter Tor data set from Dartmoor has been analysed by five different methods of similarity analysis and by information analysis. The resulting dendrograms show a high degree of similarity among the various methods and sorting strategies. The following groups emerge with a high degree of consistency:

Quadrats 1 2 9 25 and 12   Quadrats 16 and 21
Quadrats 3 18 13 20 24 and 14   Quadrats 11 17 and 22
Quadrats 8 23 19 and 15   Quadrats 4 6 and 10 (5 and 7)

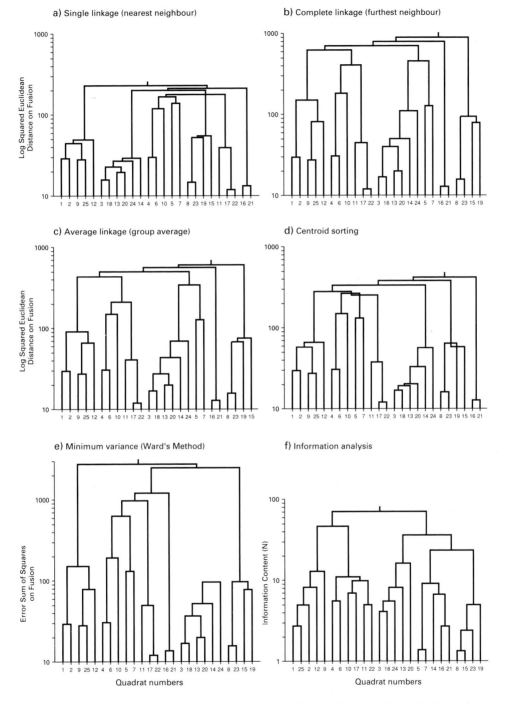

Figure 8.4 Dendrograms produced by five methods of similarity analysis and information analysis on the Gutter Tor, Dartmoor data (Table 3.4). (a) single-linkage (nearest neighbour); (b) complete-linkage (furthest neighbour); (c) average-linkage (group average); (d) centroid sorting; (e) minimum variance clustering (Ward's method); (f) information analysis

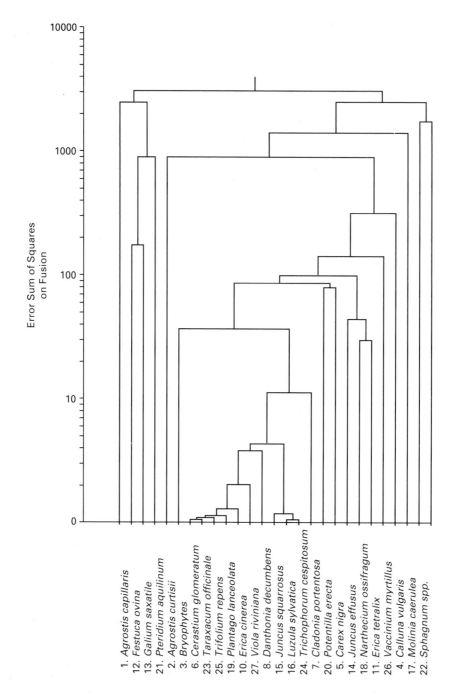

Figure 8.5 Inverse similarity analysis of the Gutter Tor data (Table 3.4) using Ward's method (minimum variance clustering)

Greater differences occur between the information analysis and the various similarity analyses, since the information analysis was performed only on presence/absence data. Nevertheless, the general patterns still remain. Another noticeable feature is the occurrence of 'reversals' in the centroid sorting dendrogram (Figure 8.4d). This is a well-known failing of the method.

For interpretation purposes, the 'best' method should be the one which makes most ecological sense. Ward's minimum-variance method is probably regarded as the optimal method of similarity analysis. Comparison of the groups from Ward's method and the subjective classification of Table 7.6 shows that virtually identical groups emerge.

The reason for the consistency between the different methods and the subjective classification is that the Gutter Tor data show a relatively clear group structure. It is likely that comparison of different methods on a data set which was closer to a continuum would result in much more variation between methods. This demonstrates the point about the subjectivity of the methodology. Even though the methods can be described as objective, the classification depends on the properties of the particular technique applied, and the resulting interpretation is still subjective, relying on the ecological knowledge and experience of the user as well as an understanding of the properties of the method.

Inverse analysis

The six dendrograms of Figure 8.4 were all **normal** analyses or quadrat classifications, where the quadrats were sorted into groups on the basis of their species content. **Inverse** analysis, or species classification, involves turning the data matrix around so that species are sorted into groups on the basis of their distribution across the set of quadrats.

In Figure 8.5, the dendrogram produced by inverse analysis of the Gutter Tor data using Ward's method is shown. Although inverse analysis was not described in the section on subjective classification in Chapter 7, the groupings of species in Table 7.6 is confirmed by the similarity analysis.

Divisive techniques

Divisive methods of classification work by taking all samples or quadrats, dividing them into two groups, those two groups into four, four into eight and so on. The earliest and most important of the divisive methods was the **monothetic** technique of **association analysis**.

Association analysis

Although association analysis is attributed to Williams and Lambert (1959; 1960; 1961), it is based on the earlier work of Goodall (1953). The method uses χ^2 calculated from 2×2 contingency tables (Chapter 3). For a normal analysis, χ^2 is computed as a measure of association between all pairs of species using equation 3:1. The χ^2 values for each species are then summed regardless of whether they are positive or negative, and that species which has the highest $\Sigma\chi^2$ value is then used as a dividing species. In the case of the Gutter Tor data (Table 3.4), the species with the highest $\Sigma\chi^2$ is *Juncus effusus*; and the quadrats are split into two groups, those containing *Juncus effusus* and those which do not. Thus the five quadrats containing *Juncus* — 1, 2, 9, 12 and 25 — form one group and the remaining 20 quadrats which do not contain *Juncus* form the other. This represents the first

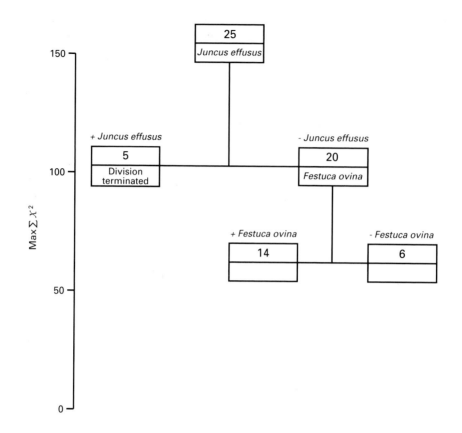

Figure 8.6 The first two levels of division of the Gutter Tor data using normal association analysis

division of the classification (Figure 8.6). The process is then repeated for each of the two new groups in turn. However, the first group containing the five quadrats with *Juncus effusus* cannot be subdivided further. Thus the group of five quadrats is finalised. Interestingly, these correspond to a group found by similarity analysis above.

The second group of 20 quadrats without *Juncus* can be divided further. The process of calculating χ^2 between all combinations of species using formula 3:1 is then repeated. However, the size of the data matrix is reduced from 20 quadrats $\times$ 27 species because four of the original 27 species (including *Juncus effusus*) only occurred in the five quadrats of the first group. Calculation of χ^2 between all combinations of the 23 species in the 20 quadrats and summing of χ^2 values for each species gives *Festuca ovina* as the species with the highest $\Sigma\chi^2$. *Festuca ovina* is thus used to split the 20 quadrats into one group of 14 quadrats containing *Festuca* and a second group of six quadrats without it. This second split is shown in Figure 8.6. Although the group of 14 quadrats could be subdivided further, this is not practicable for this very small set of data. Nevertheless, the principles of association analysis have been demonstrated. Interestingly, the six quadrats that form the group without *Festuca ovina* are 4, 6, 10, 11, 17, 22, which correspond to groupings found by similarity analysis earlier.

The method differs from the early ideas of Goodall (1953) in that he used only positive χ^2 values, whereas Williams and Lambert summed both positive and negative values regardless of sign. The effect of this was to reinforce the divisions. Williams and Lambert also proposed a set of stopping rules — the point beyond which groups would no longer be split. One method was to cease division when a group contained no significant χ^2 values and thus there were no longer any significant associations between the species. Initially, the χ^2 values of (p = 0.01) equals 6.64 and (p = 0.05) equals 3.84 were used. However, these proved to be too low for most data sets and as a result they produced too many small groups. The method only works effectively with very large data sets (hundreds or even thousands of samples), and χ^2 increases in value with the number of quadrats for a given degree of species association. It is now generally accepted that max χ^2 within a group must exceed $\sqrt{n}$, where n is the total number of quadrats in the data set, if division is to continue.

Association analysis is now only rarely used because of various problems and owing to the development of superior methods. The greatest problem is the number of misclassifications that occur in any set of data when either a quadrat happens to lack one species that is normally present in a community and which has then been used as a dividing species, or when a quadrat happens to contain a dividing species but has little else in common with the rest of the quadrats in a group. This difficulty is inevitable given the monothetic nature of the method. Also because association analysis usually only uses presence/absence data, if a set of quadrats has only a small range of community variation, the method will fail to perform effectively in the absence of abundance data. The method is also prone to 'chaining', which occurs where single quadrats are progressively split off at the start of an analysis or on formation of a group, and a well-balanced group structure does not result.

The application of association analysis is well reviewed in Ivimey-Cook and Proctor (1966), Ivimey-Cook (1972) and Coetzee and Werger (1975). Strauss (1982) also proposed a set of significance tests for species clusters. As with other methods of classification, an inverse analysis can be performed (Williams and Lambert, 1961) to give a species classification. Lambert and Williams (1962) also integrated both normal and inverse analyses to give what they described as **nodal analysis**, which represented an early form of two-way analysis.

Various criteria for calculating association and selecting dividing species have been proposed (Orlóci 1978; Wildi and Orlóci, 1990). The most commonly quoted alternative is $\sqrt{(\chi^2/N)}$ where N is the number of quadrats. This tends to give a more even split on division and this was the coefficient first recommended by Williams and Lambert.

Although association analysis is rarely used these days, examples of its application abound in the literature of the 1960s and 1970s. For this reason it is important to have an understanding of both its workings and its limitations.

Polythetic divisive methods

Polythetic divisive classification divides quadrats into groups on the basis of all the species information. Thus division is not made on the presence/absence of one species, as with association analysis, but on the basis of the species composition of the whole quadrat. Although such methods have always been seen as an optimal approach to classification, it was only in the mid-1970s that the substantial technical problems were overcome. Most polythetic divisive methods are based on ordination techniques, and one of the most important concepts in understanding them is the

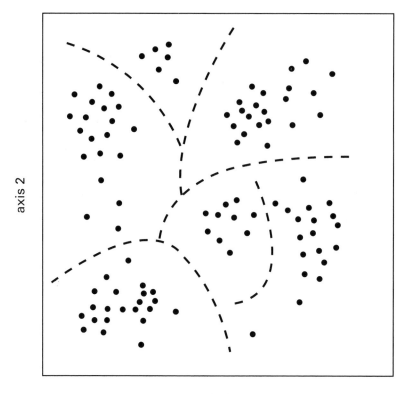

axis 1

Figure 8.7 Subjective partitioning of a two-axis ordination plot

notion of **partitioning ordination space** (Roux and Roux, 1967). This is a simple concept. On a one-, two- or three-dimensional quadrat-ordination plot, lines or partitions can be drawn through those areas of the plot which have lowest density or absence of points (Figure 8.7). This can be done subjectively, as in Figure 8.7, and if the clustering of points is marked, a satisfactory classification can be found. Thus conceptually, polythetic divisive methods are best envisaged as trying to place divisions through those least-dense areas of points representing quadrats in multidimensional space. In practice, however, they are rather more complex than this.

Two-way indicator species analysis (TWINSPAN)

Two-way indicator species analysis (TWINSPAN) (Hill, 1979b; Gauch and Whittaker, 1981) is now the most widely used technique for polythetic divisive classification. It was preceded by several developments that were similar in concept, notably AXOR (Lambert *et al.*, 1973) and POLYDIV and REMUL (Lance and Williams, 1975; Williams, 1976a,b). TWINSPAN was originally described as indicator species analysis in 1975 (Hill *et al.*, 1975). Hill (1979b) himself describes this terminology as unfortunate and confusing and suggests that perhaps the best name would have been **dichotomized ordination analysis**, since the method is based on progressive refinement of a single axis ordination from reciprocal averaging or correspondence analysis (Chapter 6).

TWINSPAN — the method — introductory principles

Pseudospecies

In phytosociology, the idea of **differential species** in defining plant communities is very important (Chapter 7). The differential species are used to make a division or a dichotomy and thus to separate one group of quadrats from another, with one set of differential species characterising one group and another set the other. However, the idea behind differential species is based on presence/absence, rather than abundance, and is thus a qualitative rather than a quantitative concept. Hill *et al.* (1975) devised the pseudospecies concept as a way of adapting abundance data to a qualitative equivalent which can then be used in the process of making a division. Thus any species abundance scale is partitioned into a series of **pseudospecies**, each of which can then be used in the process of making a dichotomy.

As an example, if percentage cover data were being analysed, these could be redefined as five pseudospecies, so that a typical species, for example *Agrostis capillaris*, could be recoded as:

Agrostis capillaris 1 percentage cover up to 2 per cent
Agrostis capillaris 2 percentage cover 3–5 per cent
Agrostis capillaris 3 percentage cover 6–10 per cent
Agrostis capillaris 4 percentage cover 11–20 per cent
Agrostis capillaris 5 percentage cover over 20 per cent

An important principle is that these classes are non-exclusive and cumulative: if only 2 per cent *Agrostis capillaris* occurred in a quadrat it would be coded as *Agrostis capillaris* 1, but if it occurred in another quadrat with 60 per cent cover all five pseudospecies would be present (*Agrostis capillaris* 1–5).

TWINSPAN output

Although the detail of the method is complex, one of the most attractive features of the technique is the computer-generated two-way table which is produced at the end of a TWINSPAN analysis. The TWINSPAN table for the garigue and maquis data from Garraf in north-east Spain (Table 3.3; Plate 3.1) is shown in Table 8.1. This represents a sorted two-way table of the original data matrix of 15 quadrats × 17 species. Quadrats have been sorted column-wise and species row-wise. Various features of this table need explanation:

(a) Quadrat numbers are written vertically across the top of the table above each column. Thus the first column is quadrat 1, the second quadrat 4, the third quadrat 5 and so on.

(b) Species names and numbers are written down the left-hand side with the species number as originally coded on data entry followed by two blocks of four letters corresponding to abbreviations of the Latin binomial. Thus the first row is species 2 *Brachypodium ramosum*, which is abbreviated to BRAC RAMO.

(c) The main block of the table presents the sorted data. Both quadrats and species have been sorted so that they form groups of both quadrats and species. The effect of this is to concentrate entries down the diagonal of the table from top-left to bottom-right. Species that do not fit easily into the overall diagonal trend may appear at the top or bottom of the table.

(d) The numbers from one to six within the table correspond to the scores for pseudospecies described above. In this instance, with percentage-cover data,

Table 8.1 The TWINSPAN table for the Garraf data (Table 3.3) (see text for explanation)

```
                    1 11    1 11
                  145206243795813

  2 BRAC  RAMO   3-4------------- 0000
  6 ERIC  MULT   433-3---------- 0000
  8 LAVA  ANGU   111------------ 0000
 15 SALV  SP.    133------------ 0000
  4 CHAM  HUMI   11------2------- 0001
 12 PIST  LENT   323-4---------- 0001
  5 CORT  SELL   ---12242-------- 0010
  7 EUPH  SP.    22-11-22-------- 0010
 10 PHIL  ANGU   23233314-------- 0010
 11 PINU  HALE   33322233-------- 0010
 13 QUER  COCC   564445451211-11 0011
 16 SEDU  MSP.   ---1-111--2----- 01
 14 ROSM  OFFI   32311-212232321 10
  1 BARE  GRND   343334356666666 11
  3 CERA  SILI   --------2211---- 11
  9 OLEA  EURO   --------3123443 11
 17 SMIL  ASPE   -------212-2312 11

                  000000001111111
                  000111110000111
                     00111
```

the scale is:

- species absent
1 less than 2 per cent cover
2 2–5 per cent cover
3 6–10 per cent cover
4 11–20 per cent cover
5 21–50 per cent cover
6 51–100 per cent cover

(e) At the bottom of the table is the dichotomized key for quadrats which shows both the group structure and the sequence of divisions. Taking the first line, the quadrats are split into the two groups indicated by the row of eight zeros (first group) and the continuation of the row as seven ones (second group). This corresponds to the first division of the quadrats into two groups of eight and seven. Each of these two groups is then split again as indicated by the second row of zeros and ones at the foot of the table. Thus the eight zeros of the first row are split in the second row into three zeros followed by five ones. Similarly, the seven ones of the first row are split into four zeros and three ones. Thus the second row summarises the second level of division, with four groups of three, five, four and three quadrats. The third row summarises the division of each of those four groups into eight and so on. In a standard analysis, division is not continued if there are four or fewer quadrats in a group. Thus only one group (Group 2) with five quadrats is split any further, as shown in the third row at the bottom of Table 8.1.

(f) On the right hand side of the table is the dichotomized key for the species. The principles are exactly the same as for quadrats, starting with the first division into two groups shown by the first column of zeros and ones. These two groups are then split into four in the second column and so on. Once again, in a standard analysis, a group is not divided further if it contains four or fewer species.

Interpretation

Interpretation of the table is subjective. For the quadrat classification, the level of subdivision at which interpretation is made depends on the size of the data set and the number of levels in the classification. In the case of the small set of data from Garraf (Table 8.1), only the second and third levels may be taken. Not all groups have to be taken from one level. If further subdivisions of a large group make more ecological sense, then it should be subdivided by looking at the next level. If two subdivisions of a small group seem to produce artificial sub-groups, then it may make sense to amalgamate them again. Similar principles apply to the species classification using the columns on the right-hand side.

The following groups emerge from the quadrat classification of the Garraf data using the second level:

Group A — Quadrats 1 4 5

Group B — Quadrats 2 10 6 12 and 14 (which could be subdivided further at level 3: into 2 10 and 6 12 14)

Group C — Quadrats 3 7 9 15

Group D — Quadrats 8 11 13

Interestingly, in Table 3.3, data on the aspect of the sites is presented as a north/south division.

Sites with a north-facing aspect = 1 2 4 5 6 10 12 14

Sites with a south-facing aspect = 3 7 8 9 11 13 15

The first division at level 1 gives a clear separation on the basis of aspect with all eight north-facing quadrats of the first side of the division separated from the remaining seven, which are south-facing (Table 8.1).

TWINSPAN — the classification method

The detailed workings of TWINSPAN are very complex. Nevertheless, it is possible to present a reasonably simplified explanation. The following is based on Hill *et al.* (1975), Hill (1979b), Jongman *et al.* (1987) and Causton (1988).

Making a dichotomy

A key concept of TWINSPAN and phytosociology is that for each division of a set of quadrats, a dichotomy can be made with a group of quadrats on one side

Table 8.2 Part of the pseudospecies table for the Garraf data (Table 3.3) — the first 6 and last species (17) are shown. There are up to six pseudospecies for each original species

Species		1	2	3	4	5	6	7	8	9	10	11	12	13	14	15	Number of pseudospecies	Cumulative pseudospecies
Bare rock (1)	1	1	1	1	1	1	1	1	1	1	1	1	1	1	1	1	6	1
	2	1	1	1	1	1	1	1	1	1	1	1	1	1	1	1		2
	3	1	1	1	1	1	1	1	1	1	1	1	1	1	1	1		3
	4			1	1		1	1	1	1		1		1	1	1		4
	5			1			1	1	1			1		1	1	1		5
	6			1			1	1	1			1		1		1		6
Brachypodium ramosum (2)	1	1				1											4	7
	2	1				1												8
	3	1				1												9
	4					1												10
	5																	
	6																	
Ceratonia siliqua (3)	1			1			1		1					1			2	11
	2			1			1											12
	3																	
	4																	
	5																	
	6																	
Chamaerops humilis (4)	1	1		1										1			2	13
	2													1				14
	3																	
	4																	
	5																	
	6																	
Cortaderia selloana (5)	1		1				1			1		1	1				4	15
	2						1			1		1	1					16
	3											1						17
	4											1						18
	5																	
	6																	
Erica multiflora (6)	1	1		1	1				1								4	19
	2	1		1	1				1									20
	3	1		1	1				1									21
	4	1																22
	5																	
	6																	
Smilax asper (17)	1		1				1	1		1		1	1	1			3	55
	2						1	1				1	1	1				56
	3							1										57
	4																	
	5																	
	6																	

Total number of species 17
Total number of pseudospecies 57

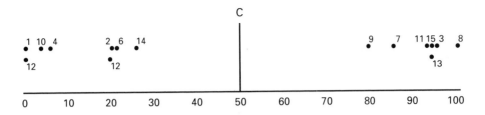

Figure 8.8 The primary ordination axis for the data from Garraf (Table 3.3), showing the positions of the 15 quadrats. The location of the centroid (the mean quadrat score on the ordination axis is also shown)

characterised by one set of differential species and a second group on the other side characterised by a second set of differential species. Ideally, species will belong to one side of the dichotomy or the other. In practice, of course, this never occurs. If it did, classification would be easy and use of numerical methods hardly necessary.

This general principle of division is applied in a series of levels, starting with the whole set of quadrats or species, dividing them into two groups, each of those into four, each of the four into eight and so on. The process of dividing a group at any level is carried out in exactly the same manner as follows:

Derivation of pseudospecies

The first stage of the analysis is the conversion of the data into pseudospecies. The concept of pseudospecies has already been introduced. Even with a small set of data, the production of pseudospecies results in a large table. Part of the pseudospecies table for the Garraf data (the first six and the last (the 17th) species across all 15 quadrats) is presented in Table 8.2. In the original data (Table 3.3), the data were coded as a six-point Domin cover scale. For the TWINSPAN analysis, the six-point scale was used to define six pseudospecies as shown in Table 8.2. Thus in quadrats 1 and 2, the first three pseudospecies of bare ground are present, while in quadrat 3, all six are present and so on. Full coding of these data gives a total of 57 pseudospecies across the 17 species, and it is these that are used in presence/absence form in the TWINSPAN analysis.

The primary ordination

This is carried out on the original raw species data using reciprocal averaging or correspondence analysis (Chapter 6). Only the **first axis ordination** is used and this **primary ordination** for the Garraf data is shown in Figure 8.8. The 15 quadrats are shown together with the centroid or mean of their scores. The quadrats can be divided into two groups on either side of the centroid — those on the left are known as the **negative** group and those on the right the **positive** group. It is important to stress that these do not refer to the presence or absence of any species but is a convenient terminology for a dichotomy or a division. In the more sophisticated version of the program produced by Hill in 1979, rarer pseudospecies are downweighted in the reciprocal averaging ordination. This is done to avoid sets of similar rare species splitting off small groups of quadrats at an early stage of the classification.

Identify preferential pseudospecies to one side or other of the dichotomy to give indicator species

When the distribution of pseudospecies in each of the two groups is examined, some pseudospecies are found to be only on one side of the division, while others are found only on the opposite side. Such pseudospecies, which occur exclusively on either side of the dichotomy, would be good **indicator species.**

Thinking of the pseudospecies data in presence/absence form, most pseudospecies will tend to be present more on one side of the division than on the other. The main exception to this would, of course, be pseudospecies which occurred in every quadrat. It thus becomes possible to work out a form of **indicator value.** For the jth pseudospecies, an indicator value (I) can be calculated as:

$$ I_j = \frac{n_j{}^+}{n_+} - \frac{n_j{}^-}{n_-} \qquad (8:1) $$

where n_+ is the total number of quadrats on the positive side of the division and n_- is the total number of quadrats on the minus side. $n_j{}^+$ is the number of quadrats on the positive side which have the jth pseudospecies and $n_j{}^-$ is the number of quadrats on the negative side which have the jth pseudospecies. Thus when a pseudospecies occurs in every quadrat on the positive side but in none on the negative side, I_j equals 1; and when a pseudospecies occurs in all quadrats on the negative side and in none on the positive, I_j equals -1. Such pseudospecies are called **perfect indicators.** If a pseudospecies occurs in every quadrat, I_j will equal 0.

In the Garraf data, *Pinus halepensis*1 and *Pinus halepensis*2 (Aleppo pine) occur in all quadrats on the negative side and in none on the positive side, while *Olea europaea*1 (Olive) occurs in no quadrats on the negative side and all quadrats on the positive side. Thus *Pinus halepensis*2, as the highest pseudospecies value for *Pinus* would be given a score of -1, while *Olea europaea*1 would be given a score of $+1$. A pseudospecies which occurs in all quadrats in the data (for example bare ground3) would obtain a score of 0.

The basic rule with the multiple pseudospecies for each species is that the pseudospecies with the highest indicator value counts as the overall indicator value for the species — for example with *Pinus halepensis*, the second pseudospecies (*Pinus halepensis*2) has a score of -1 while the third only has a score of -0.625 (5/8$-$ and 0/7$+$). *Pinus halepensis*2 is a perfect indicator while *Pinus halepensis*3 is not. Thus the final rule is that only one actual species can be an indicator species and this will be the pseudospecies number with the highest indicator value for a species.

In the TWINSPAN program, up to five indicator pseudospecies and hence species may be used to make a division. These would be the five highest indicator values regardless of whether they were positive or negative.

The refined ordination

Each quadrat is then allocated an indicator score by adding $+1$ for each positive indicator and -1 for each negative indicator that it contains. In the Garraf data, only one indicator is chosen — *Olea europaea*1, which is a positive indicator ($+$). In this very simple situation, only *Olea europaea*1 is chosen because it is a perfect indicator. *Pinus halepensis*1, *Pinus halepensis*2, *Phillyrea angustifolia*1 and *Quercus coccifera*4 could also have been chosen since they are also perfect indicators, although negative. However, there is no need to include them since they would give exactly the same result. In the refined ordination, each quadrat containing *Olea* *europaea* 1 is thus given a score of $+1$, while those without *Olea europaea* 1 are given a score of 0. This divides the quadrats into the two groups with eight on the negative side and seven on the positive.

In a more complex data set, where there are several indicator pseudospecies, which will usually not be perfect indicators but merely pseudospecies with high indicator values, each indicator is assigned the value of $+1$ or -1 depending on whether they have positive or negative values of I. Then each quadrat is given an **indicator score** based on the number of indicator pseudospecies which it contains. As an example, where there are five indicator pseudospecies, and three are positive and two are negative, then if a particular quadrat contains two of the positive indicators and two of the negative, then the indicator score is $+2 -2$ equals 0.

If a maximum of five indicator pseudospecies is to be used, then the range of indicator scores is as shown below:

Positive and negative indicators	Range of indicator scores					
5 @ -1 0 @ $+1$	-5	-4	-3	-2	-1	0
4 @ -1 1 @ $+1$	-4	-3	-2	-1	0	1
3 @ -1 2 @ $+1$	-3	-2	-1	0	1	2
2 @ -1 3 @ $+1$	-2	-1	0	1	2	3
1 @ -1 4 @ $+1$	-1	0	1	2	3	4
0 @ -1 5 @ $+1$	0	1	2	3	4	5

To make a division, an **indicator threshold** needs to be set. If with five indicator pseudospecies, there are three in the positive and two in the negative, then in the above table, a threshold of 0 will place quadrats with scores of -2, -1 and 0 in the negative group and 1, 2 and 3 in the positive. The final selection of the indicator threshold is determined by finding the position where the quadrats on the negative and positive sides, as defined by the indicator scores, agree most closely with the distribution of the quadrats on either side of the division on the original primary ordination. Thus each of the five indicator thresholds is tested in turn, and the one which gives the least discrepancy between its location on the original primary ordination and the new refined ordination is chosen.

In the small set of Garraf data, the situation is extremely simplified. Only one species, *Olea europaea* 1(+), is selected as an indicator. Thus any quadrat containing *Olea europaea* 1 has a sum of $+1$, and a quadrat without it has a score of 0. Here there is complete agreement between division on the original primary ordination and the refined ordination since *Olea europaea* 1 is a perfect indicator.

In the TWINSPAN program that Hill produced in 1979, this process of refinement of the ordination is modified and is achieved by a **transfer or iterative relocation algorithm** (Gower, 1974) with a **discriminant function** used to make the transfers. The discriminant function is derived by adding together two ordinations. The first ordination is obtained by adding together the 'preference scores' of the commoner species. A preference score of $+1$ is given to each pseudospecies that is at least **three times** more frequent on the positive side of the division than on the negative side; and vice versa, with scores of -1 being given to a pseudospecies that is three times more frequent on the negative side than the positive. Other common pseudospecies are scaled within these limits. The second ordination is derived by taking a mean of the preference scores for all pseudospecies in each quadrat. The two resulting ordinations are each scaled to an absolute maximum value of 1 and added together. The first ordination will tend to be dominant at the lower levels of

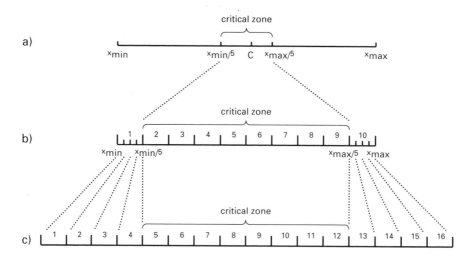

Figure 8.9 The critical zone in the refined ordination of TWINSPAN: (a) the definition of the critical zone as lying between [*xmean* − *xmin*]/5 and [*xmean* − *xmax*]/5. The critical zone thus represents 1/5th of the length of the ordination axis; (b) subdivision of the critical zone into 8 segments. In the TWINSPAN output, the areas on each side of the critical zone are also subdivided into four segments each, giving 16 segments in all; (c) in the stylised diagram which can be printed as part of TWINSPAN output (Table 8.3), the zones on either side of the critical zone are each subdivided into four segments, which, together with the eight segments of the critical zone, gives 16 segments in all

the hierarchy and the second at the higher levels. Hill (1979b) states that, in practice, this process polarizes the ordination quite strongly and reduces the number of borderline cases.

Misclassifications, borderline cases and the 'zone of indifference'

In the Garraf data, the dichotomy is very clear cut and there are no quadrats near to the central divide or axis centroid of the primary ordination axis (Figure 8.8). However, in more complex data sets, a number of quadrats will be close to or at the divide. In several cases, the location of the quadrat may be in the negative group on the primary ordination and the positive group on the refined ordination (using the indicator scores), and the possibility of **misclassification** occurs. In the original method, the assumption is made that the position of quadrats on the primary ordination is in general correct, since it is based on the complete floristic composition of the data set. In contrast, the refined ordination, being based on only 1–5 indicator pseudospecies, is less reliable. However, the purpose of the classification method is to allocate quadrats to groups on the basis of the refined ordination and the indicator scores, so the indicator score for a quadrat is preferentially used to show where a quadrat should be, while bearing in mind that the position of the quadrat on the primary ordination was different.

In order to deal with this problem, it is important to realise that most misclassifications are probably only borderline in that their primary ordination scores differ only marginally from their refined ordination scores. The exact side of the division to which they are allocated is really a matter of **indifference**. Thus a narrow **zone of indifference** is defined on the original primary ordination, where all

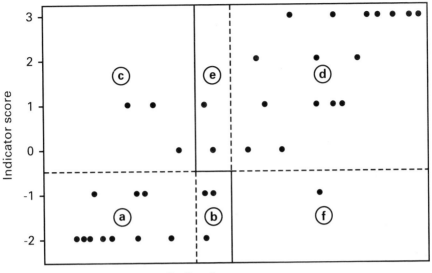

Figure 8.10 A scatter graph showing the relationship between the indicator score and the ordination score for a set of hypothetical data. Regions (a) and (d) represent the negative and positive groups respectively. Regions (c) and (f) are the misclassified negatives and positives, while the borderline quadrats lie within regions (b) and (e). The indicator threshold is set at -1. (Redrawn from Hill *et al.*, 1975; with kind permission of *Journal of Ecology*)

quadrats are classified according to their indicator score whatever their primary ordination score.

To define the zone, the primary ordination axis is divided into segments. These segments are calculated such that a **critical zone** is recognised around the centroid. Thus, using the quadrat scores (x) centred on their mean (*xmean*), the zone is defined as extending from [*xmean* $-$ *xmin*]/5 below the mean to [*xmax* $-$ *xmean*]/5 above (Figure 8.9a). This zone, which represents one-fifth of the length of the original ordination, is then further subdivided into eight segments (Figure 8.9b). The zone of indifference is then set to half the size of the critical zone (that is four segments). This zone is then gradually moved across the area of the critical zone, that is sections 2–5, 3–6, 4–7, 5–8, 6–9. In each case the indicator scores of any quadrats are re-examined and compared for any misclassifications which occur. The final optimal position for the division is that where the least number of misclassifications results.

The final division

The final point of division is thus moved so that it fits as well as possible with the indicator threshold used to produce the final **indicator ordination**. A diagram can be drawn, comparing the positions of the quadrats on the primary ordination against their positions on the refined ordination. This graph shows the six possible positions where quadrats could be located (Figure 8.10).

(a) Negative group
(b) Borderline negatives — allocated to the negative group because of their indicator scores
(c) Misclassified negatives — these are definitely placed with the negative group but their indicator scores would have placed them in the positive group
(d) Positive group
(e) Borderline positives — allocated to the positive group because of their indicator scores
(f) Misclassified positives — these are definitely placed within the positive group, but their indicator scores would have placed them in the negative group

Ideally, as is the case with the Garraf data, all quadrats would lie within groups (a) and (d). As few quadrats as possible should lie in groups (b), (c), (e) and (f).

Preferential pseudospecies

Once the division has been made, the **preferential pseudospecies** are tabulated. A pseudospecies is regarded as preferential to one side or other of the dichotomy if it is more than twice as likely to occur on one side than the other. Three categories of preferentials are recognised: positive preferentials; negative preferentials and non-preferentials. An important aspect of this is that calculation of degree of preference is dependent on group size. Thus if 60 quadrats are divided into one group of 10 and another of 50, a pseudospecies which occurs in eight out of the 10 quadrats in the first group will have a score of 0.8, while if it occurs in 18 out of the 60 in the second, it will only have a score of 0.3. The pseudospecies will therefore be deemed preferential to the smaller group even though it occurs in many more quadrats in the other larger group.

Printed output for a division

An example of the printed information relating to a division within TWINSPAN is presented in Table 8.3. This is the information produced for the first division of the Garraf data. The first line gives the division number and the number of quadrats plus the group code (*). The second line gives the information from the primary reciprocal averaging ordination. Below this are listed the indicators with their sign. In the Garraf output, only the one indicator (*Olea europaea*1) (+) is listed together with the maximum indicator score for the negative group (0) and the minimum indicator score for the positive group (1).

If requested, a stylised diagram of the pattern of the final division as shown in Table 8.3 can be printed, showing the relationship between the primary ordination axis (along the x-axis) and the indicator ordination (on the y axis). The whole axis is displayed as 16 segments, with the eight-segment critical zone extending from segments 5–12. On either side the positive and negative zones are each subdivided into four equal sub-zones to enable the pattern of quadrat distribution to be seen (Table 8.3 and Figure 8.9c). The zone of indifference is shown as lying between segment 7 and segment 10. However, given the marked polarisation of the ordination, it is not actually used. The indicator ordination shows the eight quadrats of the negative group with indicator scores of 0 clustered at the left-hand end and the seven quadrats of the positive group with indicator scores of 1 clustered to the right. An important point to realise is that the six zones of the stylised table correspond exactly to the six zones of Figure 8.10.

Below this are listed the quadrats in the positive and negative groups. In a more complex analysis borderline negatives, misclassified negatives, borderline positives and misclassified positives would all be listed here. Finally, the preferential

Table 8.3 The information printed by TWINSPAN for the first division of the Garraf data

```
Garraf vegetation data
Sample classification
```

```
DIVISION 1 (N=15) I.E. GROUP *
EIGENVALUE 0.585 AT ITERATION 1
INDICATORS, TOGETHER WITH THEIR SIGN
OLEA EURO1(+)
MAXIMUM INDICATOR SCORE FOR NEGATIVE GROUP O MINIMUM INDICATOR SCORE FOR POSITIVE GROUP 1
 1 2 3 4 5 6** 7 8 9 10** 11 12 13 14 15 16****
********************************************
 0 0 0 0 0 0** 0 0 0  0**  0  0  0  0  2  5**1
********************************************
 4 1 3 0 0 0** 0 0 0  0**  0  0  0  0  0  0**0

ITEMS IN NEGATIVE GROUP 2 (N=8) I.E. GROUP *0
QUAD 1 QUAD 2 QUAD 4 QUAD 5 QUAD 6 QUAD 10 QUAD 12 QUAD 14

ITEMS IN POSITIVE GROUP 3 (N=7) I.E. GROUP *1
QUAD 3 QUAD 7 QUAD 8 QUAD 9 QUAD 11 QUAD 13 QUAD 15

NEGATIVE PREFERENTIALS
BRAC RAMO1(2,0) CHAM HUMI1(3,0) CORT SELL1(5,0) ERIC MULT1(4,0) EUPH SP. 1(6,0) LAVA ANGU1(3,0)
PHIL ANGU1(8,0) PINU HALE1(8,0) PIST LENT1(4,0) SALV SP. 1(3,0) SEDU MSP.1(4,1) BRAC RAMO2(2,0)
CORT SELL2(4,0) ERIC MULT2(4,0) EUPH SP. 2(4,0) PHIL ANGU2(7,0) PINU HALE2(8,0) PIST LENT2(4,0)
QUER COCC2(8,1) SALV SP. 2(2,0) BRAC RAMO3(2,0) ERIC MULT3(4,0) PHIL ANGU3(5,0) PINU HALE3(5,0)
PIST LENT3(3,0) QUER COCC3(8,0) SALV SP. 3(2,0) QUER COCC4(8,0) QUER COCC5(4,0)

POSITIVE PREFERENTIALS
CERA SILI1(0,4) OLEA EURO1(0,7) SMIL ASPE1(1,6) CERA SILI2(0,2) OLEA EURO2(0,6) SMIL ASPE2(1,4)
OLEA EURO3(0,5) BARE GRND4(3,7) OLEA EURO4(0,2) BARE GRND5(1,7) BARE GRND6(0,7)

NON-PREFERENTIALS
BARE GRND1(8,7) QUER COCC1(8,6) ROSM OFFI1(7,7) BARE GRND2(8,7) ROSM OFFI2(4,6) BARE GRND3(8,7)
ROSM OFFI3(2,2)

END OF LEVEL 1
```
---

pseudospecies are listed under the three headings of negative preferentials, positive preferentials and non-preferentials. Each pseudospecies is followed by two numbers in brackets, for example BRAC RAMO1 (2, 0), which are the number of quadrats on each side of the divide in which the pseudospecies occurs. Thus *Brachypodium ramosum* 1 occurs in two quadrats on the negative side and none on the positive. Non-preferentials are shown as occurring in similar proportions of quadrats in the two groups depending on group size.

Further divisions

Further divisions are made in exactly the same way, so that each of the two groups resulting from the the first division is taken and divided in turn to give four groups. These are then each divided again to give eight and so on. In a standard analysis, division terminates when there are four or less quadrats in a group. However, this,

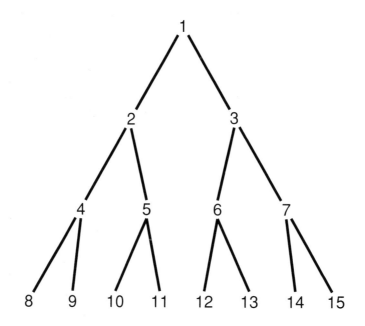

Figure 8.11 The numbering of the hierarchical group structure produced by TWINSPAN

along with most other parameters, can be varied at the start of the program.

Production of a quadrat by species two-way table

In order to produce a final table such as that shown in Table 8.2, the divisions must first be ordered and then the species must also be classified. The ordering of the quadrat groups is achieved by comparing the two site groups produced at any level with the quadrat groups at two higher levels in the hierarchy. Thus if the TWINSPAN hierarchy of divisions is numbered as in Figure 8.11, then if the positions of groups 4, 5, 6 and 7 have already been determined, each of their next divisions (8 and 9 for group 4; 10 and 11 for group 5 etc) can be swivelled and compared with the adjacent groups in the next level of the hierarchy. If 8 and 9 are swivelled and compared to group 5, then if 8 is more similar to 5 than 9, the order of groups 8 and 9 in the final table will be reversed. Again, if group 11 is more similar to group 4 than group 10 and also is less similar to group 3 than group 10, the ordering 11 10 will be used. Similar comparisons are made for all groups at the third level and switches made if necessary.

Classification of the species is achieved in a similar way to that of quadrats, except that the species classification is based on the **fidelity** (the degree to which species are confined to particular groups of quadrats) rather than on the raw data. Again a primary species ordination axis is taken out, but species are only placed in positive or negative groups with little information presented on the borderline cases between groups. The final table is then printed as in the example of Table 8.1.

The binary notation from either the bottom or the right-hand side of a TWINSPAN table can be used to produce a dendrogram of either the quadrat or the species classification. These diagrams are not usually scaled but are simply presented as series of levels showing the pattern of divisions at each level. Although

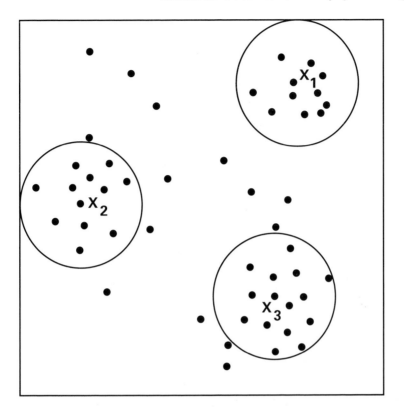

Figure 8.12 The principles of non-hierarchical clustering. X_1, X_2 and X_3 represent selected starting points, and other points within a given radius are taken in to form a group. Once circles start to overlap, rules for allocation of points to their nearest cluster have to be established or criteria have to be set for reallocation from smaller clusters to larger ones

they can be useful, they are not nearly as informative as the TWINSPAN table itself.

This explanation of the workings of TWINSPAN is still fairly complex and is difficult to grasp on a single reading. However, the really attractive feature of the method and the program is that the final table is very easily understood and relatively straightforward. Despite the complexities of the method, the end product can be seen as no more than a sorted two-way table. It is this feature which has made it so simple to use and has encouraged its widespread application.

Non-hierarchical classification

Most numerical classification of vegetation data uses hierarchical methods. However, various non-hierarchical methods exist. The simplest is the production of a constellation or plexus diagram such as Figure 3.9, where positive associations were calculated between species using χ^2 with 2×2 contingency tables. Exactly the same sort of diagram could be produced for quadrats. Another good example of this type of approach is the study of the ecology of *Juncus effusus* in North Wales by Agnew (1961).

More sophisticated methods for non-hierarchical classification have since been devised. Gauch (1979) wrote a program for composite clustering (COMPCLUS) which takes any quadrat from a set of data at random as a starting point and clusters or forms a group of all sites within a specified radius from that site (Figure 8.12). This process is then repeated until all quadrats are accounted for. At a second stage, quadrats from smaller groups or clusters are reallocated to larger groups by selecting a larger radius. Janssen (1975) devised a similar method but started by taking the first quadrat in the data set. In a computer program called CLUSLA (Louppen and van der Maarel, 1979), once a quadrat is found which lies outside of the radius of the first group, that quadrat initiates a new group. Subsequent quadrats are compared to all existing groups and allocated to that which is most similar. In all these methods, there is a strong dependence on the order in which quadrats enter the classification. FLEXCLUS (van Tongeren, 1986) is the most advanced of these methods, where a certain number of quadrats are selected at random as initiating points or are chosen by the researcher. All other sites are then allocated to the groups formed around those quadrats. A process of relocation is applied until a certain stability is achieved. A number of variations of these methods exist but non-herarchical methods are not widely used. Digby and Kempton (1987) provide useful further discussion on this topic.

Classifying large data sets

Where large sets of floristic data have been collected, it is often necessary to break the analysis down into several stages. Simply entering all data for one TWINSPAN or similarity analysis will not necessarily give satisfactory results or may not even be possible, given constraints of computer program size. The methods of both Janssen (1975) and Louppen and van der Maarel (1979), discussed in the previous section, are designed to give an initial set of divisions, within which further separate classifications may be performed. Gauch (1980) also presents a highly efficient non-hierarchical program for composite clustering of large data sets (COMPCLUS).

Van der Maarel *et al.* (1987) reviewed the problems of handling large data sets and suggested a two-step approach to classification based on stratification of the data by either geographical area or major differences in vegetation type. Numerical classification is then performed on each subset. The resulting clusters are then summarised by calculating a 'synoptic cover-abundance value' for each species in each cluster. All clusters are then subjected to further classification using these synoptic scores, and the groups resulting from these are interpreted as community types.

An alternative strategy is to sub-sample the data to produce an initial classification and then to allocate the remaining quadrats or samples to those groups. Sargent's study of railway vegetation described as a case study below is a good example of this approach.

Comparing dendrograms and group structures produced by different objective classification methods

Kent and Ballard (1988) surveyed the changing patterns of use of different classification methods over the period 1960–86. They showed the pattern of association analysis as the most widely applied technique in the late 1960s and early 1970s, similarity and information analysis in the 1970s and early 1980s with TWINSPAN becoming increasingly significant during the 1980s. Various researchers have compared the results produced by applying different techniques to the same data.

Early examples are the work of Lambert and Williams (1965) and Frenkel and Harrison (1974). More recently Gauch and Whittaker (1981) compared a range of methods of similarity analysis with TWINSPAN and partitioning of ordination space. Several simulated data sets were tested along with field data. They concluded that TWINSPAN is usually the best and most reliable technique but that there were still some situations where other techniques might be more appropriate.

Methods for comparing dendrograms and group structures have been devised, notably **cophenetic correlation** (Sneath and Sokal, 1973; Rohlf, 1974), maximisation of between or within group variance (Orlóci, 1967b) and prediction of variables from cluster membership (Gower, 1967). Digby and Kempton (1987) describe the use of two-way contingency tables to compare sets of groups resulting from different classifications of the same data set.

In general, the 'goodness' of a classification of a set of vegetation data should be assessed in terms of the ease with which ecological sense can be made of the results. Once again this involves considerable subjectivity on the part of the researcher and demonstrates the point that the interpretation of a classification remains partly an art, depending on the experience and knowledge of the user.

Characterising plant community types in terms of environmental factors

Once a set of communities have been identified by classification, they are often further characterised by associated environmental data. At its simplest, this can involve description of each group in terms of individual environmental variables such as pH, soil moisture content, nutrient status or management regime. Calculation of simple descriptive statistics (Chapter 4) such as the mean, maximum, minimum and standard deviation or their non-parametric equivalents (stem and leaf plots, medians) for continuous data is performed for the quadrats in each group. For ordinal and nominal data the construction of histograms will assist with interpretation.

More sophisticated methods for interpretation do exist, notably the use of simple discriminant functions based on a set of environmental variables to test for significant separation between groups. Ter Braak (1982, 1986b) wrote a program called DISCRIM to do this within the TWINSPAN program using only those environmental variables which optimally predict the classification. The TWINSPAN classification can be drawn as a dendrogram and the most discriminating environmental variables at each branch are shown.

Numerical classification and ordination — complementary analyses

By now it should be clear that both ordination and classification can be performed on the same data. This is known as **complementary analysis** (Kent and Ballard, 1988). One of the most effective ways of showing the relationships between the two techniques is to plot the group membership of quadrats resulting from classification on to the quadrat-ordination diagram from an ordination of the same data. In Figure 8.13a, the two-dimensional ordination plot for detrended correspondence analysis of the Garraf data is presented. In Figure 8.13b, the TWINSPAN group membership (four groups, A–D) of each quadrat has been superimposed. The relationship between the classification and the ordination should be self-evident. When used together with other diagrams for environmental factors, this approach can assist greatly with interpretation. Such plots also give some information on the group structure of the data and the extent to which the quadrats are distributed as

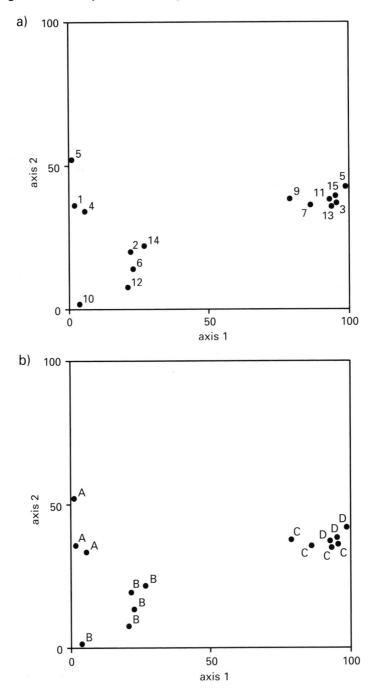

Figure 8.13   (a) Two-axis quadrat ordination plot produced by detrended correspondence analysis of the Garraf data (Table 3.3). (b) TWINSPAN groupings of the Garraf data superimposed on the ordination diagram of Figure 8.13a. Groups A–D are as in the text

a continuum or not. Unfortunately, however, the distortions inherent in most methods of ordination often tend to blur this picture.

Numerical classification and hypothesis generation

Just as with ordination, methods for numerical classification can be seen as a means of **hypothesis generation**. They are techniques for **data exploration and data reduction**. However, again, as with ordination, there has been a tendency for researchers to carry out classification as an end in itself rather than to use the results as a starting point for further work. Where the goal of the research is purely phytosociological, then this is acceptable but there are many other situations where new research ideas may be generated as a result of an initial descriptive classification.

Numerical classification and subjective classification (the Zurich–Montpellier School)

Often methods of numerical classification may be used within a phytosociological framework which is essentially that of the Zurich–Montpellier School (Chapter 7). Here the numerical methods are used to carry out the tabular rearrangement process, rather than doing it by hand and therefore subjectively. TWINSPAN produces a table which provides a useful starting point for a full Zurich–Montpellier type exercise in phytosociology. However, it is important to realise that with Zurich–Montpellier, the whole philosophy of approach is different and at the interpretation stage, the primary aim will be to fit any groupings which have been found into the established hierarchy of nomenclature and the existing abstract categorisations of the Zurich–Montpellier system.

Many other situations exist where plant community types may be recognised at the local scale for a whole range of different purposes such as attempting to understand plant-environment relationships, generate and test hypotheses concerning different management regimes for certain types of vegetation or categorisation of communities for purposes of conservation management. Such work will often contain an element of phytosociology in that one aim is to define and recognise plant-community groupings, but there may be no need to relate them to the Zurich–Montpellier classification groups for Europe or elsewhere.

This is an important point with which to end this chapter and it emphasises the need to have a clear aim and purpose to any research work in plant ecology.

Case studies

The phytosociology and mapping of the vegetation of Dartmoor, south-west England (Ward, Jones and Manton, 1972)

This research was completed by the organisation then known as the Nature Conservancy (now English Nature) in England in 1969 under the general heading of the Dartmoor Ecological Survey. Dartmoor is an area of granite upland in south-west England, characterised by moorland, acidic grassland and heath. Prior to the survey, the vegetation of Dartmoor had only been poorly described and more detailed phytosociological work had never been attempted. Also there was considerable debate over the management of the moor for grazing livestock and the effects of possible over-grazing and burning. Three main aims were presented:

(a)   A full survey of the plant communities was required;
(b)   Once the communities were defined they should be mapped using aerial photography;
(c)   The description of community types, the map and the aerial photographs would be used to provide information on the burning and grazing pressure and this could be used for planning future management.

Methods

The moor covers some 480km^2 and a transect approach with systematic random sampling was applied. Thirty-four transects were laid out across the moor to cross all major variations in the vegetation. These were followed on the ground using a compass bearing, and the vegetation was sampled at approximately every 1km. The quadrat size was 2m × 2m and species present were recorded on a 10-point Domin scale. Also relevant site characteristics and environmental data were recorded — landforms, soils and the effects of burning, grazing and drainage.

Analysis

A total of 162 quadrats were collected and these were classified using association analysis (Williams and Lambert, 1959; 1960) using χ^2 with Yates's correction and $\sqrt{\chi^2/N}$ as an index of association, where N is the number of quadrats. The dendrogram from the analysis is shown in Figure 8.14. Note the monothetic nature of the technique with the single dividing species being listed above each of the main divisions. The analysis was terminated when no single χ^2 value with Yates's correction exceeded the 0.01 probability level within a group or when there were fewer than 11 individuals within a group.

Thirteen groups emerged from the association analysis. However, the next stage of analysis was to map the vegetation using colour aerial photography. Colour photography was flown in June 1969 at a scale of 1:10,000. Markers were placed next to certain quadrats which had been described in the ground survey. From the aerial photography and the vegetation data, an air photo key was defined which characterised particular plant community types in terms of their image (colour, tone and texture) on the aerial photograph. Checking of the quadrat data against the image revealed that only eight of the 13 groups could be recognised consistently and accurately. Thus the 13 association analysis groups were reduced to eight for the purposes of mapping the vegetation.

The eight groups were: blanket bog, *Calluna-Molinia* moorland, grassland, grassland invaded by bracken, *Vaccinium* moorland, valley bog, heath and grassland with gorse. In the original paper (Ward, Jones and Manton, 1972), the typical species of each of these types is described, along with their characteristics on the aerial photograph. The vegetation map for the whole moor was then produced at a scale of 1:63,360.

A final stage of the study involved a separate analysis of grazing and burning pressure. The numbers of animals on the different community types of the aerial photographs were estimated to give a 'snapshot' sample of grazing distribution and intensity. The area of fires was also mapped, and from the different hues on the photographs it was possible to date the time since firing

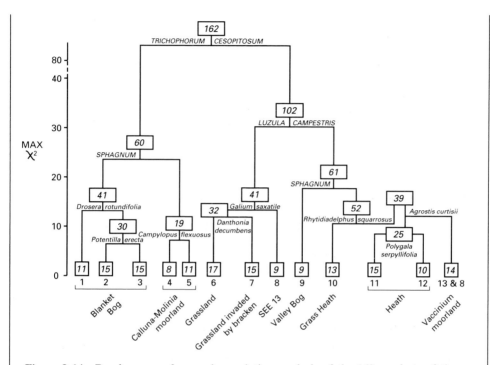

Figure 8.14 Dendrogram of normal association analysis of the 162 quadrats of the Dartmoor ecological survey (redrawn from Ward *et al.*, 1972; with kind permission of *Field Studies*)

and to determine the process of vegetation recovery. From these surveys, valuable information on the problems of grazing and burning were obtained.

This case study is of interest for several reasons. First, it is an example of the use of association analysis in a large-scale survey performed at a time when the method represented the first technique of numerical classification applied to phytosociology. Second, it shows how the recognition and definition of plant communities need not only be an end in itself but rather a starting point for vegetation mapping and a basis for further study on grazing and burning pressure. Third, although 13 major community types were recognised by association analysis, how much detail was lost by reducing these to eight categories for vegetation mapping? Should more detailed descriptions of sub-communities have been provided within each of the eight or 13 types?

Finally, the end product of this exercise was the provision of information to the various agencies charged with managing the vegetation of the moor: the National Park Authority, the Dartmoor Commoners, the local planning authority and the tenant farmers themselves. Interestingly, at the time of writing, some 20 years later, the problem of grazing is still a very controversial issue for the National Park. One last question to ask might be: 'How different would the results of the phytosociological survey have been if the data were reanalysed using TWINSPAN?'

Application of similarity analysis to a study of vegetation change on highway verges in south-east Scotland (Ross, 1986)

The vegetation of road verges is of increasing interest to plant ecologists (Way, 1969). Ross estimates that in the mid-1980s, the extent of motorway verges in Britain was increasing at the rate of 250km per year. Verges are important as reservoirs and refuges for local species which invade the original landscaping seed mix. Such species might otherwise be under threat from urbanisation, industrialisation, intensive agriculture and commercial forestry.

Surveys of verges have demonstrated the importance of both salt, which is used for de-icing, and lead from vehicle exhausts in affecting plant growth. Chow (1970), Davison (1971), Spencer *et al.* (1988) and Spencer and Port (1988) have studied such effects in some detail. As would be expected, both sodium and lead have been demonstrated to decrease as distance from the highway increases. Application of herbicides and periodic cutting for safety reasons are also commonly employed management techniques.

Ross surveyed seven sites along the A90 dual carriageway in East Lothian, Scotland at two time periods, exactly 10 years and 20 years after the original creation and seeding of the verges. The original seed mixture was as recommended by the Department of the Environment:

		%
Lolium perenne	Perennial rye grass S23	53.6
Festuca rubra	Red Fescue S59	17.9
Cynosurus cristatus	Crested dog's tail	10.7
Poa pratensis	Smooth-stalked meadow grass	8.9
Trifolium repens	White clover S100	8.9

The verges were originally sown in 1964 and surveys were carried out in 1974 and 1984. Sampling was stratified according to the verge type:

Site 1:	flat verge
Sites 2 and 5:	smooth, grassed embankment
Site 3:	rocky embankment
Sites 4 and 6:	disturbed embankment, beside lay-bys
Site 7:	rock cutting

All sites had a flat 2–3m verge adjacent to the highway.

At each site, two transects were established at right angles to the road. Percentage cover was recorded using a modified Domin scale in consecutive 1m² quadrats along the transects to the nearest field boundary. A total of 65 quadrats were sampled in 1974 and 90 in 1984.

Numerical classification was applied to the data in order to recognise species groupings and habitat types and to compare the plant communities after 10 and 20 years. Similarity analysis using two different sorting strategies (centroid sorting and single linkage) was applied within the program available in the GENSTAT package (Alvey *et al.*, 1980). The data were analysed in both abundance and presence/absence form. The dendrograms for centroid sorting of the 10- and 20-year presence/absence data are presented in Figure 8.15, with details of the major habitat groups in Table 8.4. For both the 10- and 20-year classifications, four distinct groupings were present:

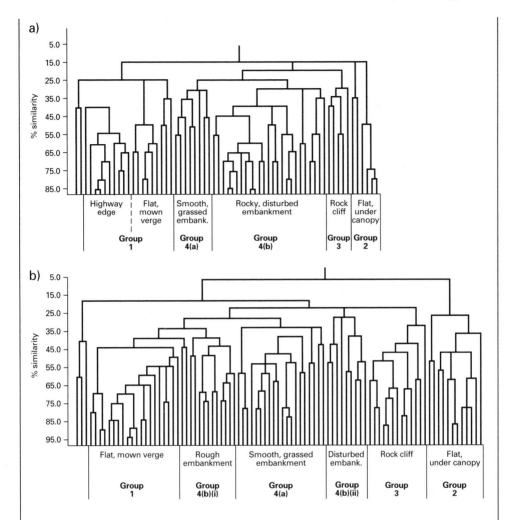

Figure 8.15 Dendrograms from similarity analysis (centroid sorting) of (A) 10-year and (B) 20-year data from road verges in East Lothian, Scotland (redrawn from Ross, 1986; with kind permission of *Journal of Biogeography*)

(1) The flat-mown verge association (1–4m from road) (*Lolium perenne, Holcus lanatus* and *Trifolium repens*);
(2) The verges under a tree-shrub canopy — *Fagus/Acer* understorey;
(3) Rock/scree cuttings;
(4) Grassed embankments — subdivided into two types at the 10-year period and three types at the 20-year period.

The results demonstrate that the grass species determine all but one of the major groups. In group 2, none of the original seed-mix species were present as dominants or frequents at either 10 or 20 years. After 20 years, *Cynosurus cristatus* and *Trifolium repens* were not even associates in any group. *Festuca rubra*, however, became the main dominant in four out of the

Table 8.4 Habitat groups from similarity analysis (centroid sorting) of (A) 10-year and (b) 20-year data on road verges in East Lothian, Scotland (Ross, 1986; reproduced with kind permission of *Journal of Biogeography*)

(A)

Group No.	No. members, 1974	Habitat	1974 (10-year survey) Dominants	Frequents	Associates
1	16	Mown verge	*Lolium perenne* *Trifolium repens* *Holcus lanatus*	*Poa annua*	*Bromus perennis* *Plantago major*
2	6	Flat under canopy	*Aegopodium podagraria* *Agropyron repens* *Galium aparine*	*Ulex europaeus* *Chamerion angustifolium*	*Dactylis glomerata* *Holcus lanatus*
3	5	Rock cliff	*Festuca rubra* *Chamaenerion angustifolium*	*Cirsium arvense* *Ulex europaeus*	*Dactylis glomerata* *Rubus fruiticosus*
4(a)	8	Smooth grassed embankment	*Festuca rubra* *Poa pratensis*	*Holcus lanatus*	*Dactylis glomerata* *Trifolium repens*
4(b)	25	Rocky disturbed embankment	*Festuca rubra* *Holcus lanatus* *Cirsium arvense*	*Trifolium repens* *Tussilago farfara* *Cynosurus cristatus*	*Ranunculus acris* *Taraxacum officinale*

Table 8.4 cont.

(B)

Group No.	No. members, 1984	Habitat	1984 (20-year survey) Dominants	Frequents	Associates
1	21	Mown verge	*Lolium perenne* *Poa pratensis* *Poa annua* *Taraxacum officinale*	*Trifolium repens* *Festuca rubra* *Bellis perennis* *Agrostis stolonifera* *Dactylis glomerata*	*Agrostis tenuis* *Plantago lanceolata* *Senecio jacobaea* *Holcus lanatus* *Pseudoscleropodium purum*
2	12	Flat under canopy	*Aegopodium podagraria* *Galium aparine* *Agrostis stolonifera*	*Agrostis tenuis* *Acer pseudoplantanus* *Fagus sylvatica*	*Lolium perenne* *Bromus ramosus*
3	13	Rock cliff	*Festuca rubra* *Chamaenerion angustifolium* *Taraxacum officinale*	*Rumex acetosella* *Dactylis glomerata* *Ulex europaeus*	*Senecio jacobaea* *Trifolium repens*
4(a)	20	Smooth grassed embankment	*Festuca rubra* *Taraxacum officinale* *Rhyitidiadelphus squarrosus*	*Lolium perenne* *Agrostis stolonifera* *Senecio jacobaea* *Ulex europaeus* *Pseudoscleropodium purum*	*Poa pratensis* *Holcus lanatus*
4(b) (i)	9	Rough embankment	*Festuca rubra* *Agrostis stolonifera* *Ryhtidiadelphus squarrosus* *Eurhynchium striatum* *Pseudoscleropodium purum*	*Dactylis glomerata* *Senecio jacobaea*	*Lophocolea heterophylla* *Lolium perenne* *Poa pratensis*
4(b) (ii)	12	Disturbed embankment	*Festuca rubra* *Agrostis stolonifera* *Bromus mollis*	*Holcus lanatus*	*Lolium perenne* *Taraxicum officinale*

six 1984 groups. Other individual species distributions are discussed in both space and time, followed by an evaluation of the effects of management practices, particularly mowing and the application of herbicides. The analysis showed that the species richness of the verges was higher in 1974 than 1984, reflecting classical successional models, where there is an initial influx of local species, followed by competition and interaction between them and the sown species and between themselves. Also the grass embankment group appeared to be diversifying in terms of habitat, with three sub-groups recognised in 1984 compared with two in 1974.

This research illustrates not only the use of similarity analysis as a means of finding pattern in a set of data. It represents a typical research project in plant ecology, where an interesting new habitat is taken and various ideas in relation to the ecology and management of that habitat are formulated. However, a necessary starting point is the recognition and definition of community types. The work is also interesting in demonstrating how repeat sets of data, in this case at 10-year intervals, can be analysed using numerical classification to enable comparison of species composition of the same quadrats at different points in time.

Finally, the work demonstrates one of the problems of similarity analyses. The results presented are for centroid sorting of presence/absence data, but the data were analysed using both centroid sorting and single linkage analysis on both presence/absence and quantitative data. This highlights the difficulties of choice in using variations of sorting strategy within similarity analysis. More recent research would suggest that Ward's method (minimum variance) or group average cluster analysis are probably the best of the sorting strategies within similarity analysis. The extent to which reanalysis using these methods would alter the interpretation of results is uncertain.

Numerical classification methods applied to the phytosociology of Britain's railway vegetation using similarity analysis and TWINSPAN (Sargent, 1984)

This project was introduced in Chapter 1 and used as an example to demonstrate problems of sampling design in Chapter 2. The purpose of the survey was to describe the nature and variation of the railway vegetation of Britain, with over 18,000km of railway being sampled in 893 track sections in 26 track classes (Table 2.8), according to the principles shown in Figure 2.21. A total of 3,502 quadrats were recorded, covering some 667 species.

One of the primary aims of the survey was a phytosociological description of the vegetation. The size of the data set meant that careful thought was required in the planning of data analysis. Simply trying to classify all 3,502 quadrats and 667 species was neither sensible nor practicable. Sargent makes the point that if TWINSPAN was applied with five cut levels to the whole data set, the analysis would require a potential pseudospecies array of $3,502 \times 667 \times 5 = 11,679,170$ elements.

A more pragmatic stepwise four-stage procedure was thus developed, whereby a stratified (by track class, Table 2.8) random subset of data was taken and classified using TWINSPAN. An attempt was made to allocate the rest of the data to the classification by using a key derived from the TWINSPAN indicator species. This failed to work and similarity analysis was used instead, giving six main groups. Finally each of the six groups was then re-sorted separately using numerical classification (TWINSPAN).

In detail, the process was carried out as follows:

(i) A subset of 937 quadrats and 442 species was selected and classified using TWINSPAN.

(ii) It was then intended that the TWINSPAN indicator species for each group be used as a dichotomised key to allocate further quadrats to the classification. The analysis of the subset showed that the maximum number of indicators allowed for in the TWINSPAN program (15) gave the least amount of misclassification (that is quadrats recognised by the program as occurring in the wrong category).

(iii) The idea of using the indicator species key was then tested by reclassifying the same 937 quadrats used to derive the TWINSPAN classification. Unfortunately, only 78 per cent of the original quadrats were allocated to their original group. Because of this, the idea of using the key was disregarded.

(iv) Instead, similarity analysis was applied with the Czekanowski coefficient as a measure of similarity and average linkage as a sorting strategy. In this case, 90 per cent of quadrats returned to their original group or next closest group, and this was the strategy finally adopted. From this, six major vegetation categories were defined:

> Heath and basifugous vegetation (noda 1–5)
> Fine-leaved grasslands (noda 6–11)
> Coarse false-oat grasslands (noda 12–18)
> Tall herb and bramble (noda 19–22)
> Scrub and secondary woodland (noda 23–27)
> Miscellaneous (noda 28–32)

(v) The rest of the data (2,565 quadrats) were then ascribed to these six groups using the Czekanowski coefficient. Each of the six groups was reanalysed separately using TWINSPAN, resulting in 32 vegetation types or noda. These noda and their relationships to Zurich-Montpellier groups are shown in Table 8.5. The 32 noda were classified as sub-communities of associations (-etum) and classes (-etea) of the Zurich-Montpellier system.

(vi) These 32 types were then characterised in terms of species constancy, track class and region and selected environmental variables.

A number of problems are highlighted by this study. The scale of the project and the size of the data matrix caused problems for data analysis. As a result, a several-stage strategy of analysis was required, whereby the analysis was based on an initial classification of a representative sub-set of the data to form six general groups, and all other quadrats were then allocated to those groups. This was a sensible strategy for such a large set of data. Also, despite using numerical methods, there is a high level of subjectivity in the analysis. The methods of numerical classification, although objective, are only used as an ordering process, once subjective decisions had been made about selection of data for input and interpretation of output.

Finally, although a major objective of the survey was to perform phytosociology and define the plant communities growing alongside British railway lines, a number of recommendations are made at the end of the report for the conservation and management of railway vegetation. Particular

Table 8.5 The 32 noda defined for British Rail vegetation by Sargent (1984). The objectively classified data (TWINSPAN) are described as sub-communities, avoiding confusion with the subjective *syntaxa* of the Zurich–Montpellier School. However, the sub-communities are placed within appropriate associations (-etum) and classes (-etea) to show their relationships with that system. Noda 29 and 32 occurred too rarely (4 and 2 stands respectively) for accurate classification

	Noda
Oxycocco-Sphagnetea Br-Bl et R Tx 1943	
Trichophoro-Callunetum McVean et Ratcliffe 1962	
Molinia caerulea subcommunity	28
Calluno-Molinietum Hill et Evans 1978	
Salix aurita subcommunity	1
Vaccinio-Piceetea Br-Bl, Siss et Vl 1939	
Vacineto-Callunetum McVean et Ratcliffe 1962	
Molinia caerulea subcommunity	2
Callunetum vulgaris McVean et Ratcliffe 1962	
Deschampsia flexuosa subcommunity	3
Agrosto-Festucetum McVean et Ratcliffe 1962	
Senecio jacobaea subcommunity	8
Molinio-Arrhenatheretea R Tx 1937	
Arrhenatheretum elatioris Br-Bl 1919	
Holcus mollis subcommunity	6
Agrostis capillaris subcommunity	7, 9
Festuca rubra subcommunity Rodwell	12, 13, 14
Brachypodium pinnatum subcommunity	11
Equisetum arvense subcommunity	15, 16
Chamaenerion angustifolium subcommunity	17, 18
Filipendula ulmaria subcommunity Rodwell	19
Urtica dioica subcommunity Rodwell	20, 22
Quercetea robori-petraeae Br-Bl et R Tx 1943	
Fago-Quercetum R Tx 1937	
Dryopteris filix-mas subcommunity	4, 5
Rhamno-Prunetea Rivas Goday et Borja Carbonell 1961	
Arrhenathero-Rosetum *assoc nov prov* Sargent	
Prunus spinosa subcommunity	21, 25
Hedera helix subcommunity	24, 27
Clematis-Viburnum subcommunity	26
Querco-Fagetea Br-Bl et Vl 1937	
Fraxino-Ulmetum Oberd 1953	
Dryopteris filix-mas subcommunity	23
Phragmitetea R Tx et Preising 1942	
Scirpo-Phragmitetum W Koch 1926	
Equisetum arvense subcommunity	30
Trifolio-Geranietea sanguinei Th Mull 1962	
Trifolio-Agrimonietum Th Mull 1962	
Arrhenatherum elatius subcommunity	10
Chenopodietea Br-Bl 1951	
Sagino-Bryetum argentei Diemont, Siss et Westhoff 1940	
Senecio viscosus subcommunity	31
Rhododendron ponticum stands	29
Asteretea tripolii Westhoff et Beeftink 1962	32

sites were identified from each railway region as being of high biological interest and the distribution of these was related to track type. Aspects of vegetation dynamics and succession were also discussed in relation to management in order to encourage species and habitat diversity. The need for a further research programme to study interaction between key railway species and certain types of activity such as ballasting, burning, grazing and the use of herbicides was also suggested.

Numerical classification as an aid to selecting lake shorelines as nature reserves in Sweden (Nilsson, 1986)

Nilsson's work was introduced in Chapter 1. As a part of a conservation evaluation of the shorelines of a number of Swedish lakes, the main aim of this project was to compare the results of a survey of a selection of shorelines for nature reserves based on methods of ecological evaluation, using criteria such as diversity and species richness, rarity, area, naturalness and typicalness (Margules and Usher, 1981), with one based on floristic composition and phytosociology. A particular concern was that the full range of community types should be represented in a nature-reserve protection programme and that with a generalised ecological evaluation focused on criteria such as diversity and rarity of plant species, this idea of floristic representativeness could be undervalued.

Twenty lakes were sampled from an area of northern Sweden (Figure 1.3), and using aerial photography, their shorelines were subdivided into 418 sections. Plants found growing along each section were identified and the results from the first 200m of each shore section were entered to numerical classification in order to define the range of community types. Two-way indicator species analysis (TWINSPAN) was the method used. The dendrogram for the first four levels of the site classification of TWINSPAN is shown in Figure 8.16. Detrended correspondence analysis (DCA) ordination was also performed on the same data, giving the two-dimensional ordination plot of Figure 8.17. The 16 groups found at level 4 of TWINSPAN (Figure 8.16) were then plotted on the ordination diagram to give a complementary analysis.

The same stretches of shore were then evaluated using the five highest ecological evaluation criteria of Margules and Usher (1981): diversity, area, rarity, naturalness and representativeness (typicalness) (see Table 3.7) Nilsson makes the point that assessment using these five criteria is highly subjective and using them, three lakes are given high conservation priority and two slightly lower, based largely on their diversity and species rarity. However, although these may be important, Nilsson believed that a proper assessment of lake shorelines for nature reserves should pay much more attention to the idea of representativeness. Thus, using the 16 plant community groups defined by TWINSPAN, six possible models for reserve selection are presented, based on their membership of particular TWINSPAN clusters. These are:

(a) At least one site from each cluster;
(b) At least a quarter of sites from each cluster;
(c) At least a half of sites in each cluster;
(d) At least three-quarters of sites in each cluster;
(e) The greatest number of the sites falling in each cluster;

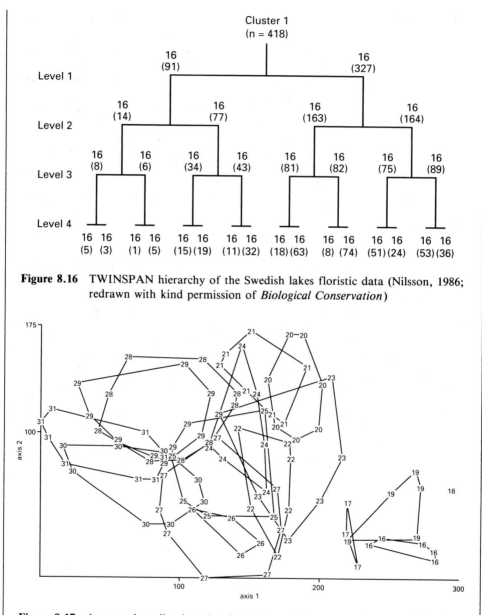

Figure 8.16   TWINSPAN hierarchy of the Swedish lakes floristic data (Nilsson, 1986; redrawn with kind permission of *Biological Conservation*)

Figure 8.17   A two-axis ordination plot from detrended correspondence analysis of the Swedish lake data. The 16 TWINSPAN groups from level 4 of the classification (Figure 8.16) are plotted as polygons using the same numbering. The polygon boundaries encompass all sites within each cluster (Nilsson, 1986; redrawn with kind permission of *Biological Conservation*)

(f)   The greatest proportion of shoreline falling within each cluster.

As the shorelines of many of the lakes had similar vegetation types, a cross-representation of all the clusters could be obtained by choosing several different series of lakes. Only one lake (Lubboträsket) is included in all of these models simply because it contains communities which do not occur elsewhere. Three lakes are outsiders in all models (Stor-Ormtjärnen, Kuträsket and Hundtjärnen), while Lakes Lubboträsket and Torrträsket together include 173 of the total 189 species found. However, Nilsson argues that further research is required to confirm the final (absolute and relative) numbers of sites which require conservation. The ordination results are not really used, but the extent to which the polygons depicted for each group in Figure 8.17 overlap could be seen as a measure of the degree of duplication of habitat among the various lakes. In the conclusion, the importance of conservation of habitat diversity as well as species diversity is stressed, while habitat diversity can only be maintained in a conservation programme by ensuring that a full range of representative sites are selected. Interestingly, Nilsson's ideas have since been applied to the different habitat of river banks (Nilsson *et al.*, 1988).

The main purpose in presenting this as a case study is to demonstrate how numerical classification (in this case TWINSPAN) can be used not only for phytosociology but also in the process of evaluation and selection of sites for conservation. Thus there is an important role for these methods in the provision of information for environmental planning.

A number of problems exist with the whole approach, however. All the models for site selection based on the classification are dependent on group size and as such are then highly dependent on the representativeness of the sampling strategy. However, allowing for this and assuming a representative sample, the number of sites in a group can be argued as being an indication of habitat rarity or scarcity, as demonstrated by Kent and Smart (1981). Also there could be difficulties in extrapolating the classification approach to a much wider area since the amount of data and the size of the analysis could become prohibitive. Adding further sites can also affect the stability of any numerical classification. Nevertheless, the approach is very interesting and shows the potential role for numerical methods of both classification and ordination beyond only phytosociology and the exploration of environmental relationships.

9

Computer programs for vegetation and environmental data analysis

Advances in computing technology have revolutionised the handling of vegetation and environmental data. Large amounts of information can now be processed very rapidly and efficiently. The major advance in recent years has been the development of the **personal computer (PC)**. Although there are many different manufacturers, most machines fitted with a reasonable amount of memory will be capable of running programs for vegetation data analysis. Many of the programs listed below were developed on mainframe machines and have since been adapted for PC.

Programs have tended to be written in either BASIC and FORTRAN. Many programs for simple statistical analysis are programmed in BASIC, while the more complex multivariate procedures for ordination and classification have largely been programmed in FORTRAN. These programs must be compiled or interpreted before they will run.

Below are listed sets of compiled programs (.EXE files) for both simple and multivariate methods, suitable for running on IBM-compatible PCs, which are available from the authors on request for a small charge. However, not all applications are available and thus other well-known packages are also described. Finally, sources of books and articles containing printed listings of programs are presented.

Programs for simple statistical analysis

The ECOSTAT package (Coker)

This is included on the disc available from the authors. The following statistical routines are covered:

Mean, variance and standard deviation
Chi-square for 2×2 contingency tables with Yates's correction
Chi-square for a $M \times N$ table
Student's t test for independent samples
Student's t test for paired samples
Mann–Whitney U test for independent samples
Wilcoxon signed ranks test for paired samples
Linear regression (least-squares)
Product moment correlation coefficient
Spearman's rank correlation coefficient
Plotless sampling calculations for the closest individual and nearest neighbour methods
Multiple correlation and regression

Other widely distributed packages

MINITAB statistical package — Release 5.1

This is only available under licence and at considerable cost. However, it is an excellent package that is easy to use. Many colleges and universities will have it on their mainframe machine under licence and thus accessible to lecturers and registered students. The PC version is very expensive, since it is only released under individual licence or to higher-education institutions in the United Kingdom through the CHEST arrangement.

Virtually all techniques for simple statistical analysis are included as well as some multivariate methods. In addition to those available in ECOSTAT, the following techniques discussed in this book are programmed:

Modes and medians, boxplots, histograms, dotplots, stem-and-leaf displays summarising categorical and ordinal data, transformations to symmetry, simple graphing of data and time series.
Multiple correlation and regression.

There are very good facilities for editing and manipulating data.
For further information see:

Ryan, B.F., Joiner, B.L. and Ryan, T.A. (1985) *MINITAB Handbook*. 2nd ed., PSW-KENT Publishing Company, Boston.

Statistical Package for the Social Sciences (SPSS)

SPSS is also only available under licence and at cost. The reason for including it here is that it is available on mainframe machines in many institutions of higher education and research. A PC version (4.0) is now available but as with MINITAB, an individual licence is expensive, although again institutions of higher education can obtain it at special rates through the CHEST arrangement.

SPSS contains programs for virtually all types of statistical analysis with the regression package being particularly useful. However, it is not as user-friendly as MINITAB, and some experience of computing is advantageous when using it.
For further information see:

Nie, N., Hadlai Hull, C., Jenkins, J.G., Steinbrenner, K. and Bent, D.H. (1975) *SPSS — Statistical Package for the Social Sciences*. 2nd ed., McGraw-Hill, New York.
Nie, N. (ed.) (1983) *SPSSX Users Guide*. McGraw-Hill, New York.

Books containing listings of programs and subroutines in BASIC and FORTRAN for simple statistical analysis

These programs will have to be entered on a PC or mainframe, checked and carefully edited before compilation. Most, but not all, include test data sets to check programs for errors. However, users will have to have some familiarity with BASIC and FORTRAN in order to get them running.

Cohen, L. and Holliday, M. (1982) *Statistics for Social Scientists — An Introductory Text with Computer Programmes in BASIC*. Harper and Row, London.

A very wide-ranging set of BASIC programs with some interesting and valuable alternative non-parametric methods.

Davies, R.G. (1971) *Computer Programming in Quantitative Biology*. Academic Press, London.

A very good set of FORTRAN programs for biology and ecology.

Davis, J.C. (1973) *Statistics and Data Analysis in Geology with FORTRAN Programs*. Wiley, New York.

A wide selection of FORTRAN programs for both simple statistical and multivariate analysis.

Lee, J.D. and Lee, T.D. (1982) *Statistical and Numerical Methods in BASIC for Biologists*. Van Nostrand Reinhold, New York.

An excellent collection of BASIC programs for most applications of simple statistics.

Programs for multivariate analysis

Most programs listed in this section are for ordination and classification.

Vegetation data analysis programs

The following programs have been compiled by the authors and are available on the disk which accompanies this book.
   The software must be installed on the computer's hard disk before use. After installation it should be possible to transfer individual programs to floppy disk but performance cannot be guaranteed.

ECOSTAT	— Simple statistics.
REFORMAT	— Takes full matrix data and converts it into the condensed form needed for the TWINSPAN and DECORANA programs. (Kovach)
DISPLAY	— Allows you to view the contents of the condensed data files used for TWINSPAN and DECORANA.
DISPLAY	— Converts condensed data files into full matrix form.

The analysis programs are:

ASSAN　　　— Association analysis;　**BRAYCURT**　— Polar ordination.

These two programs use full matrix data (see files ASSAN.DAT and BRAYCURT.DAT)

DECORANA — Hill 1979a; **TWINSPAN** — Hill 1979b.

   These programs are directly converted from the public domain versions by kind permission of Dr Mark Hill and use the Cornell condensed data format (see file CORNELL.DAT)
   The programs were compiled from edited sources Microsoft Fortran 77 v. 5.10 and the co-processor emulator and 808x code options. This means that the programs will run on any IBM-compatible PC which has 64k of base memory but they are NOT optimised for 286, 386, 486 or Pentium - based computers. A maths chip (co-processor) is not essential but speeds up processing in XT- and 286, 386 and 486SX-based computers.

MVSP (Multivariate Statistical Package) —Version 2.1
This is a shareware version and allows matrices of up to 100 x 100 items. The full version

allows users to deal with very large matrices, provides co-processors support and a manual. It is available from Dr Warren Kovach at the following address on payment of a registration fee. Contact Dr Kovach for details:

Kovach Computing Services, 85 Nant-y-Felin, Pentraeth, Inis Môn, Cymru, LL57 8UY UK

MVSP includes the following analytical programs:

— Principal component analysis (PCA)
— Principal coordinates analysis (PCO)
— Correspondence analysis/Reciprocal averaging (CA)
— Cluster analysis (Similarly analysis) -
  single-linkage/nearest neighbour
  complete-linkage/furthest neighbour
  unweighted pair group/group average
  weighted pair group
  unweighted centroid sorting
  weighted centroid sorting/median
  minimum variance (Ward's method - error sum of squares)
— (Dis)similarly matrices (10 distance measures, 4 similarly measures and 4 binary coefficients) diversity indices (Shannon, Simpson and Brillouin)
— plotting routines.

The analysis package supplied with this book also contains a number of utilities which will enable you to work easily with the analysis programs and their output. Full details of the programs are provided on the disk.

DOS VERSIONS

These programs will run on PCs which operates under Microsoft or IBM DOS 3.30 or any later versions, but the authors strongly recommend an upgrade - at least to DOS 6.0. The memory management facility in DOS 6 and later versions - called MEMMAKER - will enable you to obtain the maximum amount of memory. (below 640k), so that these programs will load and execute without running out of memory, It is possible to use other proprietary management programs to do the same job. The programs should also work with DR-DOS or Novell DOS.

Installation

Installation to a hard disk is the most straightforward way of using these programs, although once installed, individual programs can be transferred to floppy disks as required. The hard disk needs to have at least 2.5 megabytes of free space, or 4 megabytes if you want to install the MVSP program. Check this by inserting the disk into the 3.5" drive and typing:

A:FREE: (Followed by 'Enter')

This will inform you of the amount of space available on drive C.

Log on to the hard disk (suppose this is drive C) - you will get a C> prompt.

i. Type the following at the prompt.

MD VEGANAL (followed by enter)
CD VEGANAL.

Make sure that this disk is in your 3.5" drive. Suppose it is in drive A. Then type:

COPY A:*.*

(This will copy the VEGANAL compressed file and all other files).

Uncompress the files by typing VEGANAL. If disk space is at a premium, you could save nearly 700k by deleting the file VEGANAL.EXE afterwards.

Table 9.1 Sources of computer programs for ordination and numerical classification of vegetation data

Program	Copyright	CCF	M	Comment	Supplier
Ordination					
ORDIFLEX	PD	*		Flexible ordination (weighted averages, polar ordination, PCA, CA/RA)	Biological Software
DECORANA	PD	*		Detrended correspondence analysis (also CA/RA)	Biological Software
CANOCO CANODRAW CANOPLOT	C	*	M	Canonical correspondence analysis (also PCA, DCA, CA/RA) CANOPLOT and CANODRAW are plotting routines for ordination diagrams	Microcomputer Power
Numerical classification					
CLUSTAN	C			Cluster analysis/similarity analysis, association analysis, non-hierarchical clustering	Clustan Ltd.
TWINSPAN	PD	*		Two-way indicator species analysis	Biological Software
COMPCLUS	PD	*		Non-hierarchical clustering for large data sets	Biological Software
FLEXCLUS	PD	*		Flexible non-hierarchical clustering	Drs. O.F.R. van Tongeren
TABORD	PD	*		Program for tabular rearrangement	Professor E. van der Maarel
Special-purpose and combined packages					
MVSP	Sh	*		See above	Biological Software/ Warren Kovach
STATISTICAL ECOLOGY PACKAGE	B			Suite of BASIC programs for vegetation analysis — diversity indices, cluster/similarity analysis, polar ordination, PCA, CA/RA, detrended principal components, non-metric multidimensional scaling, simple and multiple regression, simple discriminant analysis	Available as diskette in Ludwig and Reynolds (1988)
MULVA-4	C		M	Specialised set of programs for analysis and simulation in phytosociology	Dr Otto Wildi

Table 9.1 cont.

Cornell Ecology Programs for DECORANA TWINSPAN COMPCLUS ORDIFLEX COMPOSE	C	*	M	These are essentially the same as the mainframe versions available on the disc from Biological Software but are optimised for PCs — expensive for individual licences	Microcomputer Power

Key

B — Diskette available under individual licence on purchase of book.

C — Commercially available program available from supplier indicated. Copyright resides with author or agents and program must not be copied or distributed except by them.

CCF — Cornell condensed data format (* — program accepts Cornell condensed data format)

M — An Apple Macintosh version is available.

PD — Public domain.

Sh — Shareware (Free use during evaluation. Register with author if program is to be permanently adopted).

Addresses

Supplier	Address
Biological Software	23 Darwin Close, Farnborough, ORPINGTON, Kent, BR6 7EP
Clustan Ltd.	16 Kingsburgh Road, EDINBURGH, Scotland, EH12 6DZ
Microcomputer Power	111 Clover Lane, ITHACA, New York, NY 14850, USA. (A catalogue of programs and prices is available on request)
Professor Eddy van der Maarel	Department of Ecological Botany, Uppsala University, Box 559, S 751 22 UPPSALA, Sweden.
Drs. O.F.R. van Tongeren	Limnological Institute, Royal Netherlands Academy of Science, Rijksstraatweg 6, 3631 AC NIEUWERSLUIS, The Netherlands
Dr Otto Wildi	Institute of Forestry Research, CH8903 BIRMENSDORF, Switzerland

ii. The MVSP program must be installed before use.

To do this, type: INSTALL

This extracts the MVSP programs and asks where the files are to be installed - on the hard disk or on the other floppy disks.

Additional information on the use of the programs on the disk is to be found by typing READIT. This enables you to page through a text file.

The disc containing both the ECOSTAT programs and those described in this section is available from:

Biological Software,
23 Darwin Close,
Farnborough,
ORPINGTON,
Kent, BR6 7EP

A fee of £8 (US$l2) is charged to cover copying, media, postage and handling and should be included with the order. The disc is only available in 3.5-inch 720k IBM format.

A number of other programs are available from Biological Software (details on request).

Other sources of programs for multivariate analysis

In Table 9.1, the most popular packages available for PCs are listed with details of copyright and supplier. Most of these packages are written in FORTRAN and will only run on IBM-compatible PCs. An increasing number of programs are available in versions which will run on Apple Macintosh computers.

Books and articles containing listings of programs for multivariate analysis (ordination and numerical classification)

Anderberg, M.R. (1973) *Cluster Analysis for Applications.* Academic Press, New York.

A detailed account of classification methods with FORTRAN programs.

Baxter, R.S. (1976) *Computer and Statistical Techniques for Planners.* Methuen.

Includes a few useful programs and subroutines in FORTRAN.

Blackith, R.E. and Reyment, R.A. (1971) *Multivariate Morphometrics.* Academic Press, New York.

A number of interesting ordination and classification programs written in FORTRAN.

Carr, J.R. (1990) 'CORSPOND: a portable FORTRAN-77 program for correspondence analysis'. *Computers and Geosciences,* **16**, 289–307.

A useful listing of a FORTRAN program for correspondence analysis.

Davies, R.G. (1971) *Computer Programming in Quantitative Biology.* Academic Press, New York.

Some elementary FORTRAN programs for multivariate analysis.

Davis, J.C. (1973) *Statistics and Data Analysis in Geology with FORTRAN programs.* Wiley, New York.

A wide selection of FORTRAN programs for multivariate analysis.

Jambu, M. (1981) 'FORTRAN IV computer program for rapid hierarchical classification of large data sets'. *Computers and Geosciences,* **7**, 297–310.

FORTRAN listing of a program for handling large amounts of data using graph theory and correspondence analysis.

Kent, M. (1977) 'BRAYCURT and RECIPRO — two programs for ordination in ecology and biogeography'. *Computer Applications New Series,* **4**, 589–647.

Two FORTRAN programs for polar ordination and correspondence analysis. Useful plotting routines.

Ludwig, J.A. and Reynolds, J.F. (1988) *Statistical Ecology — a Primer on Methods and Computing.* Wiley, New York.

An excellent suite of BASIC programs for many types of ordination and classification issued as a diskette under individual licence on purchase of the book. See Table 9.1.

Mather, P.M. (1976) *Computational Methods of Multivariate Analysis in Physical Geography.* Wiley, Chichester.

An excellent compendium of FORTRAN programs and subroutines covering most aspects of multivariate analysis. Highly recommended.

Orlóci, L. (1978) *Multivariate Analysis in Vegetation Research.* 2nd ed., Junk, The Hague.

A suite of 27 programs in BASIC — very useful but the rest of the book is rather mathematical.

Individuals and organisations

Professor John Birks, Botanical Institute, University of Bergen, BERGEN, Norway. Multivariate programs in FORTRAN, including cluster analysis and ordination techniques for IBM PCs.

Biological Software, 23 Darwin Close, Farnborough, ORPINGTON, Kent, BR6 7EP. A number of public-domain programs are available in addition to those supplied on the disc to accompany this book. Most are FORTRAN or BASIC sources but some are available as compiled executable versions which should run immediately on most IBM-compatible PCs. Write for details.

Dr Andrew Malloch, Institute of Environmental and Biological Sciences, University of Lancaster, LANCASTER, LA1 4YQ. VESPAN II package, written in FORTRAN which will handle and analyse multivariate-species data and display distributional information.

Dr Alan Morton, Blackthorn Cottage, Chawridge Lane, Winkfield, WINDSOR, Berkshire SL4 4QR. ECOLIB is a library of FORTRAN programs. A version for PC is available from Biological Software. Dr Morton has also produced DMAP, a species distribution plotting program.

Professor Eddy van der Maarel, Department of Ecological Botany, Uppsala University, Box 559, S 751 22 UPPSALA, Sweden. TABORD — a tabular arrangement program for phytosociological work based on methods of similarity analysis — widely used and effective in application.

Data format and transformations in computer programs for vegetation analysis

Data entry and formatting

In order to create data files for vegetation data to be analysed by any of these multivariate programs, the PC must have an editing program of some description. Alternatively, files can be created using a word-processing package such as Word-Perfect or WordStar which are then converted to ASCII files before being analysed.

The majority of programs (indicated by * in Table 9.1) now accept species data for ordination and classification in **condensed format**. Most vegetation data matrices are **sparse** (Chapter 3) and thus contain a large number of zero entries. These make data entry and checking prone to error. Also storage of data in complete matrix form wastes large amounts of computer memory. The condensed data format was developed to reduce these problems. With condensed data entry, only those species present in a quadrat or sample are entered, each as a couplet with a number for each species followed by its abundance score. The DATAPREP program on the disc accompanying this book converts full matrix data to condensed format.

As an example, the input file for the condensed data matrix of the Gutter Tor data (Table 3.4) is presented in a form suitable for input to the versions of DECORANA and TWINSPAN supplied on the disc available with this book:

```
27 25GUTTER TOR VEGETATION DATA - DARTMOOR - FOR VEGETATION BOOK
(I3,8(I3,F6.1))                                                      8
  1  5 15.0  9   2.0 14  5.0 17 35.0 18  20.0 22 80.0
  2  5 10.0 14  15.0 17 50.0 22 80.0 24   5.0
  3  1 80.0 12  10.0 13 10.0 20 20.0 25   3.0
  4  2 80.0  3  15.0  4 10.0  7  1.0 11   5.0 17 40.0 20  3.0
  5  3  1.0  4  35.0 12 40.0 13  1.0 26  50.0
  6  2 70.0  3   2.0 13  5.0 17 20.0 20   5.0
  7  3 15.0 12  40.0 13 35.0 26 20.0
  8 12 40.0 13  40.0 20 10.0 21 90.0 26  10.0
  9  1  1.0  5   5.0  7 10.0 13  5.0 14   5.0 20  2.0 22 75.0 24  2.0
 10  2 25.0  5  25.0  7 10.0 10  5.0 13  10.0 17 25.0 20 10.0 26 20.0
 11  4 25.0  5  10.0  7  5.0 11 20.0 17  75.0 20  5.0 26 10.0
 12  5 10.0  9   2.0 11 25.0 13  2.0 14   5.0 18 10.0 22 55.0 24 10.0
 13  1 70.0  6   1.0  8  5.0 12 30.0 13  10.0 20 10.0 27  2.0
 14  1 90.0  3   1.0  7 15.0 12 20.0 20   1.0 26 20.0
 15 12 30.0 13  20.0 20  5.0 21 60.0
 16  3  1.0  4 100.0  5  1.0  7 15.0 12   5.0 26  5.0
 17 11 20.0 13   5.0 17 90.0 20 10.0 26   5.0
 18  1 90.0  6   2.0 12 20.0 13 10.0 19   5.0 20  5.0 23  1.0 25  3.0
 19  5  5.0 12  10.0 13 25.0 20  2.0 21 100.0 27  5.0
 20  1 75.0  2  15.0  8  5.0 12 25.0 15   5.0 16  3.0
 21  4 90.0  7  20.0 12  1.0 26 20.0
 22  3  5.0  5   5.0  7 10.0 11 25.0 17  95.0 26  5.0
 23 12 50.0 13  50.0 21 95.0
 24  1 65.0 12  50.0 13 25.0 19  2.0 20  15.0 27  5.0
```

```
25   4  2.0  5  10.0  9  2.0 14 20.0 17  10.0 18  5.0 22 90.0 24  5.0
00
```

AGROCAPIAGROCURTBRYOPHYTCALLVULGCARENIGRCERAGLOMCLADPORTDANTDECUDROSROTUERICCINE
ERICTETRFESTOVINGALISAXAJUNCEFFUJUNCSQUALUZUSYLVMOLICAERNARTOSSIPLANLANCPOTEEREC
PTERAQUISPHAGNUMTARAXSP.TRICCESPTRIFREPEVACCMYRTVIOLRIVI

QUAD	1QUAD	2QUAD	3QUAD	4QUAD	5QUAD	6QUAD	7QUAD	8QUAD	9QUAD	10
QUAD	11QUAD	12QUAD	13QUAD	14QUAD	15QUAD	16QUAD	17QUAD	18QUAD	19QUAD	20
QUAD	21QUAD	22QUAD	23QUAD	24QUAD	25					

Explanation

Line 1 —   Cols. 1–5 Number of species (right-justified)
Cols. 6–10 Number of quadrats/samples (right-justified)
Cols. 11–70 A title for the data and the analysis which is reproduced throughout the output

Line 2 — Cols. 1–60 A FORTRAN FORMAT statement (I3,8(I3,F6.1)) The meaning of this statement is that the quadrat number is read in the first three columns of the row (right-justified — that is pushed to the right-hand side of the three columns). This is followed by up to eight species couplets, so that for each species present in the quadrat, there is the species number followed by the abundance of the species in that quadrat. Thus in quadrat 1 of the Gutter Tor data, species 5 (*Carex nigra*) is present with an abundance of 15 per cent, species 9 (*Drosera rotundifolia*) is present with a score of 2 per cent, species 14 (*Juncus effusus*) is present with a score of 5 per cent and so on. The species number and the abundance score within each couplet are laid out in the columns specified by the FORMAT statement, so that the species number is always written right-justified in the first three columns of each couplet and the abundance value is written in the next six columns with one decimal place (.0) with the decimal point always in the fifth column. Thus each couplet occupies nine columns and with three columns at the beginning of the line for the quadrat number, eight couplets (nine columns per couplet) can be fitted into the 80 column line of most computer VDUs. The total number of columns used in a line is then 75 ($8 \times 9 = 72 + 3 = 75$).

If there are more than eight species present in a quadrat, then the quadrat number is repeated at the start of the next line (Cols 1–3) and the remaining species are entered in the same way.

Those familiar with FORTRAN FORMAT statements will realise that the statement can be rewritten many different ways to fit particular sets of data, provided that on each line there is a quadrat number followed by a given number of species couplets.

Cols. 69–70   — The number of couplets per line (in this case eight)

Line 3 onwards — The raw vegetation data laid out as above. Note that the species couplets for each quadrat must be entered **in ascending order, as numbered in the original data matrix**.

Once all the data have been entered, the next line should have zeros instead of the quadrat number. This means that there are no further data.

Species names

After the data terminator (zeros), follow the species names, starting on a new line. These must be in the same order as the species numbering of the raw data and in the coded data. For each species name, eight columns are used, split into two groups of four letters to represent the Latin binomial, four for the genus and four for the species name — hence *Agrostis capillaris* would be AGROCAPI. Up to ten names are entered on one line and further names are continued on the next line. The result looks like a jumble of letters (see above).

Quadrat names/numbers

Following the species names, and again starting on a new line, the same principle is applied to the quadrats, samples or sites. Eight columns are used to identify each quadrat and any information on quadrat numbering in relation to transects or sampling design may be entered here. In the Gutter Tor example each sample has simply been labelled QUAD 1QUAD 2 etc. Again ten names of 80 columns can be fitted on to one line, and the identifiers should be continued at 10 per line up to the number of quadrats in the data set.

For most programs accepting this type of data entry, the species and quadrat names must be supplied.

NOTE: in later versions of TWINSPAN and DECORANA and more recently with CANOCO, the first two lines of a file may be changed and laid out in THREE lines at the start of the file as below:

```
GUTTER TOR VEGETATION DATA - DARTMOOR - FOR VEGETATION BOOK
(I3,8(I3,F6.1))
8
```

There is no need to put the species and quadrat numbers on the first line. The figure for the number of species couplets per line — eight — is moved from column 70 of line 2 and placed on its own on a new third line.

Some properties of vegetation data, species weighting and transformations

Downweighting

Raw vegetation data normally consists of a series of quadrats or samples, for each of which species have been recorded as either present or given an abundance score. Raw data may not always lend themselves immediately to analysis and it may be necessary to **weight** species. Many of the computer programs listed above have options for this, for example DECORANA and CANOCO. The commonest form of weighting is to **downweight** rare species in a set of data which will be those which occur only occasionally and randomly within the data. The reason for doing this is that rare species represent 'noise' (Chapter 3) and may confuse the major patterns and groupings within the data.

Another *ad hoc* weighting technique is suggested by van Tongeren (1986) where a lower weight is assigned to species whose abundance measurement or identification is unreliable. Another example is the presence of ecologically less relevant species such as planted trees. In such cases, downweighting will lessen the influence of the unwanted or dubious species.

However, the interpretation of the effects of downweighting, as with all interpretations from ordination and classification, will be subjective. If downweighting gives a more ecologically interpretable result, then its use will be deemed acceptable. Clearly such interpretations need to be made with care. Downweighting is discussed further in Gauch (1982b) and Jongman *et al.* (1987).

Data transformation

Transformations are applied to species or quadrat data in order to modify them in what is usually a non-linear fashion, so that parts of the measurement scale are stretched while others are shrunk. Transformation can take the form of a mathematical relationship (square root or arcsine of each value) or the direct conversion of abundance values from one scale to another (the conversion of percentage-cover data into Braun-Blanquet or Domin values).

Most data preparation programs, such as COMPOSE (Möhler, 1987) or full analysis programs such as DECORANA (Hill, 1979a) or CANOCO (ter Braak, 1987b; 1988a) provide options for a range of mathematical transformations, typically selected from the following:

Square root x	$= x^{1/2}$
Reciprocal x	$= 1/x$
Exponential x	$= a^x$
Logarithmic x	$= \log x$ [and $x = \log (x + 1)$ if the data contain zeros]
Arcsine x	$= \sin^{-1}(x)$
Inverse hyperbolic sine x	$= \sinh^{-1}(x)$

Further details of these transformations are given in van der Maarel (1979) and Jongman *et al.* (1987). Gauch (1982b) points out that untransformed data will tend to emphasise the more abundant species and that scaling of data in human mental processes is often a balance between quantitative and qualitative information with both dominant and rare species receiving some consideration.

If vegetation data are collected as part of a phytosociological survey, an automatic transformation of data will take place, since estimates of cover/abundance are converted into single digits, each of which represents a range of percentage-cover values.

In addition to the specialised scalings mentioned above, a straightforward conversion to presence/absence (1/0) is sometimes used, although this reduces information content. Some vegetation scientists would argue that the presence of a species is far more important than its abundance and this may need to be stressed in an analysis.

10

Quantitative plant ecology, vegetation science and the future

In writing this book, we have sought to demonstrate that vegetation description and analysis and quantitative plant ecology represent an exciting and challenging subject. However, in recent years considerable criticism has been levelled at the discipline. Most notable among these are the comments of some individualistic plant ecologists. Kent and Ballard (1988) drew attention to the remarks in a book edited by the English ecologist Michael Crawley, entitled *Plant Ecology*, published in 1986. In his chapter on the structure of plant communities, Crawley states:

This proliferation of multivariate techniques for the analysis of spatial variation in plant community structure has not, however, led to great advances in our understanding of the processes underlying these patterns.

He then quoted Robert May, a physicist and ecologist, who wrote in 1985:

the wilderness of meticulous classification and ordination of plant communities in which plant ecology has wandered for so long began in the pursuit of answers to questions but then became an activity simply for its own sake. [May, 1985; p. 112; Crawley, 1986; p. 33]

These same quotations were cited again and strongly rejected by van der Maarel (1989) in his wide-ranging introductory review paper for the Second Conference of the Working Group for Theoretical Vegetation Science and the International Association for Vegetation Science. While most quantitative plant ecologists would feel that these remarks are both fundamentally wrong and overstated, the criticisms nevertheless have some basis and serve their purpose, even if they only encourage plant ecologists to look carefully at and evaluate the whole philosophy and methodology of their subject.

As explained in the preface, this book represents an attempt to make vegetation description and analysis more accessible to teachers, researchers and especially to students who have an interest in this field. In particular, we have tried to provide simple but clear explanations of approaches and techniques, keeping the amount of difficult mathematics and statistics to the minimum necessary for understanding. To some mathematical ecologists, no doubt, this 'grey box' approach will seem unsatisfactory. However, we would ask them to demonstrate how they would make the subject more attractive to the majority of students and teachers who have a fascination for vegetation, plants and ecology but whose mathematical background is inevitably limited.

In seeking to answer the implied criticisms in the above quotations, we have also tried to adopt both an ecological and practical approach, showing that it is the ecological context in which vegetation description and analysis is carried out which is most important, rather than only the methods of data collection and analysis themselves. The methods are only a means to an end, and we believe that the case studies at the end of each chapter successfully emphasise this point.

The practical approach is also important. It is when students identify their own problem, design their own field programme, collect their own data, select and apply appropriate methods of description and analysis and make their own interpretations of results that the subject really comes alive. Also we are very fortunate that the practical implementation of many of the techniques of data analysis has recently become much easier with the arrival of the personal computer (PC), combined with user-friendly software, to give enormous computing power at very low cost.

Exciting new developments are already indicated. In terms of theory, the new ideas of fuzzy systems vegetation theory (Roberts, 1986; 1989) and fractals as a tool for describing spatial patterns of plant communities (Palmer, 1988), look potentially fascinating. Austin's work on the nature of the continuum, tolerance curves and environmental gradients (for example Austin and Smith, 1989) will provide a much improved understanding of plant response to environment. In terms of methodology, no doubt new techniques of ordination and classification will supersede the existing ones and in the process give Crawley, May and others justifiable further cause for concern.

Perhaps the way forward in answer to these criticisms is for us to restress the point which we have made numerous times in the course of writing this book. Methods of vegetation description and analysis should more often be seen as a prelude to more detailed research. They are inductive techniques for description and hypothesis generation. Once new ideas and questions have been generated by description at the community scale, then more detailed and specific research at the level of the individual plant, as well as at the community level, becomes possible. Both community and individualistic plant ecologists have complementary roles to play. Ultimately, together, they have the same overall goals of describing and understanding pattern and process within vegetation and ecosystems.

References

Agnew, A.D.Q. (1961) The ecology of *Juncus effusus L.* in North Wales. *Journal of Ecology* **49**, 83–102.

Allen, S.E., Grimshaw, H.M. & Rowland, A.P. (1986) Chemical analysis. In: Moore, P.D. & Chapman, S.E., (eds.) *Methods in Plant Ecology.* 2nd. ed., Blackwell Scientific, London, 285–344.

Allen, T.F.H. & Hoekstra, T.W. (1990) The confusion between scale-defined levels and conventional levels of organization in ecology. *Journal of Vegetation Science* **1**, 5–12.

Allen, T.F.H. & Wileyto, E.P. (1983) A hierarchical model for the complexity of plant communities. *Journal of Theoretical Biology* **101**, 529–540.

Alvey, N.G. *et al.* (1980) *GENSTAT: a general statistical program.* Lawes Agricultural Trust, Rothampstead Experimental Station, Numerical Algorithms Group Ltd, Oxford.

Anderberg, M.R. (1973) *Cluster Analysis for Applications.* Academic Press, New York.

Anderson, A.J.B. (1971) Ordination methods in ecology. *Journal of Ecology* **59**, 713–26.

Anderson, D.J. & Kikkawa, J. (1986) Development of concepts. In: Kikkawa, J. & Anderson, D.J. (eds.) *Community Ecology – Pattern and Process,* Blackwell Scientific Publications, Oxford, 3–16.

Auerbach, M. & Shmida, A. (1987) Spatial scale and the determination of plant species richness. *Trends in Ecology and Evolution* **2**, 238–42.

Austin, M.P. (1968) An ordination study of a chalk-grassland community. *Journal of Ecology* **56**, 739–57.

Austin, M.P. (1971) Role of regression analysis in ecology. In: *Quantitative Ecology, Proceedings of the Ecological Society of Australia* **6**, 63–75.

Austin, M.P. (1976) Performance of four ordination techniques assuming three different non-linear response models. *Vegetatio* **33**, 43–9.

Austin, M.P. (1980) Searching for a model for use in vegetation analysis. *Vegetatio* **42**, 11–21.

Austin, M.P. (1981) Permanent quadrats: an interface for theory and practice. *Vegetatio* **46**, 1–10.

Austin, M.P. (1985) Continuum concept, ordination methods and niche theory. *Annual Review of Ecology and Systematics* **16**, 39–61.

Austin, M.P. (1986) The theoretical basis of vegetation science. *Trends in Ecology and Evolution* **1**, 161–4.

Austin, M.P. (1987) Models for the analysis of species response to environmental gradients. *Vegetatio* **69**, 35–45.

Austin, M.P. & Austin, B.O. (1980) Behaviour of experimental plant communities along a nutrient gradient. *Journal of Ecology* **68**, 891–918.

Austin, M.P. & Heyligers, P.C. (1989) Vegetation survey design for conservation: Gradsect sampling of forests in North–eastern New South Wales. *Biological Conservation* **50**, 13–32.

Austin, M.P. & Orlóci, L. (1966) Geometric models in ecology. II. An evaluation of some ordination techniques. *Journal of Ecology* **54**, 217–227.

Austin, M.P. & Smith, T.M. (1989) A new model for the continuum concept. *Vegetatio* **83**, 35–47.

Austin, M.P., Cunningham, R.B. & Fleming, P.M. (1984) New approaches to direct gradient analysis using environmental scalars and statistical curve-fitting procedures. *Vegetatio* **55**, 11–27.

Bakker, P.A. (1979) Vegetation science and nature conservation. In: Werger, M.J.A. (ed.) *The Study of Vegetation*. Junk, The Hague, 249–288.

Ball, D.F. (1986) Site and soils. In: Moore, P.D. & Chapman, S.E. (eds.) *Methods in Plant Ecology*. 2nd ed., Blackwell Scientific, London, 215–284.

Ball, M.E. (1974) Floristic changes on grasslands and heaths on the Isle of Rhum after a reduction or exclusion of grazing. *Journal of Environmental Management* **2**, 299–318.

Barber, C.M. (1988) *Elementary Statistics for Geographers*. Guilford Press, New York.

Barclay-Estrup, P. (1971) The description and interpretation of cyclical processes in a heath community. III. Microclimate in relation to the *Calluna* cycle. *Journal of Ecology* **59**, 143–66.

Barclay-Estrup, P. & Gimingham, C.H. (1969) The description and interpretation of cyclical processes in a heath community. I. Vegetation change in relation to the *Calluna* cycle. *Journal of Ecology* **57**, 737–58.

Barrett, S.W. (1980) Conservation in Amazonia. *Biological Conservation* **18**, 209–35.

Bartlett, M.S. (1949) Fitting a straight line when both variables are subject to error. *Biometrics* **5**, 207–12.

Batschelet, E. (1981) *Circular Statistics in Biology*. Academic Press, London.

Baxter, R.S. (1976) *Computer and Statistical Techniques for Planners*. Methuen.

Beals, E.W. (1973) Ordination: mathematical elegance and ecological naïveté. *Journal of Ecology* **61**, 23–35.

Beals, E.W. (1984) Bray-Curtis ordination: an effective strategy for analysis of multivariate ecological data. *Advances in Ecological Research* **14**, 1–55.

Becking, R.W. (1957) The Zurich-Montpellier school of phytosociology. *The Botanical Review* **23**, 412–88.

Begon, M., Harper, J.L. & Townsend, C.R. (1990) *Ecology: Individuals, Populations and Communities*. 2nd ed., Blackwell Scientific, Oxford.

Bell, R.H.V. (1970) The use of the herb layer by grazing ungulates in the Serengeti. In: Watson, A. (ed.) *Animal Populations in relation to their Food Sources*. Blackwell, Oxford, 111–124.

Benninghoff, W.S. & Southworth, W.C. (1964) Ordering of tabular arrays of phytosociological data by digital computer. *Abstract of the 10th International Botanical Congress*, 331–332.

Benzécri, J.P. (1969) Statistical analysis as a tool to make patterns emerge from data. In: Watanabe, S. (ed.) *Methodologies of Pattern Recognition*. Academic Press, New York, 35–60.

Benzécri, J.P. (1973) *L'analyse des données*. Vol. 2. L'analyse des correspondances. Dunod, Paris.

Besag, J. (1981) On resistant techniques and statistical analysis. *Biometrika* **68**, 463–69.

Birks, H.J.B. & Birks, H.H. (1980) *Quaternary Palaeoecology*. Arnold, London.

Bishop, O.N. (1983) *Statistics for Biology*. 4th ed., Longman, London.

Blackith, R.E. & Reyment, R.A. (1971) *Multivariate Morphometrics*. Academic Press, London.

Bock, H.H. (1985) On some significance tests in cluster analysis. *Journal of Classification* **2**, 77–108.

Bourlière, F. (ed.) (1983) *Tropical Savannas*. Ecosystems of the World 13, Elsevier, Amsterdam.

Bradbury, I. (1991) *The Biosphere*. Belhaven Press, London.

Bradu, D. & Gabriel, K.R. (1978) The biplot as a diagnostic tool for models of two-way tables. *Technometrics* **20**, 47–68.

Braun-Blanquet, J. (1928) *Pflanzensoziologie. Grundzüge der Vegetationskunde*. Springer, Berlin.

Braun-Blanquet, J. (1932/1951) *Plant Sociology: the Study of Plant Communities*. (English translation), McGraw-Hill, New York.

Bray, R.J. & Curtis, J.T. (1957) An ordination of the upland forest communities of southern Wisconsin. *Ecological Monographs* **27**, 325–49.

Briggs, D.J. (1977) *Soils*. Sources and Methods in Geography, Butterworth, London.

Briggs, D.J. & Smithson, P. (1985) *Fundamentals of Physical Geography*. Hutchinson/Unwin-Hyman, London.

Brown, G.W. & Mood, A.M. (1951) On median tests for linear hypotheses. In: Neyman, J. (ed.) *Proceedings of 2nd Berkeley Symposium on Mathematical Statistics and Probability*. California University Press, Berkeley and Los Angeles, 159–66.

Brown, M.J., Ratkowsky, D.A. & Minchin, P.R. (1984) A comparison of detrended correspondence analysis and principal coordinates analysis using four sets of Tasmanian vegetation data. *Australian Journal of Ecology* **9**, 273–9.

Brown, R.T. & Curtis, J.T. (1952) The upland conifer-hardwood forest of Northern Wisconsin. *Ecological Monographs* **22**, 217–34.

Bunce, R.G.H. & Shaw, M.W. (1973) A standardised procedure for ecological survey. *Journal of Environmental Management* **1**, 129–58.

Burnett, J.D. (ed.) (1964) *The Vegetation of Scotland*. Oliver and Boyd, Edinburgh.

Burrows, C.J. (1990) *Processes of Vegetation Change*. Unwin Hyman, London.

Cain, S.A. (1932) Concerning certain phytosociological concepts. *Ecological Monographs* **2**, 475–508.

Cain, S.A. (1934a) The climax and its complexities. *American Midland Naturalist* **21**, 146–81.

Cain, S.A. (1934b) A comparison of quadrat sizes in a quantitative phytosociology study of Nash's Woods, Posey County, Indiana. *American Midland Naturalist* **15**, 529–66.

Cain, S.A. (1938) The species-area curve. *American Midland Naturalist* **19**, 573–81.

Carr, J.R. (1990) CORSPOND: a portable FORTRAN-77 program for correspondence analysis. *Computers and Geosciences* **16**, 289–307.

Carroll, J.D. (1987) Some multidimensional scaling and related procedures devised at Bell Laboratories, with ecological applications. In: Legendre, P. & Legendre, L. (eds.) *Developments in Numerical Ecology*. NATO ASI Series G14, Springer-Verlag, Berlin, 65–138.

Causton, D.R. (1988) *Introduction to Vegetation Analysis*. Unwin Hyman, London.

Chamberlin, T.C. (1890) The method of multiple working hypotheses. *Science* XV, 92–6, reprinted in *Science* **148**, 754–9.

Chatfield, C. (1985) The initial examination of data (with discussion). *Journal of the Royal Statistical Society, A* **148**, 214–53.

Chatfield, C. (1986) Exploratory data analysis. *European Journal of Operational Research* **23**, 5–13.

Chatfield, C. (1988) *Problem Solving – a Statistician's Guide*, Chapman and Hall, London.

Cherrett, J.M. (ed.) (1989) *Ecological Concepts: the Contribution of Ecology to an Understanding of the Natural World*. British Ecological Society/Blackwell Scientific Publications, Oxford.

Chow, T.J. (1970) Lead accumulation in roadside soils and grass. *Nature* **225**, 295–6.

Clapham, A.R., Tutin, T.G. & Moore, D.N. (1987) *Flora of the British Isles*. 3rd ed., Cambridge University Press, Cambridge.

Clapham, A.R., Tutin, T.G. & Warburg, E.F. (1981) *Excursion flora of the British Isles*. 3rd ed., Cambridge University Press, Cambridge.

Clapham Jr., W.B. (1983) *Natural Ecosystems*, Macmillan, New York.

Clarke, G.M. (1980) *Statistics and Experimental Design*. 2nd ed., Arnold, London.

Clements, F.E. (1916) *Plant Succession. An Analysis of the Development of Vegetation*. Carnegie Institute, Washington, Publication 242, Washington D.C.

Clements, F.E. (1928) *Plant Succession and Indicators*. H.W. Wilson, New York.

Cliff, A.D. & Ord, J.K. (1973) *Spatial Autocorrelation*. Pion, London.

Cliff, A.D. & Ord, J.K. (1981) *Spatial Processes: Models and Applications*. Pion, London.

Clifford, H.T. & Stephenson, W. (1975) *An Introduction to Numerical Classification.* Academic Press, New York.

Cody, M.L. (1989) Discussion: structure and assembly of communities. In: Roughgarden, J., May, R.M. & Levin, S.A. (eds.) *Perspectives in Ecological Theory.* Princeton University Press, Princeton, New Jersey, 227–41.

Coetzee, B.T. & Werger, M.J.A. (1975) On association analysis and the classification of communities. *Vegetatio* **30**, 201–6.

Cohen, L. & Holliday, M. (1982) *Statistics for Social Scientists – An Introductory Text with Computer Programs in BASIC.* Harper and Row, London.

Coker, P.D. (1988) *Some Aspects of the Biogeography of the Høyfjellet with Special Reference to Høyrokampen, Bøverdal, Southern Norway.* Unpublished M.Phil. thesis, University College London.

Cole, M.M. (1986) *The Savannas: Biogeography and Geobotany.* Academic Press, London.

Cole, M.M. (1987) The savannas. *Progress in Physical Geography* **11**, 334–55.

Collinson, A.S. (1977) *Introduction to World Vegetation.* George Allen and Unwin, London.

Connell, J.H. & Slatyer, R.O., (1977) Mechanisms of succession in natural communities and their role in community structure and organisation. *The American Naturalist* **111**, 1119–44.

Cormack, R.M. (1971) A review of classification. *Journal of the Royal Statistical Society – Series A* **134**, 321–67.

Cormack, R.M. (1988) Statistical challenges in the environmental sciences: a personal view. *Journal of the Royal Statistical Society – Series A* **151**, 201–10.

Cottam, G., Glenn Goff, F. & Whittaker, R.H. (1978) Wisconsin comparative ordination. In: Whittaker, R.H. (ed.) *Ordination of Plant Communities.* Junk, The Hague, 185–214.

Cox, B.C., & Moore, P.D. (1985) *Biogeography: an Ecological and Evolutionary Perspective.* 4th ed., Blackwell Scientific, Oxford.

Cox, N.J. (1989) Teaching and learning spatial autocorrelation: a review. *Journal of Geography in Higher Education* **13**, 185–190.

Crawford, R.M.M. & Wishart, D. (1967) A rapid multivariate method for the detection and classification of ecologically related species. *Journal of Ecology* **55**, 505–24.

Crawley, M.J. (ed.) (1986) *Plant Ecology.* Blackwell Scientific, Oxford.

Crowe, T.M. (1979) Lots of weeds: insular phytogeography of vacant urban lots. *Journal of Biogeography* **6**, 169–81.

Curtis, J.T. & McIntosh, R.P. (1950) The inter-relations of certain analytic and synthetic phytosociological characters. *Ecology* **31**, 434–55.

Curtis, J.T. & McIntosh, R.P. (1951) An upland forest continuum in the prairie-forest border region of Wisconsin. *Ecology* **32**, 476–96.

Czekanowski, J. (1913) *Zarys Metod Statystycznyck.* Warsaw.

Dagnelie, P. (1960) Contribution à l'étude des communautés végétales par l'analyse factorielle. *Bull. Serv. Carte. Phytogéogr., Ser. B.* **5**, 7–71; 93–195.

Dagnelie, P. (1978) Factor analysis. In: Whittaker, R.H. (ed.) *Ordination of Plant Communities.* Dr. W. Junk, The Hague, 215–238.

Dahl, E. (1956) Rondane. Mountain vegetation in South Norway and its relation to the environment. *Skr. Norske. Vid.-Akad.* I. Mat.-Naturvid. 3.

Dahl, E. (1985) A survey of the plant communities at Finse, Handangervidda, Norway. Typescript. Agricultural University of Norway. As-NLH.

Dale, M.B. (1988) Knowing when to stop: cluster concept–concept cluster. *Coenoses* **3**, 11–32.

Dale, P.E.R. (1979) A pragmatic study of Danserau's universal system for recording vegetation: application in South-East Queensland, Australia. *Vegetatio* **40**, 129–33.

Daniel, W.W. (1978) *Applied Nonparametric Statistics.* Houghton Mifflin, Boston.

Danin, A. & Orshan, G. (1990) The distribution of Raunkaier life-forms in Israel in relation to the environment. *Journal of Vegetation Science* **1**, 41–8.

Dansereau, P. (1951) Description and recording of vegetation upon a structural basis. *Ecology* **32**, 172–229.

Dansereau, P. (1957) *Biogeography: an Ecological Perspective.* Ronald Press, New York.

Dansereau, P., Buell, P.F. & Dagon, R. (1966) A universal system for recording vegetation. II. A methodological critique and an experiment. *Sarracenia* **10**, 1–64.

Dargie, T.C.D. (1986) Species richness and distortion in reciprocal averaging and detrended correspondence analysis. *Vegetatio* **65**, 95–8.

Darlington, A. (1981) *Ecology of Walls.* Heinemann Education, London.

Davies, R.G. (1971) *Computer Programming in Quantitative Biology.* Academic Press, London.

Davis, B.N.K. (1976) Wildlife, urbanisation and industry. *Biological Conservation*, **10**, 249–91.

Davis, J.C. (1973) *Statistics and Data Analysis in Geology with FORTRAN Programs.* Wiley, New York.

Davison, A.W. (1971) The effects of de-icing salt on roadside verges. I. Soil and plant analysis. *Journal of Applied Ecology* **8**, 555–61.

de Wit, C.T. (1960) On competition. *Verslagen Landbouwkundige Onderzoekingen* **66**, 1–82.

Deshmukh, I. (1986) *Ecology and Tropical Biology.* Blackwell Scientific, Oxford.

Diamond, J.M. & Case, T.J. (eds.) (1986) *Community Ecology.* Harper and Row, New York.

Dietvorst, E., van der Maarel, E. & van der Putten, H. (1982) A new approach to the minimal area of a plant community. *Vegetatio* **50**, 77–91.

Digby, P.G.N. & Kempton, R.A. (1987) *Multivariate Analysis of Ecological Communities.* Chapman and Hall, London.

Drury, W.H. & Nisbet, I.C. (1973) Succession. *Journal of the Arnold Arboretum* **54**, 331–68.

Down, C.G. (1973) Life-form succession in plant communities on colliery waste. *Environmental Pollution* **5**, 19–22.

Du Rietz, G.E. (1921) *Zur Methodologischen Grundlage der Modernen Pflanzensoziologie.* Holzhausen, Vienna.

Du Rietz, G.E. (1942a) Rishedsförband i Tarneträskomradets lagfjällbalte. *Svensk botanisk Tidskrift.* **36**, Uppsala.

Du Rietz, G.E. (1942b) De svenska fjallens vantvärld. Norrland. Natur, befolkning och naringar. *Ymer* **62**, 3–4, Stockholm.

Dunn, G. & Everitt, B.S. (1982) *An Introduction to Mathematical Taxonomy.* Cambridge University Press, Cambridge.

Dunn, R. (1989) Building regression models: the importance of graphics. *Journal of Geography in Higher Education* **13**, 15–30.

During, H.J., Werger, M.J.A. & Willems, H.J. (eds.) (1988) *Diversity and Pattern in Plant Communities.* SPB Academic Publishing, The Hague, The Netherlands.

Ebdon, D. (1985) *Statistics in Geography.* 2nd ed., Basil Blackwell, Oxford.

Eberhardt, L.L. & Thomas, J.M. (1991) Designing environmental field studies. *Ecological Monographs* **61**, 53–73.

Eden, M. (1990) *Ecology and Land Management in Amazonia.* Belhaven Press, London.

Egler, F.E. (1954) Philosophical and practical considerations of the Braun-Blanquet system of phytosociology, *Castanea* **19**, 45–60.

Elton, C.S. (1966) *The Pattern of Animal Communities.* Methuen, London.

Elton, C.S. & Miller, R.S. (1954) The ecological survey of animal communities with a practical system of classifying habitats by structural characters, *Journal of Ecology* **42**, 460–96.

Emery, M, (1986) *Promoting Nature in Cities and Towns – a Practical Guide,* Croom Helm, London.

Erickson, B.H. & Nosanchuk, T.A. (1977) *Understanding Data.* The Open University Press, Milton Keynes.

Everitt, B.S. (1980) *Cluster Analysis*. 2nd ed., Heinemann, London.

Ewusie, J.Y. (1980) *Elements of Tropical Ecology*. Heinemann, London.

Ezcurra, E. (1987) A comparison of reciprocal averaging and non-centred principal components analysis. *Vegetatio* **71**, 41–7.

Fasham, M.J.R. (1977) A comparison of non-metric multidimensional scaling, principal components analysis and reciprocal averaging for the ordination of simulated coenoclines and coenoplanes. *Ecology* **58**, 551–61.

Fearnside, P.M. (1990) The rate and extent of deforestation in Brazilian Amazonia. *Environmental Conservation* **17**, 213–26.

Feoli, E. & Orlóci, L. (1991) *Computer Assisted Vegetation Analysis*. Handbook of Vegetation Science 11, Kluwer, Dordrecht, The Netherlands.

Feyerabend, P. (1975) *Against Method*. Verso, London.

Feyerabend, P. (1978) *Science in a Free Society*. Verso, London.

Finney, D.J. (1980) *Statistics for Biologists*. Chapman and Hall, London.

Fischer, H.S. & Bemmerlein, F.A. (1989) An outline for data analysis in phytosociology: past and present. *Vegetatio* **81**, 17–28.

Fisher, R.A. (1940) The precision of discriminant functions. *Annals of Eugenics* **10**, 422–9.

Fitter, A.H. (1987) Spatial and temporal separation of activity in plant communities: prerequisite or consequence of coexistence? In: Gee, J.H.R. & Giller, P.S. (eds.) *Organization of Communities Past and Present*. British Ecological Society/Blackwell Scientific Publications, Oxford, 119–39.

Forman, R.T.T. & Godron, M. (1986) *Landscape Ecology*. Wiley, New York.

Fortin, M-J, Drapeau, P. & Legendre, P. (1989) Spatial autocorrelation and sampling design in plant ecology. *Vegetatio* **83**, 209–22.

Fosberg, F.R. (1961) A classification of vegetation for general purposes. *Tropical Ecology* **2**, 1–28.

Fowler, J. & Cohen, L. (1990) *Practical Statistics for Field Biology*. Open University Press, London.

Frenkel, R.E. & Harrison, C.M. (1974) An assessment of the usefulness of phytosociological and numerical classificatory methods for the community biogeographer. *Journal of Biogeography* **1**, 27–56.

Fridriksson, S. (1975) *Surtsey – Evolution of Life on a Volcanic Island*. Butterworth, London.

Fridriksson, S. (1987) Plant colonisation of a volcanic island, Surtsey, Iceland. *Arctic and Alpine Research* **19**, 425–31.

Fridriksson, S. (1989) The volcanic island of Surtsey, Iceland, a quarter century after it 'rose from the sea'. *Environmental Conservation* **16**, 157–62.

Gabriel, K.R. (1971) The biplot graphic display of matrices with application to principal component analysis. *Biometrika* **58**, 453–67.

Gabriel, K.R. (1981) Biplot display of multivariate matrices for inspection of data and diagrams. In: Barnett, V. (ed.) *Interpreting Multivariate Data*. Wiley, Chichester, 147–73.

Gates, M.R. & Hansell, R.I.C. (1983) On the distinctiveness of clusters. *Journal of Theoretical Biology* **101**, 263–73.

Gauch, H.G. (1979) *COMPCLUS: a FORTRAN Program for the Initial Clustering of Large Data Sets*. Cornell University, Department of Ecology and Systematics, Ithaca, New York.

Gauch, H.G. (1980) Rapid initial clustering of large data sets. *Vegetatio* **42**, 103–11.

Gauch, H.G. (1982a) Noise reduction by eigenvector ordination. *Ecology* **63**, 1643–9.

Gauch, H.G. (1982b) *Multivariate Analysis in Community Ecology*. Cambridge Studies in Ecology, Cambridge University Press.

Gauch, H.G. & Scruggs, W.M. (1979) Variants of polar ordination. *Vegetatio* **40**, 147–53.

Gauch, H.G. & Whittaker, R.H. (1972a) Coenocline simulation. *Ecology* **53**, 446–51.

Gauch, H.G. & Whittaker, R.H. (1972b) Comparison of ordination techniques. *Ecology* **53**, 868–75.

Gauch, H.G. & Whittaker, R.H. (1976) Simulation of community patterns. *Vegetatio* **33**, 13–16.

Gauch, H.G. & Whittaker, R.H. (1981) Hierarchical classification of community data. *Journal of Ecology* **69**, 135–52.

Gauch, H.G., Whittaker, R.H. & Singer, S.B. (1981) A comparative study of non-metric ordinations. *Journal of Ecology* **69**, 135–52.

Gauch, H.G., Whittaker, R.H. & Wentworth, R.T. (1977) A comparative study of reciprocal averaging and other ordination techniques. *Journal of Ecology* **65**, 157–74.

Gibson, C.W.D. & Brown, V.K. (1986) Plant succession: theory and application. *Progress in Physical Geography* **10**, 473–93.

Gilbert, O.L. (1990) *The Ecology of Urban Habitats*. Chapman and Hall, London.

Gilbertson, D.D., Kent, M. & Pyatt, F.B. (1985) *Practical Ecology for Geography and Biology*. Hutchinson/Unwin Hyman, London.

Giller, P.S. & Gee, J.H.R. (1987) The analysis of community organization: the influence of equilibrium, scale and terminology. In: Gee, J.H.R. & Giller, P.S. (eds.) *Organization of Communities Past and Present*. British Ecological Society/Blackwell Scientific Publications, Oxford, 519–42.

Gillison, A.N. & Anderson, D.J. (eds.) (1981) *Vegetation Classification in Australia*. CSIRO, Canberra, Australia.

Gimingham, C.H. (1972) *Ecology of Heathlands*. Chapman and Hall, London.

Gittins, R. (1969) The application of ordination techniques. In: Rorison, I.H. (ed.) *Ecological Aspects of Mineral Nutrition in Plants*, Blackwell Scientific, Oxford, 37–66.

Gittins, R. (1979) Ecological applications of canonical analysis. In: Orloci, L., Rao, C.R. & Stiteler, W.M. (eds.) *Multivariate Methods in Ecological Work*. Statistical Ecology, No. 7., Md. Int. Coop., Fairland, 309–535.

Gittins, R., Amir, S., Dupouey, J-L., Heiser, W.J., Meyer, M, Sokal, R.R. & Werger, M.J.A. (1987) Numerical methods in terrestrial plant ecology. In: Legendre, P. & Legendre, L. (eds.) *Developments in Numerical Ecology*. NATO ASI Series G14, Springer-Verlag, Berlin, 529–558.

Gleason, H.A. (1917) The structure and development of the plant association. *Bulletin of the Torrey Botanical Club* **43**, 463–81.

Gleason, H.A. (1926) The individualistic concept of the plant association. *Bulletin of the Torrey Botanical Club* **53**, 1–20.

Gleason, H.A. (1939) The individualistic concept of the plant association. *American Midland Naturalist* **21**, 92–110.

Goff, F.G. & Cottam, G. (1967) Gradient analysis: the use of species and synthetic indices. *Ecology* **48**, 793–806.

Golden, M.S. (1979) Forest vegetation of the Lower Alabama Piedmont. *Ecology* **60**, 770–782.

Goldsmith, F.B. (1973a) The vegetation of exposed sea cliffs at South Stack, Anglesey. I. The multivariate approach. *Journal of Ecology* **61**, 787–818.

Goldsmith, F.B. (1973b) The vegetation of exposed sea cliffs at South Stack, Anglesey. II. Experimental studies. *Journal of Ecology* **61**, 819–830.

Goldsmith, F.B. (1974) An assessment of the Fosberg and Ellenberg methods of classifying vegetation for conservation purposes. *Biological Conservation* **6**, 3–6.

Goldsmith, F.B. (1975) An evaluation of ecological resources in the countryside for conservation purposes. *Biological Conservation* **8**, 89–96.

Golley, F.B. (ed.) (1983) *Tropical Rain Forest Ecosystems*. Ecosystems of the world, Springer-Verlag, Berlin.

Goodall, D.W. (1953) Objective methods for the classification of vegetation. I. The use of positive interspecific correlation. *Australian Journal of Biology* **1**, 39–63.

Goodall, D.W. (1954) Objective methods for the comparison of vegetation. III. An essay in the use of factor analysis. *Australian Journal of Botany* **1**, 39–63.

Goodall, D.W. (1966) The nature of the mixed community. *Proceeding of the Ecological Society of America* **1**, 84–96.

Goodall, D.W. (1978) Numerical classification. In: Whittaker, R.H. (ed.) *Classification of Plant Communities*. Junk, The Hague, 247–86.

Goodman, D. (1975) The theory of diversity-stability relationships in ecology. *The Quarterly Review of Biology* **50**, 237–66.

Gordon, A.D. (1987) A review of hierarchical classification. *Journal of the Royal Statistical Society – Series A*. **150**, 119–37.

Gould, P. & White, R. (1974) *Mental Maps*. Penguin, Harmondsworth.

Gould, P. & White, R. (1986) *Mental Maps*. 2nd ed. Allen and Unwin.

Gower, J.C. (1967) A comparison of some methods of cluster analysis. *Biometrics* **23**, 623–37.

Gower, J.C. (1974) Maximal predictive classification. *Biometrics* **30**, 643–54.

Gower, J.C. (1987) Introduction to ordination techniques. In: Legendre, P. & Legendre, L. (eds.) *Developments in Numerical Ecology* NATO ASI Series G14, Springer-Verlag, Berlin, 5–64.

Gower, J.C. (1988) Classification, geometry and data analysis. In: Bock, H.H. (ed.) *Classification and Related Methods of Data Analysis*. Elsevier, North-Holland, 3–14.

Gray, A.J., Crawley, M.J. & Edwards, P.J. (eds.) (1987) *Colonisation, Succession and Stability*. 26th Symposium of the British Ecological Society, Blackwell Scientific, Oxford.

Gray, J.S. (1987) Species-abundance patterns. In: Gee, J.H.R. & Giller, P.S. (eds.) *Organization of Communities Past and Present*. British Ecological Society/Blackwell Scientific Publications, Oxford, 53–67.

Greenacre, M.J. (1981) Practical correspondence analysis. In: Barnett, M.J. (ed.) *Interpreting Multivariate Data*. Wiley, New York, 119–46.

Greenacre, M.J. (1984) *Theory and Applications of Correspondence Analysis*. Academic Press, London.

Grey, G.W. & Deneke, F.J. (1978) *Urban Forestry*. Wiley, New York.

Greig-Smith, P. (1980) The development of numerical classification and ordination. *Vegetatio* **42**, 1–9.

Greig-Smith, P. (1982) *Quantitative Plant Ecology*. 3rd ed., Studies in Ecology, Vol. 9, Blackwell Scientific, Oxford.

Greig-Smith, P. (1986) Chaos or order – organisation. In: Kikkawa, J. & Anderson, D.J. (eds.) *Community Ecology – Pattern and Process*. Blackwell Scientific, Oxford, 19–29.

Grime, J.P. (1979) *Plant Strategies and Vegetation Processes*. Wiley, Chichester.

Grubb, P.J. (1987) Global trends in species-richness in terrestrial vegetation: a view from the Northern Hemisphere. In: Gee, J.H.R. & Giller, P.S. (eds.) *Organization of Communities Past and Present*. British Ecological Society/Blackwell Scientific Publications, Oxford, 99–118.

Grubb, P.J. (1989) Towards a more exact ecology: a personal view of the issues. In: Grubb, P.J. & Whittaker, J.B. (eds.) *Towards a More Exact Ecology*. British Ecological Society/Blackwell Scientific Publications, Oxford, 3–27.

Guinochet, M. (1973) *Phytosociologie*. Masson, Paris.

Haigh, M.J. (1980) Ruderal communities in English cities. *Urban Ecology* **4**, 329–338.

Haines-Young, R. & Petch, J. (1986) *Physical Geography: its Nature and Methods*. Harper and Row, London.

Hairston, N.G. (Sr.) (1989) *Ecological Experiments. Purpose, Design and Execution*. Cambridge Studies in Ecology, Cambridge University Press, Cambridge.

Hammond, R. & McCullagh, P. (1978) *Quantitative Techniques in Geography – an Introduction*. 2nd ed., Oxford University Press, Oxford.

Hanwell, J.D. & Newson, M.D. (1973) *Techniques in Physical Geography*. Macmillan, London.

Harper, J.L. (1977) *Population Biology of Plants*. Academic Press, London.

Harrison, C., Limb, M. & Burgess, J. (1987) Nature in the city – popular values for a living world. *Journal of Environmental Management* **25**, 347–62.

Harvey, D. (1969) *Explanation in Geography*. Arnold, London.

Hatheway, W.H. (1971) Contingency-table analysis of rain forest vegetation. In: Patil, G.P., Pielou, E.C. & Waters, W.E. (eds.) *Statistical Ecology*. Vol. 3., Pennsylvania State University Press, 271–313.

Hill, A.R. (1987) Ecosystem stability: some recent perspectives. *Progress in Physical Geography* **11**, 315–33.

Hill, M.O. (1973a) Diversity and evenness: a unifying notation and its consequences. *Ecology* **54**, 427–32.

Hill, M.O. (1973b) Reciprocal averaging: an eigenvector method of ordination. *Journal of Ecology* **61**, 237–250.

Hill, M.O. (1974) Correspondence analysis. A neglected multivariate method. *Journal of the Royal Statistical Society – Series C* **23**, 340–54.

Hill, M.O. (1977) Use of simple discriminant functions to classify quantitative data. In: Diday, E., Lebart, L., Pages, J.P. & Tomassone, R. (eds.) *First International Symposium on Data Analysis and Informatics* Vol. 1, Institute de Recherche d'Informatique et d'Automatique, Le Chesnay, 181–99.

Hill, M.O. (1979a) *DECORANA – a FORTRAN Program for Detrended Correspondence Analysis and Reciprocal Averaging*. Cornell University, Department of Ecology and Systematics, Ithaca, New York.

Hill, M.O. (1979b) *TWINSPAN – a FORTRAN Program for arranging Multivariate Data in an Ordered Two Way Table by Classification of the Individuals and the Attributes*. Cornell University, Department of Ecology and Systematics, Ithaca, New York.

Hill, M.O. & Gauch, H.G. (1980) Detrended correspondence analysis, an improved ordination technique. *Vegetatio* **42**, 47–58.

Hill, M.O., Bunce, R.G.H. & Shaw, M.W. (1975) Indicator species analysis, a divisive polythetic method of classification and its application to a survey of native pinewoods in Scotland. *Journal of Ecology* **63**, 597–613.

Hinch, S.G. & Somers, K.M. (1987) An experimental evaluation of the effect of data centering, data standardization and outlying observations on principal component analysis. *Coenoses* **2**, 19–23.

Hirschfeld, H.O. (1935) A connection between correlation and contingency. *Proceedings of the Cambridge Philosophical Society* **31**, 520–7.

Hopkins, B. (1955) The species-area relations of plant communities. *Journal of Ecology* **43**, 209–26.

Hopkins, B. (1957) Pattern in the plant community. *Journal of Ecology* **45**, 451–63.

Hopkins, B. (1965) *Forest and Savanna*. Heinemann, London.

Hotelling, H. (1933) Analysis of a complex of statistical variables into principal components, *Journal of Educational Psychology* **24**, 417–41, 498–520.

Huntley, B.J. & Walker, B.H. (eds.) (1982) *Ecology of Tropical Savannas*. Ecological Studies 42, Springer-Verlag, Berlin.

Huston, M. & Smith, T. (1987) Plant succession: life history and competition. *The American Naturalist* **130**, 168–98.

Hutcheson, K. (1970) A test for comparing diversities based on the Shannon formula. *Journal of Theoretical Biology* **29**, 151–4.

Hutchings, M.J. (1983) Plant diversity in four chalk grassland sites with different aspects. *Vegetatio* **53**, 179–89.

Ivimey-Cook, R.B. (1972) Association analysis – some comments on its use. In: van der

Maarel, E. & Tüxen, R. (eds.) *Grundfragen und Methoden in der Pflanzensoziologie*. Junk, The Hague, 89–97.

Ivimey-Cook, R.B. & Proctor, M.C.F. (1966) The application of association analysis to phytosociology. *Journal of Ecology* **54**, 179–92.

Jaccard, P. (1901) Etude comparative de la distribution florale dans une portion des Alpes et du Jura. *Bulletin Soc. Vaud. Sc. Nat.* **37**, 547–79.

Jaccard, P. (1912) The distribution of the flora of the alpine zone. *New Phytologist* **11**, 37–50.

Jaccard, P. (1928) Die statistisch-floristische method als grundlage der pflanzensoziologie. *Abderhalden, Handbuch Biologisch Arbeitsmethod* **11**, 165–202.

Jackson, D.A. & Somers, K.M. (1991) Putting things in order: the ups and downs of detrended correspondence analysis. *The American Naturalist* **137**, 704–12.

Jacobs, J. (1975) Diversity, stability and maturity in ecosystems influenced by human activities. In: Van Dobben, W.H. & Lowe-McConnell, R.H. (eds.) *Unifying Concepts in Ecology*. Junk, The Hague, 187–207.

Jambu, M. (1981) FORTRAN IV computer program for rapid hierarchical classification of large data sets. *Computers and Geosciences* **7**, 297–310.

James, F.C. & McCulloch, C.E. (1990) Multivariate analysis in ecology and systematics: panacea or Pandora's box? *Annual Review of Ecology and Systematics* **21**, 129–66.

Janssen, J.G.M. (1975) A simple clustering procedure for preliminary classification of very large sets of phytosociological results. *Vegetatio* **30**, 67–71.

Johnston, R.J. (1978) *Multivariate Statistical Analysis in Geography*. Longman, London.

Jongman, R.H.G., ter Braak, C.J.G. & van Tongeren, O.F.R. (1987) *Data Analysis in Community and Landscape Ecology*. Pudoc, Wageningen.

Keddy, P.A. (1987) Beyond reductionism and scholasticism in plant community ecology. *Vegetatio* **69**, 209–11.

Keddy, P.A. (1989) *Competition*. Chapman and Hall, London.

Kenkel, N.G. & Orlóci, L. (1986) Applying metric and non-metric multidimensional scaling to ecological studies and some new results. *Ecology* **67**, 919–28.

Kenkel, N.G. & Juhász-Nagy, P. & Podani, J. (1989) On sampling procedures in population and community ecology. *Vegetatio* **83**, 195–207.

Kent, M. (1972) *A Method for the Survey and Classification of Marginal Land in Agricultural Landscapes*. Discussion Papers in Conservation, 1, University College, London.

Kent, M. (1977) BRAYCURT and RECIPRO – two programs for ordination in ecology and biogeography. *Computer Applications* **4**, 589–647.

Kent, M. (1982) Plant growth in colliery spoil reclamation. *Applied Geography* **2**, 83–107.

Kent, M. (1987) Island biogeography and habitat conservation. *Progress in Physical Geography* **11**, 91–102.

Kent, M. (1987) Reclamation of deep coal mining wastes with particular reference to Britain and Western Europe. In: Majumdar, S.K., Brenner, F.J. & Miller, E.W. (eds.) *Environmental Consequences of Energy Production*. Pennsylvania Academy of Science, 61–77.

Kent, M. & Ballard, J. (1988) Trends and problems in the application of classification and ordination methods in plant ecology. *Vegetatio* **78**, 109–24.

Kent, M. & Smart, N. (1981) A method for habitat assessment in agricultural landscapes. *Applied Geography* **1**, 9–30.

Kent, M. & Wathern, P. (1980) The vegetation of a Dartmoor catchment. *Vegetatio* **43**, 163–72.

Kershaw, K.A. (1968) Classification and ordination of Nigerian savanna vegetation. *Journal of Ecology* **56**, 467–82.

Kershaw, K.A. & Looney, J.H.H. (1985) *Quantitative and Dynamic Plant Ecology.* 3rd ed., Arnold, London.

Kikkawa, J. (1986) Complexity and stability. In: Kikkawa, J. & Anderson, D.J. (eds.) *Community Ecology: Pattern and Process.* Blackwell Scientific, 41–62.

Knox, R.G. (1989) Effects of detrending and rescaling on correspondence analysis: solution stability and accuracy. *Vegetatio* **83**, 129–36.

Kormondy, E.J. (1984) *Concepts of Ecology.* Prentice-Hall, Englewood Cliffs, New Jersey.

Krebs, C.J. (1978) *Ecology: the Experimental Analysis of Distribution and Abundance.* Harper & Row, New York.

Kruskal, J.B. (1964a) Multidimensional scaling by optimizing goodness of fit to a nonmetric hypothesis. *Psychometrika* **29**, 1–27.

Kruskal, J.B. (1964b) Nonmetric multidimensional scaling: a numerical method. *Psychometrika* **29**, 115–29.

Kruskal, J.B. & Landwehr, J.M. (1983) Icicle plots: better displays for hierarchical clustering. *The American Statistician* **37**, 162–8.

Küchler, A.W. (1967) *Vegetation Mapping.* Ronald Press, New York.

Kuhn, T.S. (1962) *The Structure of Scientific Revolutions.* University of Chicago Press, Chicago.

Kuhn, T.S. (1970) Logic of discovery or the psychology of research? In: Lakatos, I. & Musgrave, A. (eds.) *Criticism and the Growth of Knowledge.* Cambridge University Press, Cambridge, 1–24.

Kunick, W. (1982) Comparison of the flora of some cities of the central European lowlands. In: Bornkamm, R., Lee, J.A. & Seaward, M.R.D. (eds.) *Urban Ecology.* 2nd European Ecological Symposium, Blackwell Scientific, 13–22.

Lambert, J.M. & Williams, W.T. (1962) Multivariate methods in plant ecology. IV. Nodal analysis. *Journal of Ecology* **50**, 775–802.

Lambert, J.M. & Williams, W.T. (1965) Multivariate methods in plant ecology. VI. Comparison of information analysis and association analysis. *Journal of Ecology* **54**, 635–64.

Lambert, J.M., Meacock, S.E., Barrs, J. and Smartt, P.F.M. (1973) AXOR and MONIT: Two new polythetic-divisive strategies for hierarchical classification. *Taxon* **22**, 173–6.

Lance, G.N. & Williams, W.T. (1966) A generalised sorting strategy for computer classification. *Nature* **211**, 218.

Lance, G.N. & Williams, W.T. (1967) A general theory of classification sorting strategies. I. Hierarchical systems. *Computer Journal* **9**, 373–80.

Lance, G.N. & Williams, W.T. (1975) REMUL: a new divisive polythetic classificatory program. *Australian Computer Journal* **7**, 109–12.

Law, R. & Watkinson, A.R. (1989) Competition. In: Cherrett, J.M. (ed.) *Ecological Concepts: the Contribution of Ecology to an Understanding of the Natural World.* British Ecological Society/Blackwell Scientific Publications, Oxford, 243–84.

Lee, J.D. & Lee, T.D. (1982) *Statistics and Numerical Methods in BASIC for Biologists.* Van Nostrand Reinhold, New York.

Legendre, P. & Fortin, M-J. (1989) Spatial pattern and ecological analysis. *Vegetatio* **80**, 107–138.

Legendre, P. & Legendre, L. (eds.) (1987) *Developments in Numerical Ecology.* NATO ASI Series G14, Springer-Verlag, Berlin.

Levin, S.A. (ed.) (1976) *Ecological Theory and Ecosystem Models.* Institute of Ecology, Indianapolis.

Louppen, J.M.W. & van der Maarel, E. (1979) CLUSLA: a computer program for the clustering of large phytosociological data sets. *Vegetatio* **40**, 107–14.

Ludwig, J.A. & Reynolds, J.F. (1988) *Statistical Ecology: a Primer on Methods and Computing.* Wiley, New York.

Luken, T.O. (ed.) (1990) *Directing Ecological Succession.* Chapman and Hall.

Mabberley, D.J. (1991) *Tropical Rain Forest Ecology.* 2nd ed., Blackie, Glasgow.

MacArthur, R.H. & Wilson, E.O. (1967) *The Theory of Island Biogeography.* Princeton University Press, Princeton, New Jersey.

Magurran, A. (1988) *Ecological Diversity and its Measurement.* Croom Helm, London.

Margalef, R. (1975) Diversity, stability and maturity in natural ecosystems. In: Van Dobben, W.H. & Lowe-McConnell, R.H. (eds.) *Unifying Concepts in Ecology.* Junk, The Hague, 152–60.

Margules, C. & Usher, M.B. (1981) Criteria used in assessing wildlife conservation potential: a review. *Biological Conservation* **21**, 79–109.

Marsh, C. (1988) *Exploring Data: an Introduction to Data Analysis for Social Scientists.* Polity Press, Cambridge.

Mather, P.M. (1976) *Computational Methods of Multivariate Analysis in Physical Geography.* Wiley, London.

May, R.M. (1973) *Stability and Complexity in Model Ecosystems.* Princeton University Press, Princeton, N.J.

May, R.M. (1981) Patterns in multi-species communities. In: May, R.M. (ed.) *Theoretical Ecology: Principles and Applications.* Blackwell Scientific, Oxford, 197–227.

May, R.M. (1984) An overview: real and apparent patterns in community structure. In: Strong, D.R., Simberloff, D, Abele, L.G. & Thistle, A.B. (eds.) *Biological Communities: Conceptual Issues and the Evidence.* Princeton University Press, Princeton, N.J., 3–18.

May, R.M. (1985) Evolutionary ecology and John Maynard Smith. In: Greenwood, P.J., Harvey, P.H. & Slatkin, M. (eds.) *Essays in Honour of John Maynard Smith.* Cambridge University Press, Cambridge, 107–116.

May, R.M. (1989) Levels of organization in ecology. In: Cherrett, J.M. (ed.) (1989) *Ecological Concepts: the Contribution of Ecology to an Understanding of the Natural World.* British Ecological Society/Blackwell Scientific Publications, Oxford, 339–63.

McIntosh, R.P. (1967a) An index of diversity and the relation of certain concepts of diversity. *Ecology* **48**, 392–404.

McIntosh, R.P. (1967b) The continuum concept of vegetation. *Botanical Review* **33**, 130–187.

McIntosh, R.P. (1980) The background and some current problems of theoretical ecology. *Synthese* **43**, 195–255.

McIntosh, R.P. (1986) *The Background of Ecology – Concept and Theory.* Cambridge University Press, Cambridge.

McNaughton, S.J. (1983) Serengeti grassland ecology: the role of composite environmental factors and contingency in community organisation. *Ecological Monographs* **53**, 291–320.

McNaughton, S.J. (1985) Ecology of a grazing ecosystem: the Serengeti. *Ecological Monographs* **55**, 259–94.

McPherson, G. (1989) The scientist's view of statistics – a neglected area. *Journal of the Royal Statistical Society – Series A* **152**, 221–40.

McVean, D.N. & Ratcliffe, D.A. (1962) *Plant Communities of the Scottish Highlands.* Monographs of the Nature Conservancy, No. 1., H.M.S.O., London.

Meades, W.J. (1983) Heathlands. In: South, G.R. (ed.) *Biogeography and Ecology of the Island of Newfoundland.* Junk, The Hague, 267–318.

Mellinger, M.V. & McNaughton, S.J. (1975) Structure and function of successional vascular plant communities in central New York. *Ecological Monographs* **45**, 161–82.

Milne, G. (1935) Some suggested units of classification and mapping, particularly for East African soils. *Soil Research* **4**, 183–98.

Milner, C. (1978) Shetland ecology surveyed. *Geographical Magazine* **50**, 730–753.

Minchin, P.R. (1987a) An evaluation of the relative robustness of techniques for ecological ordination. *Vegetatio* **69**, 89–107.

Minchin, P.R. (1978b) Simulation of multidimensional community patterns: towards a comprehensive model. *Vegetatio* **71**, 145–56.

Mitchell, J. (1977) The effect of bracken distribution on moorland vegetation and soils. Unpublished Ph.D. thesis, University of Glasgow.

Möhler, C.L. (1987) *COMPOSE – a program for formatting and editing data matrices.* Microcomputer Power, Ithaca, New York.

Mojena, R. (1977) Hierarchical grouping methods and stopping rules: an evaluation. *The Computer Journal* **20**, 359–63.

Moore, G.W., Benninghoff, W.S. & Dwyer, P.S. (1967) A computer method for the arrangement of phytosociological tables. *Proceedings of the Association for Computer Machinery* **20**, 297–99.

Moore, J.J. (1962) The Braun-Blanquet system – a reassessment. *Journal of Ecology* **50**, 701–9.

Moore, J.J., Fitzsimons, P., Lambe, E. & White, J. (1970) A comparison and evaluation of some phytosociological techniques. *Vegetatio* **20**, 1–20.

Moravec, J. (1971) A simple method for estimating homogeneity of sets of phytosociological relevés. *Folia Geobotanica Phytotaxonomie* **6**, 147–70.

Mortimer, A.M. (1974) *Studies of germination and establishment of selected species with reference to the fates of seeds.* Unpublished Ph.D.thesis, University of Wales.

Moss, R.P. (1977) Deductive strategies in geographical generalisation. *Progress in Physical Geography* **1**, 23–39.

Mucina, L. & van der Maarel, E. (1989) Twenty years of numerical syntaxonomy. *Vegetatio* **81**, 1–15.

Mueller-Dombois, D. & Ellenberg, H. (1974) *Aims and Methods of Vegetation Ecology.* Wiley, New York.

Myers, N. (1984) *The Primary Source – Tropical Forests and Our Future.* Norton, London.

Nature Conservancy Council (1979) *Nature Conservation in Urban Areas: Challenge and Opportunity.* Nature Conservancy Council.

Naveh, Z. & Lieberman, A.S. (1984) *Landscape Ecology – Theory and Practice.* Springer-Verlag, New York.

Nelder, J.A. (1986) Statistics, science and technology: the address of the president (with proceedings). *Journal of the Royal Statistical Society – Series A* **149**, 109–21.

Newbould, P. (1965) Production ecology and the International Biological Programme. *Geography* **49**, 98–104.

Nie, N., Hadlai Hull, C., Jenkins, J.G., Steinbrenner, K. & Bent, D.H. (1975) *SPSS – Statistical Package for the Social Sciences.* 2nd ed., McGraw-Hill, New York.

Nie, N.H. (ed.) (1983) *SPSSX Users Guide.* McGraw-Hill, New York.

Nilsson, C. (1986) Methods of selecting lake shorelines as nature reserves. *Biological Conservation* **35**, 269–91.

Nilsson, C., Grelsson, G. Johansson, M. & Sperens, U. (1988) Can rarity and diversity be predicted in vegetation along river banks? *Biological Conservation* **44**, 201–12.

Noble, I.R. & Slatyer, R.O. (1980) The use of vital attributes to predict successional changes in plant communities subject to recurrent disturbances. *Vegetatio* **43**, 5–21.

Noest, V. & van der Maarel, E. (1989) A new dissimilarity measure and a new optimality criterion in phytosociological classification. *Vegetatio* **83**, 157–65.

Noy-Meir, I. (1973) Data transformations in ecological ordination. I. Some advantages of non-centring. *Journal of Ecology* **61**, 329–41.

Noy-Meir, I. & van der Maarel, E. (1987) Relations between community theory and community analysis in vegetation science: some historical perspectives. *Vegetatio* **69**, 5–15.

Noy-Meir, I., Walker, D. & Williams, W.T. (1975) Data transformation in ecological ordination. II. On the meaning of standardization. *Journal of Ecology* **63**, 779–800.

Odum, E.P. (1989) *Ecology and Our Endangered Life-Support Systems*. Sinauer Associates, Inc., Sunderland, Massachusetts.

Økland, R.H. & Bendiksen, E. (1985) The vegetation of the forest-alpine transition in the Grunningdalen area, Telemark, Southern Norway. *Sommerfeltia* **2**, 1–224.

Oksanen, J. (1983) Ordination of boreal heath-like vegetation with principal components analysis. correspondence analysis and multidimensional scaling. *Vegetatio* **52**, 181–9.

Oksanen, J. (1987) Problems of joint displays of species and site scores in correspondence analysis. *Vegetatio* **72**, 51–7.

Oksanen, J. (1988) A note of the occasional instability of detrending in correspondence analysis. *Vegetatio* **74**, 29–32.

O'Neill, R.V. (1989) Perspectives in hierarchy and scale. In: Roughgarden,, J., May, R.M. & Levin, S.A. (eds.) *Perspectives in Ecological Theory*. Princeton University Press, Princeton, New Jersey, 140–56.

Orlóci, L. (1966) Geometric models in ecology. I. The theory and application of some ordination methods. *Journal of Ecology* **54**, 193–215.

Orlóci, L. (1967a) Data centering: a review and evaluation with reference to component analysis. *Systematic Zoology* **16**, 208–12.

Orlóci, L. (1967b) An agglomerative method for the classification of plant communities. *Journal of Ecology* **55**, 193–206.

Orlóci, L. (1972) On information analysis in phytosociology. In: van der Maarel, E. & Tüxen, R. (eds.) *Grundfragen und Methoden in der Pflanzensoziologie*. Junk, The Hague, 75–88.

Orlóci, L. (1974) Revision for the Bray and Curtis ordination. *Canadian Journal of Botany* **52**, 1773–6.

Orlóci, L. (1978) *Multivariate Analysis in Vegetation Research*. Junk, The Hague.

Orians, G.H. (1975) Diversity, stability and maturity in natural ecosystems. In: Van Dobben, W.H. & Lowe-McConnell, R.H. (eds.) *Unifying Concepts in Ecology*. Junk, The Hague, 139–50.

Palczynski, A. (1984) Natural differentiation of plant communities in relation to hydrological conditions of the Biebrza valley. *Polish Ecological Studies* **10**, 347–85.

Palmer, M.W. (1988) Fractal geometry: a tool for describing spatial patterns of plant communities. *Vegetatio* **75**, 95–102.

Pawlowski, B. (1966) Review of terrestrial plant communities. A. Composition and structure of plant communities and methods of their study. In: Szafer, W. (ed.) *The Vegetation of Poland*. Pergamon Press/PWN Polish Scientific Publishers, Warsaw, 241–81.

Pears, N. (1985) *Basic Biogeography*. 2nd ed., Longman, London.

Pearson, K. (1901) On lines and planes of closest fit to systems of points in space. *Philosophical Magazine, Sixth Series* **2**, 559–72.

Peet, R.K., Knox, R.G., Case, J.S. & Allen, R.B. (1988) Putting things in order: the advantages of detrended correspondence analysis. *The American Naturalist* **131**, 924–34.

Peterken, G.P. (1967) *Guide to the Check Sheet for IBP Areas*. IBP Handbook No. 4, Blackwell Scientific, Oxford.

Pickett, S.T.A. & Kolasa, J. (1989) Structure of theory in vegetation science. *Vegetatio* **83**, 7–15.

Pielou, E.C. (1969) *An Introduction to Mathematical Ecology*. Wiley, New York.

Pielou, E.C. (1975) *Ecological Diversity*. Wiley, New York.

Pignatti, S. (1980) Reflections on the phytosociological approach and the epistomological basis of vegetation science. *Vegetatio* **42**, 181–5.

Pimm, S.L. (1984) The complexity and stability of ecosystems. *Nature* **307**, 321–6.

Podani, J. (1984) Spatial processes in the analysis of vegetation: theory and review. *Acta Botanica Hungarica* **30**, 75–118.

Podani, J. (1989) Comparisons of ordinations and classifications of vegetation data. *Vegetatio* **83**, 111–128.

Poole, R.W. (1974) *An Introduction to Quantitative Ecology*. McGraw-Hill, Tokyo.

Poore, M.E.D. (1955a,b,c) The use of phytosociological methods in ecological investigations. I. The Braun-Blanquet system. *Journal of Ecology* **43**, 226–44. II. Practical issues involved in trying to apply the Braun-Blanquet system. *Journal of Ecology* **43**, 245–69. III. Practical applications. *Journal of Ecology* **43**, 606–51.

Poore, M.E.D. (1956) The use of phytosociological methods in ecological investigations. IV. General discussion of phytosociological problems. *Journal of Ecology* **44**, 28–50.

Poore, M.E.D. & McVean, D.N. (1957) A new approach to Scottish mountain vegetation. *Journal of Ecology* **45**, 401–39.

Popper, K.R. (1972a) *The Logic of Scientific Discovery*. 6th revised impression, Hutchinson, London.

Popper, K.R. (1972b) *Objective Knowledge*. Oxford University Press, Oxford.

Popper, K.R. (1976) *Unended Quest: an Intellectual Autobiography*. Fontana, London.

Prance, G.T. (1977) The phytogeographical subdivisions of Amazonia and their influence on the selection of biological reserves. In: Prance, G.T. & Elias, T.S. (eds.) *Extinction is for ever*. The New York Botanic Garden, New York, 193–213.

Prance, G.T. (1978) Conservation problems in the Amazon Basin. In: Schofield, E.A. (ed.) *Earthcare: Global Protection of Natural Areas*. 14th Proc. Biennial Wilderness Conference, Westview Press, Colorado, 191–207.

Pratt, D.J., Pratt, D.J., Greenway, P.J. & Gwynne, M.D. (1966) A classification of East African rangeland with an appendix on terminology. *Journal of Applied Ecology* **3**, 369–82.

Prentice, I.C. (1977) Non-metric ordination models in ecology *Journal of Ecology* **65**, 85–94.

Prentice, I.C. (1980) Vegetation analysis and order invariant gradient models. *Vegetatio* **42**, 27–34.

Proctor, J. (1985) Tropical rain forest: ecology and physiology. *Progress in Physical Geography* **19**, 402–3.

Proctor, J. (1990) Tropical rain forests. *Progress in Physical Geography* **14**, 251–69.

Putnam, R.D. & Wratten, S.D. (1984) *Principles of Ecology*. Croom Helm, London.

Pyšek, P. & Lepš, J. (1991) Response of a weed community to nitrogen fertilization: a multivariate analysis. *Journal of Vegetation Science* **2**, 237–44.

Quenouille, M.H. (1959) *Rapid Statistical Calculations*. Griffin, London.

Rahel, F.J. (1990) The hierarchical nature of community persistence: a problem of scale. *The American Naturalist* **136**, 328–44.

Ramensky, L.G. (1930) Zur Methodik der vergleichenden Bearbeitung und Ordnung von Pflanzenlisten und anderen Objekten, die durch mehrere, verschiedenartig wirkende Factoren bestimmt werden. *Beiträge zur Biologie der Pflanzen* **18**, 269–304.

Randall, R.E. (1978) *Theories and Techniques in Vegetation Analysis*. Oxford University Press, Oxford.

Ratkowsky, D.A. (1984) A stopping rule and clustering method of wide applicability. *Botanical Gazette* **145**, 518–23.

Raunkaier, C. (1928) Dominansareal artstaethed of formationsdominanter. *Kgl. Danske Vidensk Selsk. Biol. Meddel.* **7**, 1.

Raunkaier, C. (1934) *The Life Forms of Plants and Statistical Plant Geography*. Clarendon Press, Oxford.

Raunkaier, C. (1937) *Plant Life Forms*. Clarendon Press, Oxford.

Redfield, G.W. (1988) Holism and reductionism in community ecology. *Oikos* **53**, 276–8.

Richards, P.W. (1952) *The Tropical Rain Forest*. Cambridge University Press, Cambridge.

Rieley, J.O. & Page, S.E. (1990) *Ecology of Plant Communities – a Phytosociological Account of the British Vegetation*. Longman.

Ricklefs, R.E. (1979) *Ecology*. 2nd ed., Nelson, London.

Roberts, D.W. (1986) Ordination on the basis of fuzzy set theory. *Vegetatio* **66**, 123–31.

Roberts, D.W. (1987) A dynamical systems perspective on vegetation theory. *Vegetatio* **69**, 27–33.

Roberts, D.W. (1989) Fuzzy systems vegetation theory. *Vegetatio* **83**, 71–80.

Robotnov, T.A. (1979) Concepts of ecological individuality of plant species and of the continuum in the works of L.G. Ramenskii. *Soviet Journal of Ecology* **9**, 417–22.

Rorison, I.H. (1969) Ecological inferences from laboratory experiments on mineral nutrition. In: Rorison, I.H. (ed.) *Ecological Aspects of the Mineral Nutrition of Plants*. British Ecological Society Symposium, Blackwell Scientific Publications, 155–75.

Rohlf, F.J. (1974) Methods of comparing classifications. *Annual Review of Ecology and Systematics* **5**, 101–13.

Ross, S.M. (1986) Vegetation change on highway verges in south-east Scotland. *Journal of Biogeography* **13**, 109–17.

Roughgarden, J. (1989) The structure and assembly of communities. In: Roughgarden, J., May, R.M. & Levin, S.A. (eds.) *Perspectives in Ecological Theory*. Princeton University Press, Princeton, New Jersey, 203–226.

Roughgarden, J., May, R.M. & Levin, S.A. (eds.) (1989) *Perspectives in Ecological Theory*. Princeton University Press, Princeton, New Jersey.

Roux, G. & Roux, M. (1967) A propos de quelques méthodes de classification en phytosociologie. *Revue de Statistique Appliquée* **15**, 59–72.

Rowe, J.S. (1961) The level of integration concept and ecology. *Ecology* **42**, 420–7.

Ryan, B.F., Joiner, B.L. & Ryan, T.A. (1985) *MINITAB Handbook*. 2nd ed., PWS-Kent, Boston, USA.

Sargent, C. (1984) *Britain's Railway Vegetation*. Institute of Terrestrial Ecology, Abbots Ripton, Huntingdon.

Schoener, T.W. (1989) The ecological niche. In: Cherrett, J.M. (ed.) *Ecological Concepts: the Contribution of Ecology to an Understanding of the Natural World*. British Ecological Society/Blackwell Scientific Publications, Oxford, 79–113.

Shaw, G. & Wheeler, D. (1985) *Statistical Techniques in Geographical Analysis*. Wiley, London.

Shenstone, J.C. (1912) The flora of London building sites. *Journal of Botany* **50**, 117–24.

Shepard, R.N. (1962) The analysis of proximities: multidimensional scaling with an unknown distance function. *Psychometrika* **27**, 125–39; 219–46.

Shepard, R.N. & Carroll, J.D. (1966) Parametric representation of non-linear data structures. In: Krishnaiah, P.F. (ed.) *Multivariate Analysis*. Academic Press, New York, 561–92.

Shimwell, D.W. (1971) *Description and Classification of Vegetation*. Sidgewick and Jackson, London.

Shipley, B. & Keddy, P.A. (1987) The individualistic and community-unit concepts as falsifiable hypotheses. *Vegetatio* **69**, 47–55.

Shmida, A. & Wilson, M.V. (1985) Biological determinants of species diversity. *Journal of Biogeography* **12**, 1–20.

Sibley, D. (1987) *Spatial Applications of Exploratory Data Analysis*. Concepts and Techniques in Modern Geography, No. 49. Geo Books, Norwich.

Siegel, S. (1956) *Non-parametric Statistics for the Social Sciences*. McGraw-Hill/ Kogakusha, New York and Tokyo.

Silk, J. (1979) *Statistical Concepts in Geography*. George Allen and Unwin, London.

Silvertown, J.W. (1987) *Introduction to Plant Population Ecology*. 2nd ed., Longman, London.

Smartt, P.F.M. (1978) Sampling for vegetation survey: a flexible systematic model for sample location. *Journal of Biogeography* **5**, 43–56.

Smartt, P.F.M., Meacock, S.E. & Lambert, J.M. (1974) Investigations into the properties of quantitative vegetational data. I. Pilot study. *Journal of Ecology* **62**, 735–59.

Smartt, P.F.M., Meacock, S.E. & Lambert, J.M. (1976) Investigations into the properties of quantitative vegetational data. II. Further data type comparisons. *Journal of Ecology* **64**, 41–78.

Smith, R.J., & Atkinson, K. (1975) *Techniques in Pedology: a Handbook for Environmental and Resource Studies*. Elek Science, London.

Smith, T. & Huston, M. (1989) A theory of the spatial and temporal dynamics of plant communities. *Vegetatio* **83**, 49–69.

Sneath, P.H.A. & Sokal, R.R. (1973) *Numerical Taxonomy*. Freeman, San Francisco.

Sobolev, L.N. & Utekhin, V.D. (1978) Russian (Ramensky) approaches to community systematization. In: Whittaker, R.H. (ed.) *Ordination of Plant Communities*. Junk, The Hague, 71–98.

Sokal, R.R. (1974) Classification: purposes, principles, progress and prospects. *Science* **185**, 1115–23.

Sokal, R.R. & Michener, C.D. (1958) A statistical method for evaluating systematic relationships. *University of Kansas Science Bulletin* **38**, 1409–38.

Sokal, R.R. & Thompson, J.D. (1987) Applications of spatial autocorrelation in ecology. In: Legendre, P. & Legendre, L. (eds.) *Developments in Numerical Ecology*. NATO ASI Series, Vol. G 14, Springer-Verlag, Berlin, 431–66.

Sørensen, T. (1948) A method of establishing groups of equal amplitude in plant sociology based on similarity of species content. *Det Kongelige Danske Videnskabernes Selskab, Biologiske Skrifter*, Bind V, Nr. 4, Copenhagen.

South, G.R. (1983) (ed.) *Biogeography and Ecology of Newfoundland*. Junk, The Hague.

Southwood, T.R.E. (1978) *Ecological Methods*. Chapman and Hall, London.

Southwood, T.R.E. (1987) The concept and nature of the community. In: Gee, J.H.R. & Giller, P.S. (eds.) *Organization of Communities Past and Present*. British Ecological Society/Blackwell Scientific Publications, Oxford, 3–27.

Spencer, H.J. & Port, G.R. (1988) Effects of roadside conditions on plants and insects. II. Soil conditions. *Journal of Applied Ecology* **25**, 709–15.

Spencer, H.J., Scott, N.E., Port, G.R. & Davison, A.W. (1988) Effects of roadside conditions on plants and insects. I. Atmospheric conditions. *Journal of Applied Ecology* **25**, 699–707.

Sprugel, D.G. (1991) Disturbance, equilibrium and environmental variability: what is 'natural' vegetation in a changing environment? *Biological Conservation* **58**, 1–18.

Stace, C. (1991) *New Flora of the British Isles*. Cambridge University Press.

Strauss, R.E. (1982) Statistical significance of species clusters in association analysis. *Ecology* **62**, 634–9.

Sukopp, H., Hejný, S. & Kowarik, I. (eds.) (1990) *Urban Ecology: Plants and Plant Communities in Urban Environments*. SPB Academic Publishing, The Hague, The Netherlands.

Tansley, A.G. (1920) The classification of vegetation and the concept of development. *Journal of Ecology* **8**, 118–49

Tansley, A.G. (1935) The use and abuse of vegetational concepts and terms. *Ecology* **16**, 284–307.

ter Braak, C.J.F. (1982) *DISCRIM – a modification to TWINSPAN (Hill, 1979) to construct simple discriminant functions and to classify attributes, given a hierarchical classification of samples*. Report C82 ST10756. TNO Institute of Mathematics, Information Processing and Statistics, Wageningen.

ter Braak, C.J.F. (1983) Principal components biplots and alpha and beta diversity. *Ecology* **64**, 454–62.

ter Braak, C.J.F. (1985) Correspondence analysis of incidence and abundance data: properties in terms of a unimodal response model. *Biometrics* **41**, 859–73.

ter Braak, C.J.F. (1986a) Canonical correspondence analysis: a new eigenvector technique for multivariate direct gradient analysis. *Ecology* **67**, 1167–79.

ter Braak, C.J.F. (1986b) Interpreting a hierarchical classification with simple discriminant function: an ecological example. In: Diday, E. *et al.* (eds.) *Data Analysis and Informatics* **4**, North-Holland, Amsterdam, 11–21.

ter Braak, C.J.F. (1987a) Ordination. In: Jongman, R.H.G., ter Braak, C.J.F. & van Tongeren, O.F.R. (eds.) *Data Analysis in Community and Landscape Ecology*. PUDOC, Wageningen, 91–173.

ter Braak, C.J.F. (1987b) The analysis of vegetation-environment relationships by canonical correspondence analysis. *Vegetatio* **64**, 69–77.

ter Braak, C.J.F. (1988a) *CANOCO – a FORTRAN Program for Canonical Community Ordination by [Partial] [Detrended] [Canonical] Correspondence Analysis (Version 2.0)*. TNO Institute of Applied Computer Science, Wageningen.

ter Braak, C.J.F. (1988b) CANOCO – an extension of DECORANA to analyse species-environment relationships. *Vegetatio* **75**, 159–60.

ter Braak, C.J.F. (1988c) Partial canonical correspondence analysis, In: Bock, H.H. (ed.) *Classification and Related Methods of Data Analysis*, Elsevier, North Holland, 551–8.

ter Braak, C.J.F. & Prentice, I.C. (1988) A theory of gradient analysis, *Advances in Ecological Research* **18**, 271–317.

Tilman, D. (1988) *Plant Strategies and the Dynamics and Structure of Plant Communities*. Princeton University Press, Princeton, New Jersey.

Tivy, J. (1982) *Biogeography: a Study of Plants in the Ecosphere*. 2nd ed., Longman, London.

Tomlinson, R. (1981) A rapid sampling technique suitable for expedition use, with reference to the vegetation of the Faroe Islands. *Biological Conservation* **20**, 69–81.

Tong, S.T.Y. (1989) On non-metric multidimensional scaling ordination and interpretation of the Mattoral vegetation in lowland Murcia. *Vegetatio* **79**, 65–74.

Tukey, J.W. (1969) Analysing data: sanctification or detective work. *American Psychologist* **24**, 83–91.

Tukey, J.W. (1977) *Exploratory Data Analysis*. Addison-Wesley, Reading, Mass.

Tutin, T.G., Heywood, V.H., Burges, N.A., Moore, D.M., Valentine, D.H., Walters, S.M. & Webb, D.A. (1964–80) *Flora Europaea*. 5 vols, Cambridge University Press, Cambridge.

Usher, M.B. (ed.) (1986) *Wildlife Conservation Evaluation*. Chapman and Hall, London.

van der Maarel, E. (1975) The Braun-Blanquet approach in perspective. *Vegetatio* **30**, 13–19.

van der Maarel, E. (1979) Transformation of cover-abundance values in phytosociology and its effects on community similarity. *Vegetatio* **39**, 97–114.

van der Maarel, E. (1984a) Vegetation science in the 1980s. In: Cooley, J.H. & Golley, F.B. (eds.) *Trends in Ecological Research for the 1980s*. Plenum Press, New York, 89–110.

van der Maarel, E. (1984b) Dynamics of plant populations from a synecological viewpoint. In: Dirzo, R. & Sarukhan, J. (eds.) *Perspectives on Plant Ecology*. Sinauer Associates Inc., Sunderland, Massachusetts.

van der Maarel, E. (1988a) Vegetation dynamics: patterns in time and space. *Vegetatio* **77**, 7–19.

van der Maarel, E. (1988b) Species diversity in plant communities in relation to structure and dynamics. In: During, H.J., Werger, M.J.A. & Willems, J.H. (eds.) *Diversity and Pattern in Plant Communities*. SPB Academic Publishing, The Hague, The Netherlands, 1–14.

van der Maarel, E. (1989) Theoretical vegetation science on the way. *Vegetatio* **83**, 1–6.

van der Maarel, E. (1990) Ecotones and ecoclines are different. *Journal of Vegetation Science* **1**, 135–8.

van der Maarel, E., Janssen, J.G.M. & Louppen, J.M.W. (1978) TABORD – a program for structuring phytosociological tables. *Vegetatio* **38**, 143–56.

van der Maarel, E., Espejel, I. & Moreno-Casasola, P. (1987) Two-step vegetation analysis based on very large data sets. *Vegetatio* **68**, 139–43.

van Tongeren, O.F.R. (1986) FLEXCLUS: an interactive program for classification and tabulation of ecological data. *Acta Botanica Nederlandia* **35**, 137–42.

van Tongeren, O.F.R. (1987) Cluster analysis. In: Jongman, R.H.G., ter Braak, C.F.J. & van Tongeren, O.F.R. (eds.) *Data Analysis in Community and Landscape Ecology.* PUDOC, Wageningen, 174–212.

von Post, H. (1862) *Forsök till en systematik uppstallning af vextstallena i mellersta Sverige.* Bonnier, Stockholm.

Walker, D. (1989) Diversity and stability. In: Cherrett, J.M. (ed.) *Ecological Concepts: the Contribution of Ecology to an Understanding of the Natural World.* British Ecological Society/Blackwell Scientific Publications, Oxford, 115–45.

Walker, J., Sharpe, P.J.H., Penridge, L.K. & Wu, H. (1989) Ecological field theory: the concept and field tests. *Vegetatio* **83**, 81–95.

Ward, J.H. (1963) Hierarchical grouping to optimize an objective function. *American Statistical Association Journal* **58**, 236–44.

Ward, S.D., Jones, A.D. & Manton, M. (1972) The vegetation of Dartmoor. *Field Studies* **3**, 505–33.

Wardlaw, A.C. (1985) *Practical Statistics for Experimental Biology.* Wiley, Chichester.

Waring, R.H. (1989) Ecosystems: fluxes of matter and energy. In: Cherrett, J.M. (ed.) *Ecological Concepts: the Contribution of Ecology to an Understanding of the Natural World.* British Ecological Society/Blackwell Scientific Publications, Oxford, 17–41.

Wartenberg, D., Ferson, S. & Rohlf, F.J. (1987) Putting things in order: a critique of detrended correspondence analysis. *The American Naturalist* **129**, 434–48.

Wathern, P. (1976) *The Ecology of Development Sites.* Unpublished Ph.D. thesis, Department of Landscape Architecture, University of Sheffield.

Watt, A.S. (1947) Pattern and process in the plant community. *Journal of Ecology* **35**, 1–22.

Wassen, M.J., Barendregt, A., Palczynski, A., de Smidt, J.T. & de Mars, H. (1990) The relationship between fen vegetation gradients, groundwater flow and flooding in an undrained valley mire at Biebrza, Poland. *Journal of Ecology* **78**, 1106–1122.

Way, J.M. (ed.) (1969) *Road Verges: their Function and Management.* Nature Conservancy, Monks Wood Experimental Station, Abbots Ripton, Huntingdon.

Webb, L.J. (1978) A structural comparison of New Zealand and South-East Australian rain forests and their tropical affinities. *Australian Journal of Ecology* **3**, 7–21.

Webb, L.J., Tracey, J.G. & Williams, W.T. (1976) The value of structural features in tropical forest typology. *Australian Journal of Ecology* **1**, 3–28.

Webb, L.J., Tracey, J.G., Williams, W.T. & Lance, G.N. (1970) Studies in the numerical analysis of complex rain forest communities. V. a comparison of the properties of floristic and physiognomic-structural data. *Journal of Ecology* **58**, 203–32.

Webber, P.J. (1978) Spatial and temporal variation and its productivity. In: Tieszen, L.L. (ed.) *Vegetation and Production Ecology of an Alaskan Arctic Tundra.* Ecological Studies 29, Springer-Verlag, New York, 37–112.

Werger, M.J.A. (1974a) The place of the Zurich-Montpellier method in vegetation science. *Folia Geobotanica et Phytotaxonomica* **9**, 99–109.

Werger, M.J.A. (1974b) On concepts and techniques applied in the Zurich-Montpellier method of vegetation survey. *Bothalia* **11**, 309–23.

Werger, M.J.A. & Sprangers, J.T.C. (1982) Comparison of floristic and structural classification of vegetation. *Vegetatio* **50**, 175–83.

Westfall, R.H., Dednam, G, van Rooyen, N. & Theron, G.K. (1982) PHYTOTAB – a program package for Braun-Blanquet tables. *Vegetatio* **49**, 35–7.

Westhoff, V & van der Maarel, E. (1978) The Braun-Blanquet approach. In: Whittaker, R.H. (ed.) *Classification of Plant Communities.* Junk, The Hague, 289–374.

Wheeler, B.D. (1980) Plant communities of rich-fen systems in England and Wales. I. Introduction. Tall sedge and reed communities. *Journal of Ecology* **68**, 365–96.

White, R.E. (1979) *Introduction to the Principles and Practice of Soil Science.* Blackwell, Oxford.

Whitmore, T.C. (1990) *An Introduction to Tropical Rain Forests.* Clarendon Press, Oxford.

Whitney, G.G. (1985) A quantitative analysis of the flora and plant communities of a representative midwestern U.S. town. *Urban Ecology* **9**, 143–60.

Whitney, G.G. & Adams, S.D. (1980) Man as a maker of new plant communities. *Journal of Applied Ecology* **17**, 431–48.

Whittaker, R.H. (1948) *A Vegetation Analysis of the Great Smokey Mountains,* Unpublished Ph.D. thesis, University of Illinois, Urbana.

Whittaker, R.H. (1951) A criticism of the plant association and climatic climax concepts. *NorthWest Science* **25**, 17–31.

Whittaker, R.H. (1953) A consideration of climax theory: the climax as a population and pattern. *Ecological Monographs* **23**, 41–78.

Whittaker, R.H. (1956) Vegetation of the Great Smoky Mountains. *Ecological Monographs* **26**, 1–80.

Whittaker, R.H. (1960) Vegetation of the Siskiyou Mountains, Oregon and California. *Ecological Monographs* **30**, 279–338.

Whittaker, R.H. (1962) Classification of plant communities. *Botanical Review* **28**, 1–239.

Whittaker, R.H. (1965) Dominance and diversity in land plant communities. *Science* **147**, 250–60.

Whittaker, R.H. (1967) Gradient analysis of vegetation. *Biological Review* **42**, 207–64.

Whittaker, R.H. (1972a) Evolution and measurement of species diversity. *Taxon* **21**, 213–51.

Whittaker, R.H. (1972b) Convergences of ordination and classification. In: van der Maarel, E. & Tüxen, R. (eds.) *Grundfragen und Methoden in der Pflanzensoziologie.* Junk, The Hague, 39–57.

Whittaker, R.H. (1975) *Communities and Ecosystems.* 2nd ed., Macmillan, London.

Whittaker, R.H. (ed.) (1978a) *Ordination of Plant Communities.* Junk, The Hague.

Whittaker, R.H. (ed.) (1978b) *Classification of Plant Communities.* Junk, The Hague.

Whittaker, R.H. & Levin, S.A. (1977) The role of mosaic phenomena in natural communities. *Theoretical Population Biology* **12**, 117–39.

Whittaker, R.H., Levin, S.A. & Root, R.B. (1973) Niche, habitat and ecotope. *The American Naturalist* **107**, 321–38.

Whittaker, R.H. & Gauch, H.G. (1978) Evaluation of ordination techniques. In: Whittaker, R.H. (ed.) *Ordination of Plant Communities.* Junk, The Hague, 277–336.

Whittaker, R.J., Bush, M.B. & Richards, K. (1989) Plant recolonization and vegetation succession on the Krakatau Islands, Indonesia. *Ecological Monographs* **59**, 59–123.

Wiegert, R.G. (1988) Holism and reductionism in ecology: hypotheses, scale and systems models. *Oikos* **53**, 267–9.

Wiegleb, G. (1989) Explanation and prediction in vegetation science. *Vegetatio* **83**, 17–34.

Wildi, O. & Orlóci, L. (1983) *Management and Multivariate Analysis of Vegetation Data.* Swiss Federal Institute of Forestry Research, CH8903, Birmensdorf.

Wildi, O. & Orlóci, L. (1990) *Numerical Exploration of Community Patterns.* SPB Academic Publishing, The Hague, The Netherlands.

Williams, A.G., Kent, M. & Ternan, J.L. (1987) Quantity and quality of bracken throughfall, stemflow and litterflow in a Dartmoor catchment. *Journal of Applied Ecology* **24**, 217–30.

Williams, G.J., Harris, J.R., Kemp, P.R., Cartwright, D. & Gerwick, B.C. (1979) Introducing the principles of vegetation sampling in the laboratory. *American Biology Teacher* **41**, 14–7.

Williams, W.T. (1971) Principles of clustering. *Annual Review of Ecology and Systemetics* **2**, 303–326.

Williams, W.T. (ed.) (1976a) *Pattern Analysis in Agricultural Science.* Elsevier, New York.

Williams, W.T. (1976b) Hierarchical divisive strategies. In: Williams, W.T. (ed.) *Pattern Analysis in Agricultural Science.* Elsevier, New York.

Williams, W.T. (1981) Underlying assumptions in numerical classification. In: Gillison, A.N. & Anderson, D.J. (eds.) *Vegetation Classification in Australia*. CSIRO, Canberra, Australia, 117–29.

Williams, W.T. & Lambert, J.M. (1959) Multivariate methods in plant ecology. I. Association analysis in plant communities. *Journal of Ecology* **47**, 83–101.

Williams, W.T. & Lambert, J.M. (1960) Multivariate methods in plant ecology. II. The use of an electronic digital computer for association analysis. *Journal of Ecology* **48**, 689–710.

Williams, W.T. & Lambert, J.M. (1961) Multivariate methods in plant ecology. III. Inverse association analysis. *Journal of Ecology* **49**, 717–29.

Williams, W.T., Lambert, J.M. & Lance, G.N. (1966) Multivariate methods in plant ecology. V. Similarity analysis and information analysis. *Journal of Ecology* **54**, 427–45.

Wilson, M.V. (1981) A statistical test of the accuracy and consistency of ordinations. *Ecology* **62**, 8–12.

Wishart, D. (1987) *CLUSTAN User Manual (CLUSTAN 3)*. 4th ed., Computing Laboratory, University of St. Andrews, Scotland.

Index